With Best Wishes
For Great

T.D. Shahan

BIOCHEMISTRY OF FOODS AND THE BIOCATALYSTS

by

Ishak I. Shahied, Ph.D.

Department of Chemistry
University of Missouri
Kansas City, Missouri

VANTAGE PRESS
New York   Washington   Atlanta   Hollywood

And ye shall serve the Lord your God, and he shall bless thy bread, and thy water; and I will take sickness away from the midst of thee. He that cometh to me shall never hunger; and he that believeth on me shall never thirst. Thy flesh shall be fresher than a child's: ye shall return to the days of thy youth, as I satisfy thy mouth with good things; so that thy youth is renewed like the eagle's; and thine health shall spring forth speedily. And the Lord shall satisfy thy soul in drought, and make fat thy bones: and thou shalt be like a watered garden, and like a spring of water, whose waters fail not. That I will give you the rain of your land in his due season, the first rain and the latter rain, and the land shall yield her increase, and the trees of the field shall yield their fruit, and your threshing shall reach unto the vintage, and the vintage shall reach unto the sowing time: and ye shall eat your bread to the full; as I will also bless thy corn, and thy wine, and thine oil, the increase of thy kine, and the flocks of thy sheep.

FIRST EDITION

All rights reserved, including the right of reproduction in whole or in part in any form.

Copyright © 1977 by Ishak I. Shahied, Ph.D.

Published by Vantage Press, Inc.
516 West 34th Street, New York, New York 10001

Printed in the United States of America

Standard Book Number 533-02164-2

CONTENTS

PREFACE

ACKNOWLEDGMENTS

PART I THE CHEMISTRY OF FOODS

Chapter

1. WATER IN FOODS

    Forms of Water in Foods   1
        Free water
        Hydrate forms
        Bound water
        Imbibed water as gels
        Adsorbed water
    Characteristics of Water as Life-Sustaining System   1
        Water as temperature stabilizer
        Its specific heat
        Its latent heat of vaporization
        Its latent heat of fusion
        Its specific gravity
        Its heat conductance
        Its solvent property
    Some Unique Properties of Water   2
        Its transparency to light
        Its high surface tension
        Its hydrogen ion activity
    Structure of Water   3
    Effects of Water on Foods   3

2. LIPIDS

    Definition   5
    Classification and Functions   5
        Saponifiable lipids

Fatty acids
Complex lipids
Nonsaponifiable
Steroids
Fatty Acids   7
Complex Lipids   11
Fat Soluble Vitamins   12
    Vitamin A
    Vitamin D
    Plant sterols
    Vitamin E
    Vitamin K
Digestion and Absorption of Lipids   17
Physical Properties of Fats and Oils   19
    Crystalization
    Hydrolysis of fats and relation to foods
    Double bond rearrangement
    Pyrrolysis
    Hydrogenation
    Terpenoids
Processing of Edible Oils and Fats   24
Refining of Oils   24
Autoxidation of Unsaturated Fatty Acids   27
Methods of Determining Lipid Oxidation in Foods   29
    Peroxide test
    TBA test
    Carbonyls
Factors That Cause Oxidation of Fats   30
Factors That Inhibit Oxidation of Fats   30
Possible Ways of Reducing Fat Oxidation   31
    Antioxidants
    Mechanism of action of antioxidants
    Synergists
Sources of Oxidation   33
Lipid Metabolism   34
    β-oxidation of fatty acids
    Role of carnitine
    Fatty acids synthesis
    Synthesis of triglycerides
    Biosynthesis of cholesterol

3. **CARBOHYDRATES**
   Introduction 44
   Monosaccharides 44
   Various Forms of Glucose in Solution 50
   Oligosaccharides 51
       Cellibiose
       Lactose
       Sucrose
       Trehalose
       Maltose
   Trisaccharides 54
   Aldonic Acid 55
   Uronic Acid 56
   Saccharic Acid 56
   Mucopolysaccharides 57
       Hyaluronic acid
       Chitin
       Chondriotin sulfate
       Hemicellulose
   Polysaccharides 59
   Enzymatic Action on Starch 61
   Degree of Sweetness of Sugars 62
   Pectic Substances 62
       Pectin
       Pectic acid
   Gums and Mucilages 65
   Types and Sources of Carbohydrates in the American Diet 66
   Analytical Procedures for Detection of Carbohydrates 68
   Enzymatic Browning 71
       Properties of the phenolase enzymes
       Optimum temperature range of activity of phenolases
       Possible mechanism of enzyme inhibition
       Possible mechanism of melanoidin formation
   Nonenzymatic Browning 76
       Aldol condensation
       Carmilization
   Prevention of Maillard Browning 84

Carbohydrate Metabolism	85
Catabolism	
Anabolism	
Glycolytic system	
Control mechanisms	
The Tricarboxylic Acid (TCA) Cycle	91
Energy production	
The Pentose Shunt	94
Anaplerotic reactions	
Metabolism of galactose	

4. PROTEINS

Introduction	104
Amino Acids in Foods	104
Formation of Peptide Linkages Between Amino Acids	109
Acid-Base Characteristics of Amino Acids	109
Classification of Proteins	111
Simple proteins	
Albumins	
Globulins	
Glutelins	
Prolamines	
Albuminoids	
Histones	
Protamines	
Conjugated proteins	
Nucleoproteins	
Glycoproteins	
Phosphoproteins	
Chromoproteins	
Lipoproteins	
Derived proteins	
Shape and Structure of Proteins	113
Fibrous	
Globular	
Milk Proteins	114
Fractionation of milk proteins	
Casein fractionation	
Role of casein in cheese-making	

Egg Proteins   116
- *Egg Composition*
  - *The shell*
  - *Outer shell membrane*
  - *Inner shell membrane*
  - *Albumen Layer*
  - *Vitelline membrane*
  - *Egg yolk*
- *Egg White Proteins*
  - *Ovalbumin*
  - *Conalbumin*
  - *Lysozyme*
  - *Ovomucin*
  - *Ovomucoid*
  - *Ovoinhibitor*
  - *Flavoprotein*
  - *Avidin*
  - *Minor proteins*
- *Proteins in the Egg Yolk*

Cereal Proteins   121
- *Glutelins*
- *Prolamines*
- *Albumins*
- *Globulins*
- *Proteoses*
- *Amino acids*

Wheat Proteins   122
- *Composition of hard wheat*
- *Morphological composition of wheat protein*
  - *Glutin protein*
- *Soluble proteins*
  - *Albumins*
  - *Globulins*

Vegetable Proteins   123
Meats   124
- *Types of muscles*
- *Composition of skeletal muscle*
- *Protein composition of skeletal muscle*
- *Types of proteins of skeletal muscle*
- *Connective tissue proteins*

Keratins
Connective tissue membranes of skeletal muscle
Muscle contraction
Lipid component of muscle
Water extractives
Connective tissue
Composition of cooked meats
Meat salts
Amino acid content of meat
Rigor mortis
Meat color
   Structure of myoglobin and hemoglobin
   Green pigments
Meat curing
   Function of smoking of meats
   Meat tenderness
Enzymes in meat
   Enzymes of bacteria and fungi
Changes brought in meat by cooking
Meat emulsions
   Definition of emulsions
   Percent of soluble proteins

Metabolism of Amino Acids   146
   Making use of nitrogen
   Nitrogen fixation
   Processes of NH_3 fixation
   The essential amino acids
   Amino acids requirement
   Digestion and absorption of proteins
   General reactions of amino acids
   Transamination reactions of amino acids
   Deamination reactions of amino acids
   Dehydrogenation reactions
   Amino and keto acids interconversions
   Decarboxylation of some amino acids
   Important reactions of specific amino acids
   Glycine
   Lysine
   Tryptophan

      Valine, Leucine, Isoleucine
      Interconversion of sulfur–amino acids
      Urea cycle
      Relationship between urea and TCA cycles
      Formation of creatine

5. FLAVONOIDS

    Classification   171
    Flavones   174
    Flavanone   174
        Anthocyanins
        Anthoxanthins
        Catechols and tannins
    Occurrence of Anthoxanthins   175
    General Properties of the Flavonoids   177
    Biological Activity of the Bioflavonoids   178
    Bitterness of Some Flavonoids   178
    Tannins   178
    Changes That Occur During Processing   180

6. FLAVORS IN FOODS

    Introduction   182
    Sensory Evaluation of Flavor   185
    Flavor Profile   185
    Flavors in Various Foods   186
    Off-Flavor Development   187
    Milk Off-Flavors   188
    Vegetable-Off-Flavor   189

7. VEGETABLES AND FRUITS

    Introduction   192
    Chemical Composition of Fruits and Vegetables   192
    Classification   193
    Structure of Fruits and Vegetables   193
    Turgor of Fruits and Vegetables   195
    General Changes That Occur During Cooking of
        Fruits and Vegetables   195
    Volatile Sulfur Compounds and Flavor   197

8. MILK

    Introduction   200

Vitamins A, D, E, and K Content	201
Protein Content	201
Noncombustible Ash Content	203
Vitamin Content of Milk	203
Enzymes in Milk	203

9. CEREAL CHEMISTRY

Introduction	205
Vitamins and Minerals	206
Gas Production in Dough-Making	206
Gas Retention in Dough	207

PART II THE CHEMISTRY OF VITAMINS

10. VITAMIN A

Introduction	213
Vitamin A—Precursors and Forms	214
Factors in Conversion of Carotenes to Vitamin A	215
Tests for Vitamin A	216
Absorption and Uptake of Vitamin A	216
Biological Function of Vitamin A	217
Mechanism of Vitamin A Involvement in Vision	219
Vitamin A's Role in Bone	220
Vitamin A versus Thyroxine	221
Vitamin A and Glycogen Synthesis	221
Symptoms of Vitamin A Deficiency	222
Summary of Deficiency Symptoms of Vitamin A	223
Summary of Excessive Vitamin A Symptoms	224

11. VITAMIN D

Introduction	227
Structure of the Various Forms of Vitamin D	229
Functions of Vitamin D	230
Mechanism of Ca^{2+} Transport	231
Sources of Vitamin D	232
Relationship between Vitamin D and Other Chemicals	233
Relationship of Vitamin D with Other Hormones	233
Deficiency Symptoms of Vitamin D	233
Summary of Vitamin D Deficiency Symptoms	233

    Toxicity Symptoms of Vitamin D   234

12. **VITAMIN E**
    Introduction 236
    Various Forms of Vitamin E 236
    Sources of Vitamin E 237
    Structures of the Various Forms of Vitamin E 237
    Absorption and Utilization of Vitamin E 237
    Functions of Vitamin E 237
    Summary of the Metabolic Functions of Vitamin E 240
    Vitamin E Deficiency Symptoms 241
    Factors Determining Vitamin E's Intake 241

13. **VITAMIN K**
    Introduction 244
    Various Forms of Vitamin K 246
    The Intrinsic System of Vitamin K Action 247
    Factors Affecting Poor Clotting Time or
        Vitamin K Deficiency 248
    Vitamin K Antagonists 248
    Deficiency Symptoms of Vitamin K 248

14. **VITAMIN C**
    Introduction 251
    Oxidation-Reduction System of Ascorbic Acid 253
    Other Biochemical Functions of Vitamin C 254
    Effect of Ascorbic Acid on Metal Ions 255
    Ascorbic Acid and Adrenal Functions 256
    Relationship Between Ascorbic Acid and Cholesterol 257
    Various Functions of Vitamin C 257
    Factors Affecting Ascorbic Acid Requirements 258
    Symptoms of Scurvy 258
    Biosynthesis of Ascorbic Acid 259

15. **THIAMIN (B_1)**
    Introduction 263
    Structure 264
    Functions of Thiamin 264
    Specific Reactions Catalyzed by TPP 265
        Nonoxidative decarboxylation

      Oxidative decarboxylation
      Carbon transfer reactions
      Transketolase reactions

Possible Mechanism of the Nonoxidative Decarboxylation Reaction	267
Possible Mechanism of Decarboxylation Reaction	268
Deficiency Symptoms of Thiamin	269

16. RIBOFLAVIN (B_2)

Introduction	271
Structure and Forms	271
Role of Coenzymes in Electron Transport	273
Functions of FAD and FMN	273
Enzymes Activated by FMN	274
Enzymes Activated by FAD	274
Metallo-Flavoproteins	274
Deficiency Symptoms of Riboflavin	274

17. NIACIN

Introduction	275
Function of the Coenzymes	275
Source of the Vitamin	276
Structure and Forms of Niacin	276
Deficiency Symptoms of Niacin	278

18. PANTOTHENIC ACID

Introduction	279
Structure	279
Types of Reactions Catalyzed by Co-A	280
Nucleophillic attack	
Head to tail condensation	
Summary of Functions of Pantothenic Acid	281
Deficiency Symptoms of Pantothenic Acid	282

19. BIOTIN

Introduction	283
Structure	284
Possible Mechanism of Biotin's Binding of CO_2	284
Sources of Biotin	285
Biochemical Functions of Biotin	285

      Role in carbohydrate metabolism
      Role in propionate metabolism
      Role in lipid metabolism
      Role in urea synthesis
      Role in purine synthesis
   Summary of Possible Biotin Roles in Biochemical
      Systems   287
      Carbohydrate metabolism
      Lipid metabolism
      Amino acid and nucleic acid metabolism
      Enzymes Activated with Biotin
      Role in Protein Synthesis
   Deficiency Symptoms of Biotin   288

20. PYRIDOXINE (B_6)

   Introduction   289
   Various Forms of B_6 290
   Roles of B_6 291
      Decarboxylation of amino acids
      Transamination of amino acids
      Racemization of amino acids
   Possible Reaction Mechanisms of B_6 291
   Role of B_6 in Tryptophan Metabolism   294
   Metabolism of B_6 294
   Deficiency Symptoms of B_6 295

21. VITAMIN B_{12}

   Introduction   296
   Structure   296
   Absorption of B_{12} 298
   General Functions of B_{12} 298
   Coenzyme Functions of B_{12} 298
   Factors Affecting B_{12} Requirement   300
   Relationship between B_{12} and Folic Acid   301
   Deficiency Symptoms of B_{12} 301
   Summary of B_{12} Deficiency Symptoms   301
      In man
      In chicks
      In swine
      In ruminants

22. **FOLIC ACID**

    Introduction   303
    Structure and Forms   304
    Coenzyme Forms   304
    Functions of Tetrahydrofolic Acid   306
    Examples of Reactions Catalyzed by THF   307
    Metabolism of Histidine and the Role of THF   308
    Symptoms of Folic Acid Deficiency   309

PART III THE CHEMISTRY OF MINERALS

23. **MINERALS**

    Introduction   315
    Minerals and the Biological System   315
    The Essential Minerals   319
        Macroelements
        Borderline elements
        Microelements
    Calcium   319
        Source
        Role of calcium
        Deficiency symptoms of calciun.
        Toxicity symptoms of calcium
        Requirement
        Human daily calcium requirements
        Factors affecting calcium requirement
    Phosphorus   322
        Sources
        Biochemical functions
        Toxicity symptoms
    Sodium   322
        Biochemical functions
        Deficiency symptoms
        Toxicity symptoms
    Potassium   323
        Toxicity symptoms
    Magnesium   323
        Functions
        Symptoms of Mg Deficiency
        Role in enzyme activation

     Summary of Functions of Mg^{++}
The Trace Elements   325
Iron   326
     Functions
     Requirements
     Sources
     Absorption of iron
     Biochemical functions of iron
     Iron-copper relationship
Copper   327
     Functions
     Sources of Cu
     Summary of Functions of Cu
     Deficiency symptoms of copper
     Toxicity symptoms of copper
Zinc   329
     Requirement
     Sources
     Biochemical functions of Zinc
     Deficiency symptoms of Zinc
     Toxicity symptoms of Zinc
Manganese   331
     Functions
     Biological roles
Molybdenum   332
     Functions
     Requirement
     Toxic level
     Sources
     Toxicity symptoms
Cobalt   334
     Requirement
     Toxic level
     Source
     Biochemical functions
     Deficiency symptoms
     Toxicity symptoms
Selenium   335
Fluorine   336
Chromium   336

PART IV PHYSIOLOGY AND BIOCHEMICAL REACTIONS OF THE HORMONES

24. CATECHOLAMINES
- Hormones of the Adrenal Medulla — 343
- Regulation of Catecholamine Secretion — 344
- Origin, Storage, and Release of Catecholamines — 345
- Secretion and Release of Catecholamines — 345
- Biosynthesis of Catecholamines — 346
- Physiological Effects of Adrenal Medullary Hormones — 347
- Action of Catecholamines on the Various Organs and Tissues — 348
- Receptors — 348
- Biochemical Action of Catecholamines — 349
- Biochemical Mechanism of Action of Epinephrine — 350
    - *Hyperglycemia*
    - *Lipolysis*
- Control of Secretion — 351
- Internal Negative Feedback — 352
- Metabolism of Catecholamines — 352
- Distribution of Urinary Metabolites of Epinephrine — 353
- Estimation — 353
- Action of Catecholamines on Adipose Tissue — 354

25. HORMONES OF THE POSTERIOR PITUITARY
- Oxytocin — 356
    - *Synthesis and Storage*
    - *Release of Oxytocin*
    - *Actions of Oxytocin*
    - *Assay of Oxytocin*
- Antidiuretic hormone (ADH) — 357
    - *Synthesis and Storage*
    - *Functions of ADH*
    - *Control of ADH Secretion*
    - *Assay of ADH*

26. HORMONES OF THE ANTERIOR PITUITARY
- Introduction — 360
- Histology of Pituitary Gland — 360
- Major Groups of Pituitary Hormones — 361

27. ADRENOCORTICOTROPIC HORMONE (ACTH)
   - Chemical Structure — 365
   - Biological Activity — 366
   - Mechanism of Action of ACTH on Cell Membrane — 366
   - Secretion and Feedback of ACTH — 369
   - Assay of ACTH — 369

28. GROWTH HORMONE (GH)
   - Chemical Structure — 371
   - Biological Effects of Growth Hormone — 371
   - Secretion and Feedback Control of Growth Hormone — 373
   - Factors Influencing Growth Hormone Secretion — 373
   - Assay of Growth Hormone — 374

29. PROLACTIN
   - Introduction — 376
   - Structure — 376
   - Biological Effects — 377
   - Assay Methods — 377
   - Metabolism — 377
   - Control of Prolactin Secretion — 378

30. THYROID HORMONES
   - Introduction — 380
   - Thyroxine (T_4) — 381
   - Structure of the Thyroid Hormones — 382
   - Formation of the Hormone — 383
   - Control of the Release of the Thyroid Hormones — 386
   - Functions of the Thyroid Hormones — 387
   - Theories of Thyroid Hormone Action — 388
   - Action of Thyroxine Deficiency in Youth — 388
   - Specific Actions of T_3 and T_4 — 389
   - Relation of Thyroxine to Other Hormones — 389
       - *Relationship between adrenal cortex and thyroid hormone*
       - *Effect of thyroid hormone on gonads*
       - *Relation to parathyroid gland*
   - Characteristics of Thyrotoxicosis — 391
   - Goitrogenic Compounds — 391

Examples of Goitrogenic Compounds　　391
　　　Associated Diseases and Dysfunctions　　393
　　　　Graves' Disease
　　　　Plumner's Disease
　　　　Hashimoto Thyroiditis
　　　　Myexedemia
　　　　Cretinism
　　Lats　　394

31. PARATHYROID HORMONE
　　Introduction　　398
　　Function of the Parathyroid Gland　　399
　　Control of Calcium Metabolism　　399
　　Interaction of Parathyroid-Calcitonin-Vitamin D　　400

32. INSULIN
　　Introduction　　404
　　Synthesis　　405
　　Action on Cell Membranes　　406
　　Factors that Influence Insulin's Release　　406
　　Biological Action　　406
　　Insulin and Carbohydrate Metabolism　　407
　　Influence of Insulin on Skeletal Muscle Cells　　408
　　Insulin and Fat Metabolism　　409
　　Insulin and Protein Metabolism　　410
　　Miscellaneous Actions of Insulin　　410
　　Symptoms and Causes of Diabetes Mellitus　　411
　　Bioassay for Insulin　　412
　　Orally Effective Agents for Control of Diabetes
　　　　Mellitus　　412
　　Types of Insulin　　413

33. GLUCAGON
　　Introduction　　416
　　Function　　416
　　Action vs. Insulin　　416
　　Glucagon and Lipolysis　　418

34. GONADOTROPIC AND SEX HORMONES
　　Steroid Structures　　419
　　Naming　　421

Steroid Biosynthesis 422
Biosynthesis and Metabolism of Sex Hormones 423
Follicle-Stimulating Hormone 423
Hormones of the Ovary 424
 Estrogen
 Action of estrogen
 Types of estrogens
 Sources of the naturally occurring estrogens in the body
 Solubility
 Biological effects of estrogen
 Bioassay for estrogens
 Synthetic estrogens
 DES, MGA
 Plant estrogens
Progesterone 431
 Separation of progesterone
 Synthetic Pathway
 Action of progesterone
Chorionic Gonadotropin 434
Pregnant Mare Serum Gonadotropin 434
Hormone of the Testes 435
Action of Androgen 435
Functions of Testes 436
Adrenal Output of Sex Hormones 437
Feedback Control of Gonadal Hormones (male) 438
Feedback Control of Gonadal Hormones (female) 439
Causes of Infertility 440
The Pill 440
Activity of Steroids 441
Assays for Sex Hormones 441

35. ADRENAL CORTEX HORMONES

Introduction 443
Structure of the Adrenal Cortex 443
Zones and Their Secretion 443
Regeneration and Hypertrophy of the Adrenal Cortex 444
Blood Supply 444
Nerve Supply of the Adrenal Cortex 444

Biosynthesis of the Adrenal Cortex Hormones	444
Metabolism of Glucocorticoids	446
Binding to Proteins in Circulation	446
Hormones of the Adrenal Cortex	447
Aldosterone	
Control of aldosterone secretion	
Action of aldosterone	
Cortisol	
Corticosterone	
Desoxycorticosterone	
Control of cortisol and corticosterone secretion	
Effects of Adrenalectomy	452
Miscellaneous Functions of the Glucocorticoids	453
Scheme of Biosynthesis of the Adrenal Cortex Hormones	453

36. GASTROINTESTINAL HORMONES

Introduction	455
Gastrin	455
Secretin	456
Enterogastrone	456
Cholecystokinin	457
Villikinin	457
Enterocrinin	457
Index	461

PREFACE

This book was designed and written so it may serve as a textbook for students taking advanced undergraduate and graduate courses in Nutrition, Food Chemistry, Vitamins and Minerals, Endocrinology, and Applied Biochemistry. This is not a biochemistry text as such, however, a student of biochemistry will find it useful.

Chapters covering lipids, carbohydrates, and proteins are fairly extensive. They will not only provide the student taking a Nutrition or a Food Chemistry course with the information he needs, but, also, a biochemistry student may find most of the material he might need for coverage of these three subjects. The chemistry and biochemistry of lipids, carbohydrates, and proteins are adequately covered, including the metabolism of fatty acids, sugars, and amino acids. I chose not to go into details of lipid metabolism as this can be easily found in most biochemistry texts. The discussions on carbohydrate and amino acid metabolism are of more detail, however. I have almost completely avoided discussion of the hormonal influence on the metabolism of the three above nutrients, as this was fairly adequately discussed in the section on Hormones in this text. For an example, if and when the student requires to learn of the influence of the hormone *insulin* on carbohydrate metabolism, he should refer to the chapter on "Insulin," where he may find a detailed list of enzymes that might be either activated or inactivated by insulin in due process of sugar metabolism. The same principle should apply to fat and protein metabolism.

Chapter 4, on "Proteins," contains a section on "Meat Chemistry" which, in my opinion, provides a student taking an advanced course in Meat Science or Meat Chemistry with all the information he may need on the subject.

Both the fat-soluble and the B vitamins are discussed in this text. There is hardly any discussion of historical events, as my interest has been mainly in their biochemical roles and reactions. The role of the fat-soluble vitamins in control reactions, and the role of the B vitamins as cofactors for the enzymatically catalyzed reactions, in my opinion, have been adequately discussed and outlined.

My discussion of "Minerals" is not as detailed as that of the vitamins; however, a student of nutrition may find all the important points that he requires on the subject.

It is my impression that the section on "Hormones" in this text will adequately provide the student taking any Endocrinology course with practically all the information that he may need for his course coverage.

Ishak I. Shahied, Ph.D.

ACKNOWLEDGMENTS

I am most grateful to Dr. K. C. Back, to Major E. Van Stee, to Dr. A. A. Thomas and to Col. J. A. Winstead of the Toxic Hazards Division of the Aerospace Medical Research Laboratory, Wright-Patterson Air Force Base, Ohio, for all the encouragement they rendered me during the preparation of this manuscript.

I also owe Cindy Truman and Barbara Allen many thanks for typing the manuscript.

Of course, I shall always be most thankful to Drs. D. A. Cramer, J. F. Masken, L. Charkey, E. Kienholz and the many others for the training and good stay that I was privileged to enjoy at Colorado State University, Fort Collins, Colorado.

Thanks are also extended to Miss Lois Ann Nadolny for the moral support she provided.

Ishak I. Shahied, Ph.D.

Part I

THE CHEMISTRY OF FOODS

Chapter 1

WATER IN FOODS

FORMS OF WATER IN FOODS

Water is the most abundant chemical compound. It exists in the free form as free liquid in which other substances are dissolved or dispersed, such as in milk, cellular fluids, or cytoplasm. There is free water present in most foods.

The hydrate form of water is manifested in proteins, starches and some salts. Here H-bonding between hydrogen ions and oxygen or nitrogen is abundant.

The imbibed form of water is not much different from the hydrate form in most cases.

The bound form of water is very hard to remove from foods, even after removal of all the free water present. Plants' not freezing in winter is mainly due to the presence of bound water.

The adsorbed form of water is usually in the solid forms of food where the solid surface adsorbs water to it. The finely divided forms of foods usually have a great water adsorption capacity.

WATER AS LIFE-SUSTAINING SYSTEM

The high specific heat of water (exceeded only by ammonia), its high latent heat of vaporization, its high latent heat of fusion (higher than most organic compounds)—all of these properties make water an excellent temperature stabilizer. As a matter of fact, life in its present form would immediately collapse if it were not for to the blessing of ocean's water which absorbs the radiant energy falling from the sun on the earth's surface. This prevents us from burning alive at daytime, or freezing to death during the night. The high specific heat of water causes the oceans, seas, and lakes to act as heat reservoirs, where heat is maintained and moderate temperature is the result. Water, thus, acts as a buffer against heat changes.

Another remarkable property of water is its specific gravity. Water has lower specific gravity at 0°C than at 4°C, which causes ice formation on the surface of body water, thus serving as an insulator and preventing freezing of the interior of body.

Water is a better heat conductor than any other liquid known. Water is also the best solvent known; it dissolves more substances than other solvents. The dipole moment of water causes electrolytes ionization and thus their solubility. This high dipole moment of water (78.5, vs. 24.5 for ethanol; vs. 2.5 for benzene) renders water a very polar material.

The transparency of water permits the process of photosynthesis to proceed underneath the water surface, which supports the growth of plants underneath the surface of water.

The high surface tension of water allows it to move near the surface of soil, thus becoming available to plants.

STRUCTURAL PROPERTIES OF WATER

Some of the unique properties of water are predicted from its structure. Water exists in a molecular form; in an aggregate, through hydrogen-bonding; or in a crystalline structure, depending on the temperature and pressure of the aqueous system. The water molecule itself is of the tetrahedron form with oxygen in its center. In the ice form, the whole molecule of water is tied up in a hydrogen-bonded crystalline form. Upon melting of ice, however (from 0°C to 4°C), there is a loss of 15 percent of the hydrogen-bond formation. The partial breakage of the H-bonds results in partial loss of the crystalline structure and in a closer packing of the molecule. Meanwhile, as the temperature goes up, extension of the bond angles takes place, and, consequently, there is a decrease in the water molecule's density due to the greater bond distance. The maximum density, however, is reached at 4°C and then starts declining.

OTHER PROPERTIES OF WATER

Freezing of water may occur at a temperature higher than 0°C if hydrates are formed. A very unique property of water is due to the H-bonding. This even becomes of significant importance in the properties of proteins, carbohydrates, and other polar materials.

As temperature increases, the H-bonding is gradually broken till water starts boiling, a point where most of the H-bonds become broken. Thus, the H-bonding is the reason for the high boiling point of water.

EFFECTS OF WATER ON FOODS

The presence of free water in food, gives it a sensation of juiciness and tenderness as the free water is pressed out when food is being chewed. This free water also contributes to the viscosity of foods. Surface stickiness of certain foods such as hard candies is due to interaction of water with structural components of the food, such as with protein or with carbohydrate.

Water content of cells of fruits and vegetables gives the crispness of those fruits or vegetables.

Turger, or the fullness or rigidity of cells of fruits or vegetables, depends on the water content. If water is lost from cells, turgidity is likewise lost. Tenderness and juiciness of emulsified products are directly related to their moisture and fat contents, while the texture of such products is totally dependent on their water content. There are cases where the water centent of the product is increased by adding tetrasodium pyrophosphates, which increase the product's ability to bind water for better emulsion formation.

REFERENCES

Dick, D. A. 1966. *Structure, properties, movement, and control of cellular water.* Washington, DC: Butterworth, Inc.,

Edsall, J. T., and Wyman, J. 1958. Detailed treatment of water and solutions of electrolytes, including amino acids. In *Biophysical Chemistry*. Vol. 1. New York. Academic Press, Inc.

Eisenberg, D., and Kauzmann, W. 1969. *The structure and properties of water.* Fair Lawn, N.J: Oxford University Press.

Henderson, L. J. 1958. *The fitness of the environment.* New York: The Macmillan Co.

Kavanau, J. L. 1965. Water. In *Structure and function in biological membranes.* P. 170.

Klotz, I. M. 1962. Water. In *Horizons in biochemistry*. New York: Academic Press Inc., P. 523.

Montgomery, R., and Swenson, C. A. 1969. *Quantitave problems in biochemical sciences.* San Francisco: W. H. Freeman Co.

Whipple, H. E. 1965. Forms of water in biological systems. *Ann. N.Y. Acad. Sci.* 125:249.

Wicke, E. 1966. Structure, formation, and molecular mobility in water and aqueous solutions. *Angew. Chem. Intern. Ed.* 5:106.

Chapter 2

LIPIDS

DEFINITION

Lipids are defined as large heterogeneous compounds soluble in nonpolar organic solvents as chloroform, ether, benzene, etc., and insoluble in water.

CLASSIFICATION AND FUNCTIONS

The general function of lipids is their excellent source of calories, as they provide more calories per gram than other foods. Lipids are qualitatively and quantitatively well stored and require no water of hydrolysis, thus representing an enormous potential for a source of energy far superior to glycogen. In adipose tissue, for example, storage of fuel is 90 percent represented by lipids. Meanwhile the complete degradation of palmitic acid to $CO_2 + H_2O$ provides over 900 Kcal, which is an abundance of energy.

Lipids being non-water soluble, play a major structural role in cell membranes, cell walls of plants and bacteria, etc.

In animals, lipids serve as insulators, and in the protection of organs as the liver, spleen, and heart, etc. They also have a major metabolic role, as some vitamins, hormones, cholesterol, and polyunsaturated fatty acids are all lipids.

Saponifiable Lipids

Those that are soap-making when warming up with water and alkali such as fatty acids which are the simplest and most important form.

Complex lipids

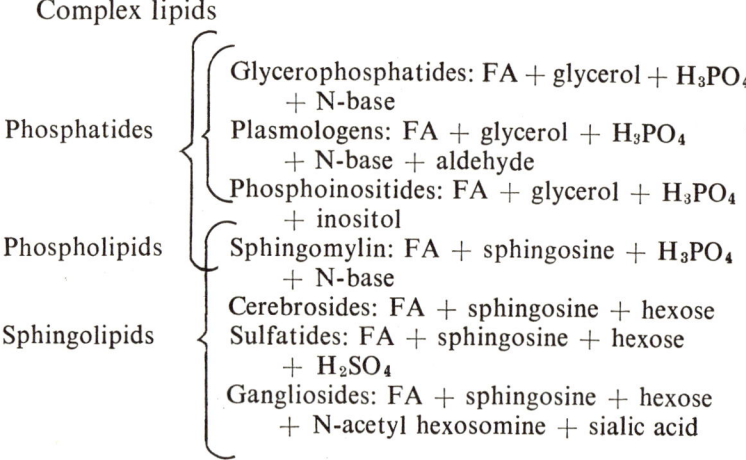

Nonsaponifiable

These are long-chained, usually with methylene in middle, are more common in plants and bacteria, and are less common in animals. Unlike the saponifiable type, these are not saturated alkanes but may be derivatives of an isoprene building block.

$$CH_2 = C - CH = CH_2$$
$$|$$
$$CH_3$$

Steroids

The parent of steroids is cholesterol.

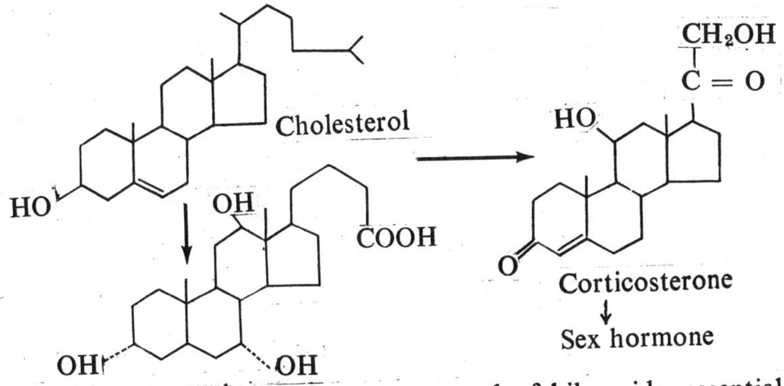

[Structural diagrams of Cholesterol, Cholic acid, and Corticosterone → Sex hormone]

Cholic acid, which is a parent compound of bile acids, essential for absorption of lipids.

FATTY ACIDS

The fatty acids are the common derivatives of lipids. They are straight-chain, saturated, or unsaturated (the slight majority is unsaturated). They are almost always of even number carbon atoms.

Nomenclature of fatty acids

No. of Carbons	Common Name	Systemic Name
2	acetic	(completely soluble in water)
4	butyric	(6% soluble in water)
6	caproic	(Hexanoic, less than 1% soluble in H_2O)
8	caprylic	octanoic
10	capric	decanoic
12	lauric	dodecanoic
14	myristic	tetradecanoic
16	palmitic	hexadecanoic
18	stearic	octadecanoic
20	arachidonic	eicosanoic

| 22 | behenic | docosanoic |
| 24 | lignoceric | tetracosanoic |

Branched chain fatty acids are usually found in microorganisms, the rumen, etc. Such form may have CH_3 group branching upward or downward of the chain:

$$CH_3 - \underset{\underset{CH_3}{|}}{CH} - CH_2 \longrightarrow \text{Iso-form}$$

$$CH_3 - CH_2 - \underset{\underset{CH_3}{|}}{CH} - CH_2 \longrightarrow \text{Anteiso-form}$$

The three most common unsaturated fatty acids are: Oleic (9-octadecanoic) with one double bond at carbon-9; Linoleic (9, 12-octadecadienoic) with two double bonds, at C-9, and 12; Linolenic (9, 12, 15-octadecatrienoic) with three double bonds, at C-9, 12, and 15; and Arachidonic (5, 8, 11, 14-eicosatetranoic) with four double bonds at C-5, 8, 11, and 14. The double bond numbering is from the carboxyl group, and there are two kinds of isomers of the double bonds, trans and cis.

trans

cis

Most fatty acids in nature are cis, but some are trans. Eight percent of fatty acids of cow's milk is trans. Some fatty acids may have an OH group such as cerebronic acid, a 24-C acid with an OH on C-2 and is found in brain glycolipids.

Neutral fats or triglycerides

Triglycerides are composed of glycerol with three fatty acids on each of three OH groups. The three fatty acids may be a mixture or may be the same.

$$\begin{aligned}
&1.\ CH_2-O-\overset{\displaystyle O}{\underset{\displaystyle \|}{C}}-R_1\\
&2.\ CH\ -O-\overset{\displaystyle O}{\underset{\displaystyle \|}{C}}-R_2\\
&3.\ CH_2-O-\overset{\displaystyle O}{\underset{\displaystyle \|}{C}}-R_3
\end{aligned}$$

A Typical Triglyceride

The specific gravity of triglycerides is less than that of water and they are less polar than fatty acids; thus, they are more soluble in fat solvents than fatty acids. Triglycerides are odorless and tasteless. Their composition in adipose tissue varies with species and with the diet. There are usually more unsaturated fatty acids in plants than in animals and vice versa. An animal's triglyceride, such as lard, is solid where at least half the fatty acids are saturated. Castor oil is rich in recinoic acid which possesses an OH at the C-12 position. Most plants and animals have a sizeable portion of the unsaturated fatty acids. Coconut oil has high concentration of the saturated fatty acids, the property which makes it serve as substitute for milk fat.

The most abundant fatty acids of plants and animals are the 16- and 18-carbon fatty acids. Most of the fatty acids of the marine animals, on the other hand, are of the 20- to 22- carbons fatty acids.

Chemical properties

Hydrolysis of the ester linkage with alkali or with acid: If alkali is used, this is saponification where triglycerides $+ H_2O +$ heat $+$ alkali $\rightarrow$ glycerol $+$ three fatty acids' salt of sodium or potassium. The potassium salts are milder and more water soluble than the sodium salts. The saponification number, thus, is the mg of potassium hydroxide or alkali required to hydrolyze a gram of fat. A high molecular weight fat means it has fewer molecules per gram and, thus, has a lower saponification number, and vice versa.

Formation of methyl esters of the fatty acids has been used as a tool for gas liquid chromatographic (GLC) analysis of the fatty acids. Here, methyl alcohol in acid is applied to the fat under heat, resulting in the production of glycerol + fatty acid methyl esters. These methyl esters of fatty acids are then volatilized and detected through GLC.

Either fatty acids or the triglycerides with fatty acids may be subject to catalytic hydrogenation, where saturation of the double bonds is done partially or completely. Nickel is usually used as a catalyst in this process.

Oxidation of fatty acid, for example, with potassium permanganate and alkaline, will break the double bond of the fatty acid giving rise to a carboxyl group on each side of the two fragments produced.

Example: Oleic acid or oxidation of oleic acid where

$$CH_3 - (CH_2)_7 - CH = CH (CH_2)_7 COOH \xrightarrow{\text{Alkaline}}_{\substack{\text{potassium} \\ \text{permanganate}}}$$

$$CH_3 - (CH_2)_7 - COOH + HOOC (CH_2)_7 COOH$$

attack on the double bond results in the production of aldehydes and semialdehydes.

$$\text{Oleic acid} \xrightarrow{O_3} \underset{\text{aldehyde}}{CH_3 - (CH_2)_7 CH = O} + \underset{\text{semialdehyde}}{O = CH_2 (CH_2)_7 COOH}$$

Oxidation of the double bond may tell about its position in the fatty acid chain.

COMPLEX LIPIDS

Phospholipids

Phospholipids are present in all of animal cells, and, while triglycerides are the storage units, phospholipids are the structural units of fat in the body. Phospholipids are an integral part of the membranes of animals, microorganisms, mitochondrion, multienzyme complexes, brain, liver, etc. Phospholipids are more polar than simple lipids, and they are precipitated with acetone. The common unit of phospholipids is phosphatidic acid which is a derivative of L-α-glycerophosphate. The most common phospholipid is lecithin.

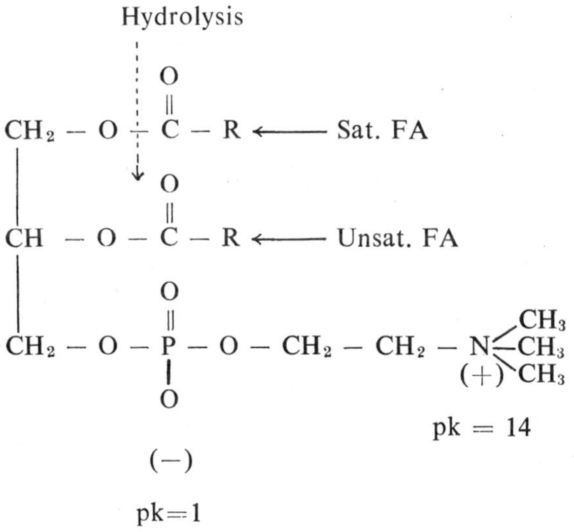

$$\begin{array}{l} \text{Hydrolysis} \\ \\ CH_2 - O \;\vdots\; \overset{O}{\overset{\|}{C}} - R \longleftarrow \text{Sat. FA} \\ \quad\quad\quad\;\; \downarrow \\ CH \;\, - O - \overset{O}{\overset{\|}{C}} - R \longleftarrow \text{Unsat. FA} \\ \\ CH_2 - O - \underset{\underset{(-)}{O}}{\overset{O}{\overset{\|}{P}}} - O - CH_2 - CH_2 - N\overset{CH_3}{\underset{(+)\;CH_3}{\lessgtr CH_3}} \\ \\ \quad\quad\; pk = 1 \quad\quad\quad\quad\quad\quad\quad\quad pk = 14 \end{array}$$

The fact that lecithin is a good detergent or emulsifier, finds it a wide use in the food industry, especially in the stabilization of emulsions. Cephalin is the same molecule, but instead of a choline base it has ethanolamine.

Plasmologen: 10% of the phospholipids of human brain is mainly plasmologen.

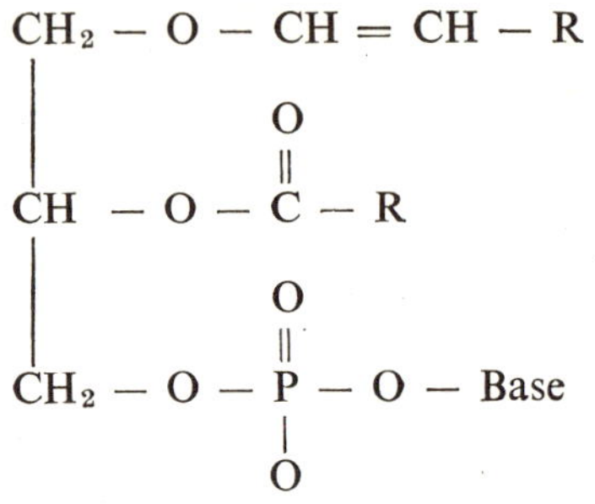

$$\begin{array}{l} CH_2 - O - CH = CH - R \\ | \\ CH - O - \overset{O}{\underset{\|}{C}} - R \\ | \\ CH_2 - O - \overset{O}{\underset{\|}{P}} - O - \text{Base} \\ | \\ O \end{array}$$

A polyglycerophosphatide in lower species is cardiolipin.

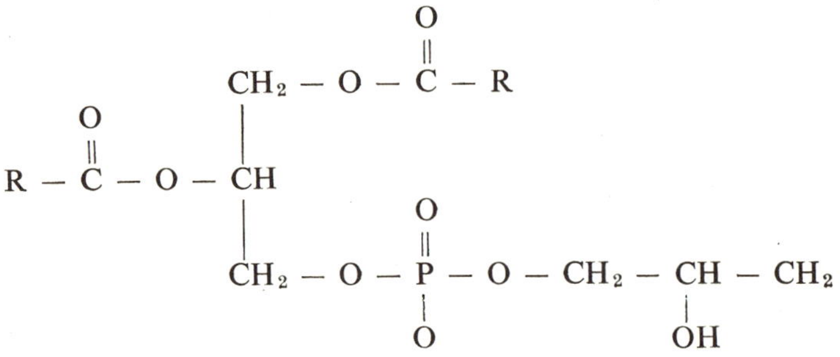

$$\begin{array}{l} CH_2 - O - \overset{O}{\underset{\|}{C}} - R \\ | \\ R - \overset{O}{\underset{\|}{C}} - O - CH \\ | \\ CH_2 - O - \overset{O}{\underset{\|}{P}} - O - CH_2 - \underset{\underset{OH}{|}}{CH} - CH_2 \\ | \\ O \end{array}$$

FAT SOLUBLE VITAMINS

These are vitamins A, D, E, and K. They are soluble in lipid solvents, and their role is less well defined than the B vitamins. They are, however, stored more efficiently than the water-soluble vitamins.

Vitamin A

β-carotene is the major input in the diet for vitamin A. β-carotene is oxidized or split in the middle by the enzyme β-carotene dioxygenase in the intestinal mucosa into trans-retinal. This, in turn, is converted to the alcohol form through the action of a dehydrogenase enzyme. The alcohol form is esterified, and then is hydolyzed in the liver into the form of protein-bound transretinol. This form acts on tissue, as shown in following Figure 1.

Vitamin A_1 with the trans-double bond is the key vitamin in vision. It is not found in the provitamin form in abundancy.

Functions of Vitamin A

In sulfation of chondriotin-SO_4 which preserves the integrity of epithelial tissue.

The major role of vitamin A is in the visual cycle, where its influence is exerted on the retinal rods and also on the cones, as shown in the following Figure 2.

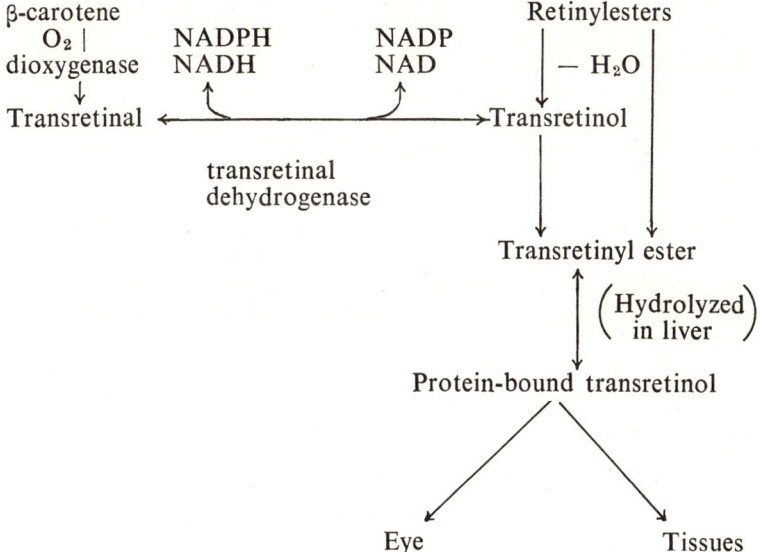

Figure 1

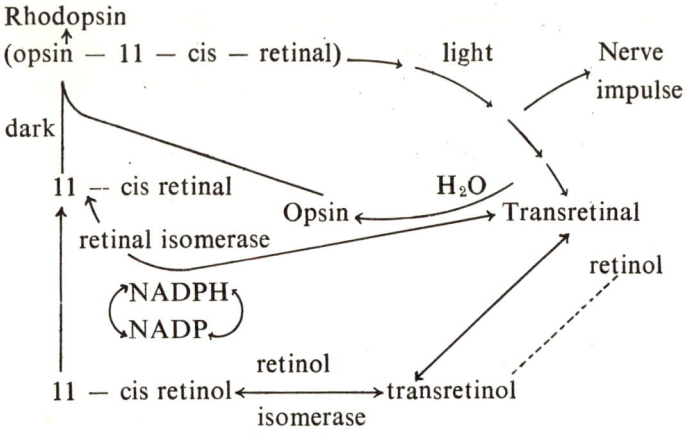

Figure 2

Nonsaponifiable material

Alcohols, sterols, hydrocarbons, fat-soluble vitamins, and pigments are all nonsaponifiable material.

Vitamin D

Vitamin D is made from the provitamin form, the 7-dehydrocholesterol, as ring B opens up with U.V. light.

Plant sterols (phytosterols)

Sitosterol, is like cholesterol in structure, except it has C_2H_5 group at C-24, but, unlike cholesterol, sitosterol is not absorbed in the intestines.

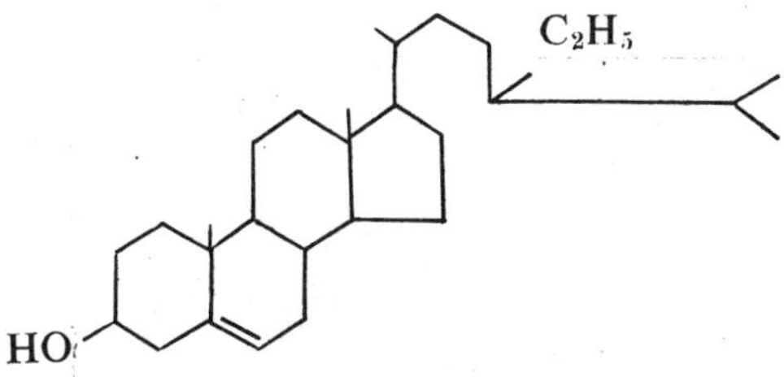

Stigmasterol is a 24-carbon plant sterol found in high concentration in soybean oil.

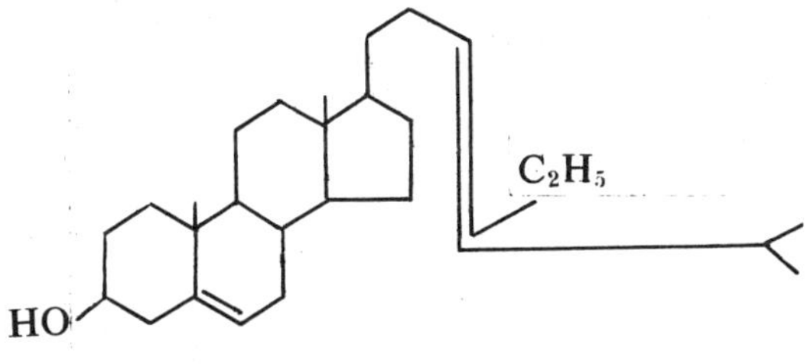

The analog of vitamin D_3 is ergostrol, vitamin D_2, found in yeast.

Vitamin D's major function is in calcium and phosphate metabolism, as it helps in the active uptake of Ca^+ and promotes the reabsorption of PO_4 in the renal tubule.

Vitamin E

Vitamin E is a true vitamin, as far as experimental evidence shows, only in animals. Its function in man is unknown, other than its antioxidant role. This role of vitamin E stems from its characteristic of being easily oxidized. Thus, vitamin E, especially in vegetable fat, serves as a natural antioxidant, where it spares other vitamins, such as vitamin A, and prevents the toxicity of high doses of unsaturated fatty acids.

Vitamin E serves a reproduction function in rodents and lower animals, but this is not known to occur in man. Deficiency of vitamin E causes muscular dystrophy in animals, but not in man.

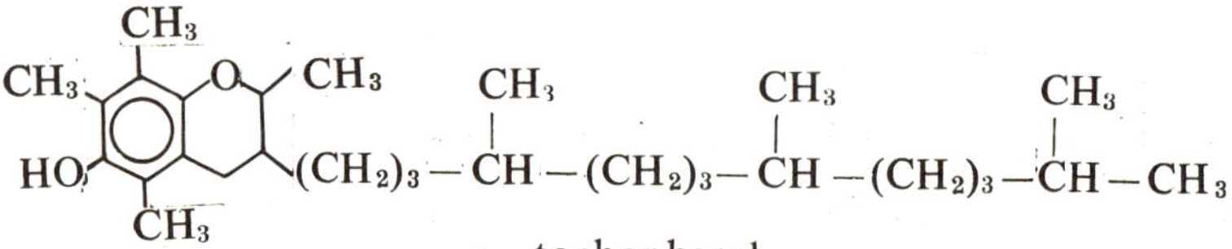

$$\text{HO}\underset{\text{CH}_3}{\overset{\text{CH}_3}{\bigodot}}\underset{\text{CH}_3}{\overset{\text{O}}{}}\text{CH}_3\ (\text{CH}_2)_3-\underset{|}{\overset{\text{CH}_3}{\text{CH}}}-(\text{CH}_2)_3-\underset{|}{\overset{\text{CH}_3}{\text{CH}}}-(\text{CH}_2)_3-\underset{|}{\overset{\text{CH}_3}{\text{CH}}}-\text{CH}_3$$

α—tochopherol

Lettuce, wheat germ, corn oil, and soybean oil are rich sources of vitamin E.

If cows are fed vitamin E, it will prevent oxidation of their milk and production of off-flavor.

Vitamin K

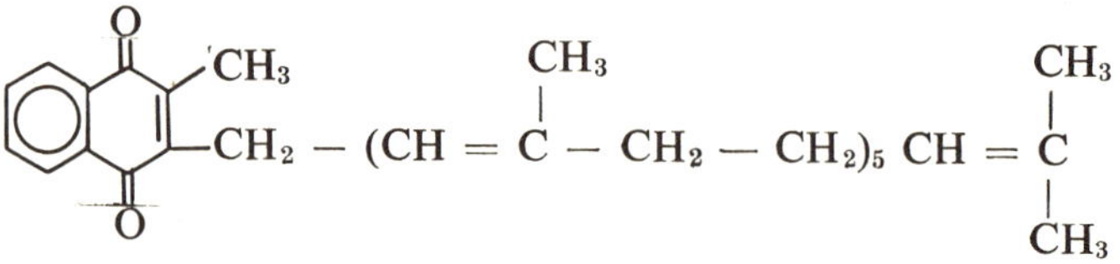

$$\text{[naphthoquinone]}-\text{CH}_2-(\text{CH}=\underset{|}{\overset{\text{CH}_3}{\text{C}}}-\text{CH}_2-\text{CH}_2)_5\ \text{CH}=\underset{|}{\overset{\text{CH}_3}{\underset{\text{CH}_3}{\text{C}}}}$$

K_2—antihemorrhagic

Vitamin K is necessary for blood clotting because of its antihemorrhagic function.

Vitamin K is found in spinach, cabbage, cauliflower, and chestnuts.

DIGESTION AND ABSORPTION OF LIPIDS

Uptake of lipids, mainly in the form of triglycerides, may range from 30-50 grams per day in man. There is no enzymatic action on lipid at the mouth level, and the first lipolytic action occurs at the stomach, where gastric lipase partly produces fatty acids and glycerol from triglycerides. Mostly, however, it is the pancreatic lipases and bile secretion from the liver that cause the digestion of lipids in the small intestines.

Bile consists mainly of bile salts, where the bile acids are conjugated with tyrosine or choline producing, respectively, taurocholic and glycocholic acids. These are excellent anionic detergents that contain minor amounts of cholesterol and phospholipids.

When the triglycerides and other lipids reach the small intestines, they usually are large, 500-1000 mu spherical diameter. This spherical particle has a charge on the outside and is hydrophobic in the inside. The charge units on the outside cause the interaction of that part with water. The triglyceride is coated with specific protein and partly emulsified with phospholipids, forming the chilomicron which goes to the lymph and is spilled into the blood. The chilomicron is 90 percent lipids and 0.2 to 0.3 percent protein in noncovalent linkage, is fairly water soluble, and is carried into plasma. It is about half hydrophillic (on the outside) and half hydrophobic (on the inside). The chilomicron, thus, may get to tissues, such as liver and adipose tissues. Thus, the function of chilomicron formation is to bring the triglycerides into tissues. The triglycerides are then acted upon in the liver by the lipase enzyme, which causes their hydrolysis to glycerol + three fatty acids (FA). These fatty acids are taken apart, two carbons at a time, inside the mitochondria, where they are oxidatively degraded for the production of energy (β-oxidation).

The fatty acid is activated with a thiokinase enzyme into the acyl-coenzyme form, which with the aid of carnitine can penetrate the mitochondria readily. The carnitine inside the mitochondria splits away from the acyl-carnitine, leaving the isolated fatty acid inside and the carnitine moving to the outside of the mitochondria.

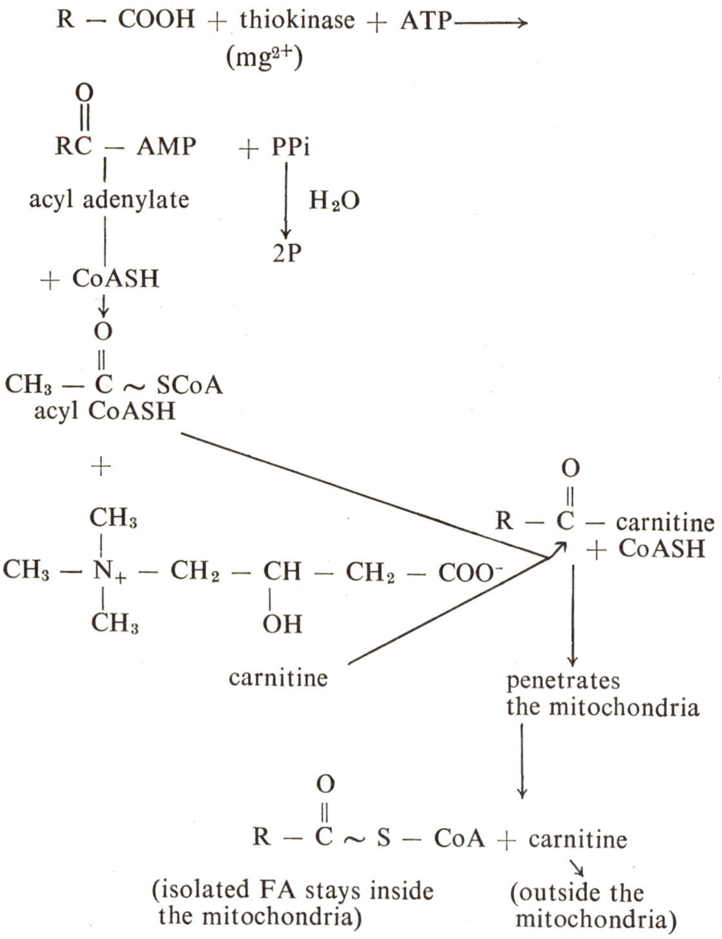

$$R - COOH + \text{thiokinase} + ATP \xrightarrow{(mg^{2+})}$$

$$\underset{\text{acyl adenylate}}{\overset{O}{\underset{\|}{RC}} - AMP} + PPi \xrightarrow{H_2O} 2P$$

$$+ \text{CoASH} \downarrow$$

$$\underset{\text{acyl CoASH}}{CH_3 - \overset{O}{\underset{\|}{C}} \sim SCoA}$$

$$+$$

$$\underset{\underset{CH_3}{|}}{\overset{CH_3}{|}} CH_3 - N_+ - CH_2 - \underset{\underset{OH}{|}}{CH} - CH_2 - COO^-$$

carnitine

$$R - \overset{O}{\underset{\|}{C}} - \text{carnitine} + \text{CoASH}$$

↓

penetrates the mitochondria

↓

$$R - \overset{O}{\underset{\|}{C}} \sim S - CoA + \text{carnitine}$$

(isolated FA stays inside the mitochondria)   (outside the mitochondria)

The CoA form of the fatty acid inside the mitochondria undergoes β-oxidation as follows:

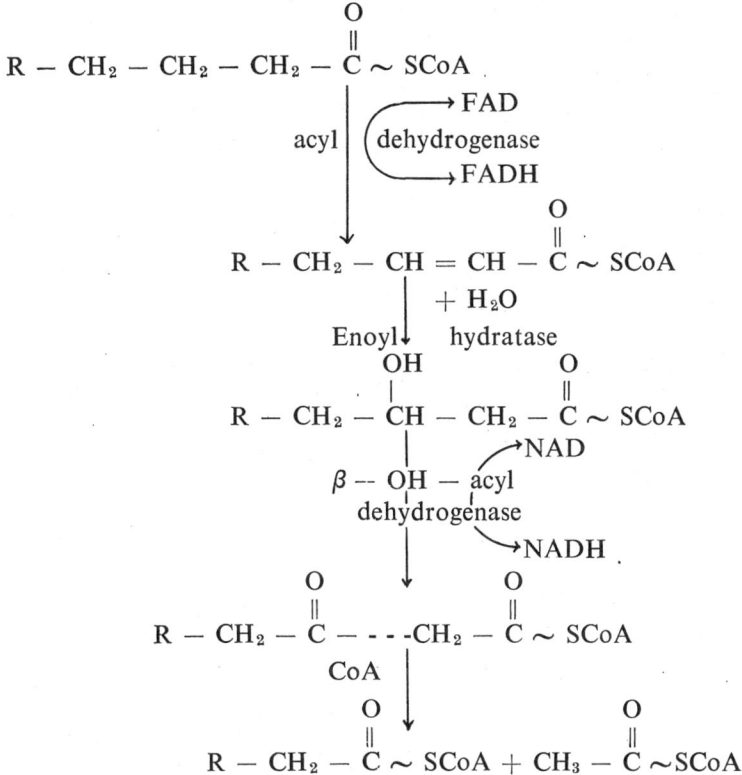

This occurs only with the even-numbered carbon fatty acids, ending with acetyl-CoA.

PHYSICAL PROPERTIES OF FATS AND OILS

Crystalization

Fats may crystallize in different forms under various conditions, thus exhibiting polymorphism as crystalization occurs at different melting points and according to the history of the specific treatments.

Based on their solubility variations, fats may be separated by applying polar solvents which will dissolve the unsaturated fatty acids' salts. A mixture of two immiscible solvents will partition the salts of the unsaturated FA into the more polar solvent, and the salts of the more saturated FA into the less polar solvent.

The Reichert Meissl number determines the amount of water-soluble volatile FA. All of C-4 and part of C-6 are water soluble, steam distillable. As milk fat is the only fat that contains 10% butyric acid, if a foreign fat is introduced to milk fat, this will lower the Reichert Meissl number. The Reichert Meissl number is, thus, the number of ml of 0.1 N alkali necessary to neutralize the steam-distillable fatty acids of 5.0 grams of fat.

The iodine number is a measure of the degree of unsaturation of a fat sample. Highly unsaturated vegetable fats have higher iodine numbers than the more saturated fats.

Characterization of fats now depends more on gas-liquid chromatography.

Hydrolysis of fats and relation to foods

Hydrolysis through the action of the enzyme lipase and water causes the hydrolytic cleavage of fat and the production of fatty acids that become objectionable flavor-wise. This type of lipolysis is called hydrolytic rancidity.

$$C_3H_5(OOCR)_3 \xrightarrow[H_2O]{Lipase} C_3H_5(OH)_3 + 3\ RCOOH$$

Oxidative rancidity, meanwhile, gives rise to carbonyls.

Double bond rearrangement

Most natural fats have a cis configuration and the first double bond between C-9 and C-10, second double bond between C-12 and C-13, and third between C-15 and C-16. A shift of double bonds in either direction may occur, resulting in a cis-cis 9, 11 FA; or cis-cis 12, 10 FA; or cis-trans 9, 11; or cis 12-trans 10. The switch from cis to trans is feasible.

Pyrrolysis

When fat is heated to high temperatures, it breaks into fragments. Acrolein is produced when fats are heated to high temperatures. As one molecule of water splits from glycerol, the dehydration of glycerol thus produces acrolein, which gives the heated fat a strongly pungent odor.

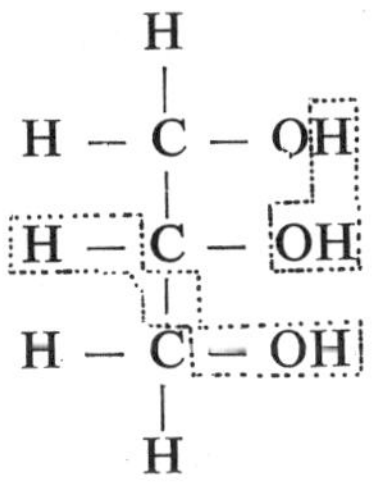

$$\begin{array}{c} H \\ | \\ H - C - OH \\ | \\ H - C - OH \\ | \\ H - C = OH \\ | \\ H \end{array}$$

One way of distinguishing between saturated and unsaturated fats is by making lead salt or urea aducts of the two fats and then partitioning them between polar and non-polar solvents. The urea aducts of the unsaturated fat salt will be more soluble in the polar solvent.

Hydrogenation

With a suitable catalyst, like Ni, hydrogen gas may be added to the double bonds of the unsaturated fat. Partial hydrogenation of the fat may give rise to isometric and geometric forms such as iso-oleic acid. During the hydrogenation process, however, the fatty acids with the largest number of double bonds are selectively hydrogenated first. The selectivity in hydrogenation becomes more remarkable when manipulating with the temperature and pressure. At higher temperatures, of course, the rate of the reaction is increased and more of the unnatural or the iso-forms, etc., are produced.

Most natural fats have saturated and unsaturated fatty acids. Plants have an abundance of the unsaturated fatty acids. Oleic acid composes 34% of edible fats and oils, linoleic 29%, and palmitic 11%. Although in nature most fatty acids exist in the cis-form, trans-fatty acids are, to a good extent, found in the rumen.

Terpenoids

Terpenoids are made up of isoprene units; examples are carotene, steroids, and rubber. Squalene is a high-chain hydrocarbon $C_{30}H_{50}$, and is the precursor of cholesterol. It is found in milk, fish liver oil, and rice bran.

$$(CH_5)-\underset{\underset{CH_3}{|}}{C}=CH-(CH_2)_2-\underset{\underset{CH_3}{|}}{C}=CH-(CH_2)_2-\underset{\underset{CH_3}{|}}{C}-----$$
$$-----=CH-(CH_2)_2$$

β-carotene:

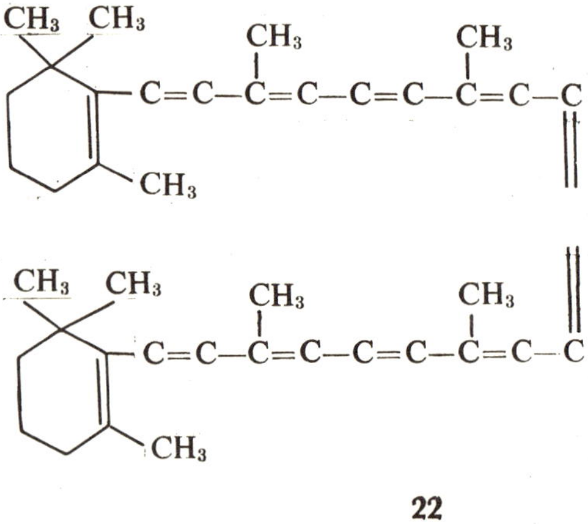

[Structure showing cyclohexene ring with two CH₃ substituents and CH₃ side group, connected to C=C–C(CH₃)=C–C=C–C(CH₃)=C–C chain]

[Second structure showing cyclohexene ring with two CH₃ substituents and CH₃ side group, connected to C=C–C(CH₃)=C–C=C–C(CH₃)=C–C chain]

Hydrolysis of β-carotene produces two moles of Vitamin A. Most land animals do not have an appreciable quantity of Vitamin A or D stored. Several fats, however, contain β-carotene, i.e., the provitamin form of A. Such fats are palm oil, olive oil, and butter. These carotenoids also are the main color pigments of fats and oils. The animal's carotenoids are assumed to have come from plant origin after being consumed by the animal and deposited in his body lipid.

There is thought to be a relationship between carotenoids and photosynthesis, as they are found in cells alongside with chlorophyll. There are the α, β, and γ carotenes which give rise to varying vitamin A activity.

The β-carotene on hydrolysis gives two moles of Vitamin A activity, the α-carotene gives one mole of vitamin A activity, while γ-carotene does not have vitamin A activity.

Hydrolysis of lycopene, which is the red colored pigment in tomatoes, watermelons, and paprika, gives no vitamin A activity.

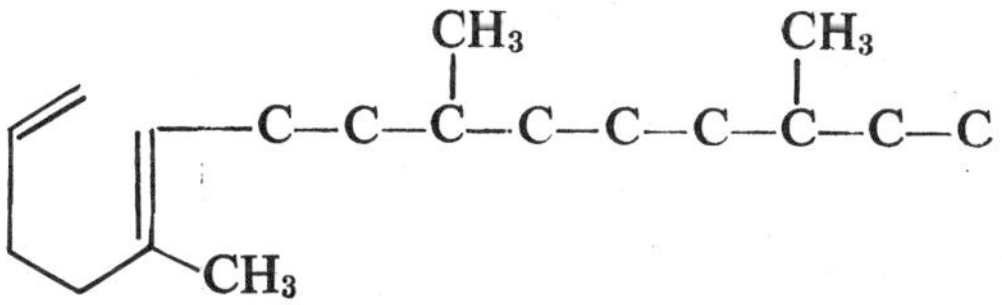

Xanthophylls contain OH groups and upon hydrolysis give one mole of vitamin A activity.

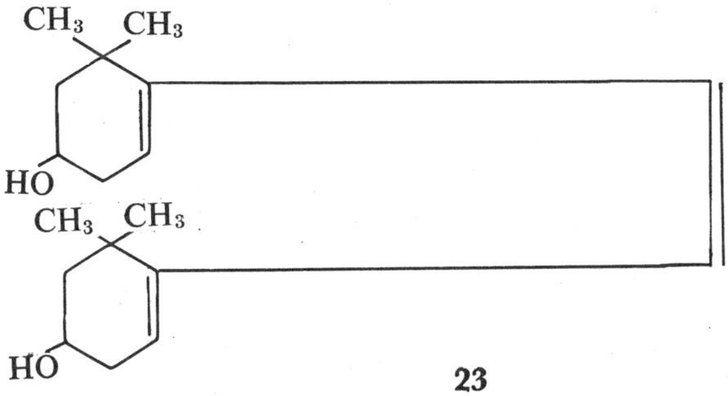

PROCESSING OF EDIBLE OILS AND FATS

Some products require that extraneous materials be removed before extraction of the fat. In the case of cereals, germ cells should be separated before pressing. For the best way of making oil or fat available, a combination of heating and pressing procedures may become necessary for extraction of the fat. Solvent extraction of fat is widely used for its separation. Common solvents usually are petroleum ether, chloroform, benzene, carbon disulfide, and others. Solvent extraction at times may prove expensive to use unless some method of recovering the solvent is applied. In some tissues where only small amounts of residual fat is expected, solvent extraction becomes about the only method to successfully extract the minute amount of fat present.

REFINING OF OILS

The first step should be the removal of cellular material and derivatives of protein and carbohydrate, usually through centrifugation. Undesirable free fatty acids are removed by steam or alkali treatment. This may also remove undesirable resins. The steam refining is usually done by blowing steam into heated oil, usually under vacuum (this treatment removes C-14 from the fat). Steam refining, then, may be followed by alkali refining. When the hot oil is treated with hot alkali, soap is formed and is removed by washing with hot water, and centrifugation is applied till the oil is free from alkali.

Pigments are removed through their adsorption to Fuller's Earth or by addition of a small amount of charcoal, a process that is called "bleaching," where oil is heated to between 105 and 115°C, agitated, and passed through the filtering bed where there is the Fuller's Earth. If high temperature is used, a vacuum should be applied to prevent any rancidity that might be caused by the high temperature. Oxygen should, of course, be eliminated from the oil. Chemical bleaching is used for industrial but not for edible oils.

Odor is removed from fat by heating at 220 to 250°C under vacuum of 6 mmHg. After volatile odors are removed, oils are usually very bland.

Many salad oils, such as cottonseed oil, are winterized, where the oil is cooled at various low temperatures, and each time the clear oil is removed by filtration. This process is important for salad oils as it helps maintain them in a clear state at refrigeration temperature. 0.05% of soybean lecithin may be used for clarification of oils.

Sometimes monoglycerides and antioxidants are added to oils. Air may be whipped into a product to give it a white, crystal appearance.

Butter is a water-in-oil emulsion that contains 80% fat. Cream is kept at 50°F for forty minutes to break the emulsion, after which the fat is continually worked to get water worked into the butter. The composition of butter is 2.5% salt, 1.5% curd, 16% water, and usually starter distillate that contains diacetyl and volatile fatty acids is added to add flavor to butter.

Beef oils, two of which are made from the internal fat tissue of beef: After rendering at low temperature, the beef tissue is held at 90°F for several days. The liquid portion is the "oleo oil," while the solid, crystalline portion is the "oleo stearin." The oleo oil is used in the manufacturing of some margarines, and may get modified into a shortening. Oleo stearin is used for compound-shortening making.

Lard is processed from hog fat. Lard made from kidney fat is graded as the best in quality, back-fat lard is intermediate in quality, while the lowest quality lard is made from intestinal fat. Lard may be wet or dry rendered. In the U.S., it is usually by steam-rendering. A disadvantage that lard has is a physical characteristic where it develops large crystals, or graining. This renders it difficult to cream in batter and doughs, as it gives it a narrow melting point. It is possible continuously to interesterify lard by metering in a measureable amounts of catalysts which will induce trisaturated instead of disaturated fatty acid formation. When esterification is completed, of course, soaps are formed, but are removable through washing and centrifugation. The trisaturation will convert the otherwise liquid disaturate into the solid form through the process of interesterification that adds another saturated fatty acid to the other two saturated fatty acids. The trisaturate form crystallizes out and has a wider range of melting points which, consequently, gives the fat more plasticity. The process of esterification may be stopped through poisoning the Na^+ or K^+ catalyst by adding H_2O or CO_2.

All salad oils in the U.S. are highly refined oils except olive oil, which does not undergo deodorization. Cottonseed oil must be winterized, while soybean oil may or may not undergo winterization. Corn oil should be winterized as well, or lecithin may instead be added to it for clarity. Soybean oil possesses the problem of undergoing reversion, which is a mild type of oxidation, while marine oils may have fishy aromas which may be stopped by deodorization, but usually will return on prolonged storage. The first occurring fishy odor is due, however, to the development of free amines, while the second is due to carbonyls which develop with autoxidation of the fat. Hydrogenated vegetable oils have become superior to lard as they possess better plasticity, better stability, are more bland in flavor, of better emulsification, and cream better than lard. Oils used in hydrogenated shortenings are of the same salad oils such as cottonseed, peanut, or soybean oils. For making special hydrogenated shortenings that must have long shelf lives, elimination of linolenic acid (the source of oxidative deterioration) becomes necessary.

To make batter incorporate more water and more sugar, thus cakes may become sweeter, superglycerated shortening was developed where mono- and di-glycerides are incorporated as emulsifying agents in the batter. These, in turn, will induce increased incorporation of water and sugar and by changing the sugar rate of incorporation to flour from a 1 : 1 into a 1¼ : 1 ratio.

AUTOXIDATION OF UNSATURATED FATTY ACIDS

For example, a pentadiene system, i.e., unsaturated fatty acid with two double bonds has a very labile H on the adjacent methylene group which is very easily lost as a free radical. The mechanism of autoxidation is, thus, a free radical formation where a free H radical is lost. If O_2 is present, it will be picked up in our example at one of any of three points.

(Linoleic acid)

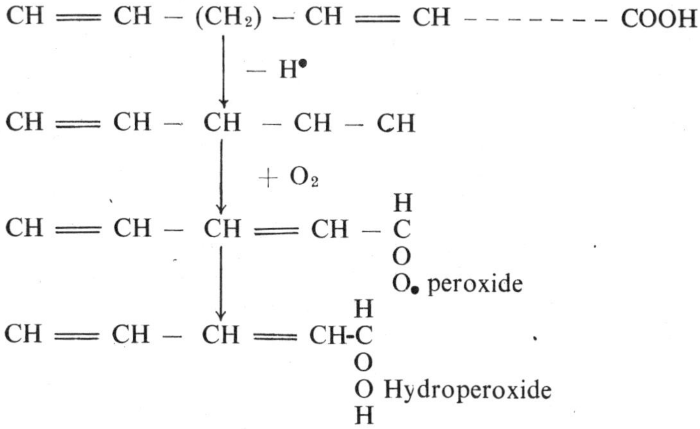

$$CH = CH - (CH_2) - CH = CH -------- COOH$$

$$\downarrow -H^{\bullet}$$

$$CH = CH - CH - CH - CH$$

$$\downarrow + O_2$$

$$CH = CH - CH = CH - \overset{H}{\underset{\underset{O_{\bullet}}{O}}{C}} \quad \text{peroxide}$$

$$\downarrow$$

$$CH = CH - CH = CH\text{-}\overset{H}{\underset{\underset{H}{\overset{O}{\underset{O}{|}}}}{C}} \quad \text{Hydroperoxide}$$

This hydroperoxide may undergo dismutation.

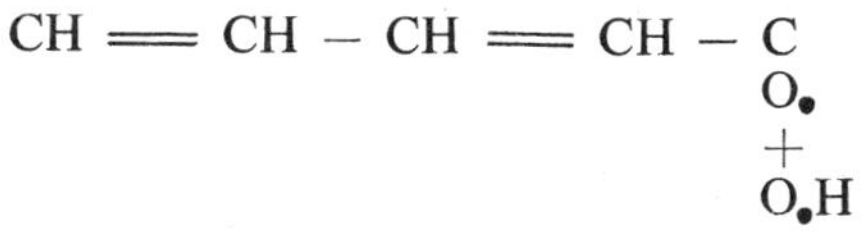

$$CH = CH - CH = CH - \underset{\underset{O_{\bullet}H}{+}}{\overset{C}{O_{\bullet}}}$$

Dismutation takes place as the first reaction after peroxides start breaking down. The free radical formed at the peroxide level may react with a H or abstract a H from the fatty acid chain, forming another free radical.

A hydrogen free radical may react with an oxygen free radical, forming an alcohol.

$$RC-H-R_1 \xrightarrow{\text{Dismutation}} \text{aldehyde} + R^\bullet$$
$$\underset{O^\bullet}{|}$$

or
$$R-\underset{\underset{O^\bullet}{|}}{C} + RH^\bullet \longrightarrow R-\underset{\underset{\underset{H}{|}}{O}}{C} + R^\bullet$$

or ... $R^\bullet \longrightarrow R - \underset{\underset{O}{\|}}{C} - R_1 + RH$

or ... $RO^\bullet \longrightarrow R - \overset{\overset{O}{\|}}{C} + ROH$

Once peroxides are formed, there is subsequently a shift in the double bond's position. This spontaneous oxidation is not limited to just foods, but it exists in any other natural system. Products of this oxidation reaction may be the following: lipoperoxides, aldehydes, acids, keto aldehydes, epoxy compounds, and polymers. The free radicals, upon joining together, form polymers.

Secondary reactions, such as oxidation of substrate materials such as pigments and vitamins, with the peroxides formed, results in the subsequent loss of color or vitamins, or the destruction of aroma.

A system low in neutral fat but rich in phospholipids is very susceptible to oxidation due to the comparatively high level of unsaturated fatty acids present in phospholipids. This may be true of such foods as milk and meat.

Oxidation of lipids becomes a keen problem in causing defects in particular foods, like stored marine products, meat, vegetable oils, dehydrated foods, and dairy products. Oxidation of the lipid part of a food product often determines its shelf life.

METHODS OF DETERMINING LIPID OXIDATION IN FOODS

Peroxide test

Where treating fat with potassium iodide and where the peroxides are liberated and the iodine is titrated with sodium thiosulfate,

$$R - OOH + KI \longrightarrow I_2 \text{ (titrated with } Na_2 S_2 O_3).$$

In this test, the fat product is exposed to mild heating (37 to 70°C), and air is bubbled through it for a certain period of time.

Thiobarbituric Acid (TBA) test

This test is a reaction of two TBA molecules with malonaldehyde producing a yellow-reddish color which is measured spectrophotometrically at a wavelength of 532 mu. This reaction is produced in an acid media at 60°C. The TBA test is widely used in determining the early stages of oxidation in fat. Here, the origin of the malonaldehyde is the pentadiene system that arises from the polyunsaturated fatty acid.

The TBA test can be directly applied to foods. In milk, malonaldehyde in the trichloroacetic acid (TCA) filtrate reacts with TBA for sixty minutes, giving reddish-yellow color, the intensity of which is an indication of milk-fat oxidation.

$$\text{S}=\text{C}\begin{smallmatrix}\text{NH}-\text{C}\\\text{N}-\text{C}\\\text{H}\end{smallmatrix}\begin{smallmatrix}\text{O}\\\text{CH}_2\\\text{O}\end{smallmatrix} \;+\; \begin{smallmatrix}\text{O}\\\text{C}-\text{CH}-\text{C}\\\text{H}\quad\quad\text{H}\end{smallmatrix} \longrightarrow$$

TBA   malonaldehyde

$$\text{S}=\text{C}\begin{smallmatrix}\text{H}\\\text{N}-\text{C}\\\text{N}-\text{C}\\\text{H}\end{smallmatrix}\begin{smallmatrix}\text{O}\\\\\text{O}\end{smallmatrix}=\text{CH}-\text{CH}=\text{CH}-\begin{smallmatrix}\text{O}\quad\text{H}\\\text{C}-\text{N}\\\text{C}-\text{N}\\\text{O}\quad\text{H}\end{smallmatrix}\text{C}=\text{S}$$

(reddish-yellow color)

Distillation at a reduced pressure is usually applied in recovery of the volatiles of fat. At least some of these volatiles do produce an off-flavor when fat is oxidized.

Carbonyls

An important volatile component of oxidized fat is the carbonyls. Of these carbonyls, the volatile unsaturated ones contribute mostly to the off-flavor produced. The carbonyls generally produced in an oxidized fat may take several forms, such as alkanals, alkanones, alkenals, alkenones, or alkdienans.

FACTORS THAT CAUSE OXIDATION OF FAT	FACTORS THAT INHIBIT OXIDATION OF FAT
High temperature	Refrigeration
Light; U.V. and visible	Opaque packaging
Ionizing radiation (α, β, γ, x-rays)	Exclusion of oxygen (nitrogen packing)
Peroxides (from partially oxidized fat)	Blanching (to destroy lipoxidase enzymes)
Lipoxidase enzymes	Metal deactivation (chelation of metals)
Trace metals (Cu, Fe, etc.) Organic ions	Use of antioxidants

POSSIBLE WAYS OF REDUCING FAT OXIDATION

By selectively hydrogenating the fat. This will reduce the double bonds which contribute to the oxidation.

By giving animals certain feeds, for example, pigs consuming barley produce fat that is relatively more resistant to oxidation than those consuming alfalfa. Feeding whiting to fish trouts produces more resistance to oxidation than when feeding herring.

Low moisture level in foods usually provides more protection against oxidation than high levels.

Antioxidants

Antioxidants may exert their effect in various ways such as:

a) Chain breaking or breaking of the free radical chain mechanism;
b) Deactivation of the metal catalysts;
c) Use of antioxidants, such as butylated hydroxyanisole (BHA), butylated hydroxy toluene (BHT), hydroxyquinone, galic acid, or N-propylgalate, which are all phenolic types of compounds.

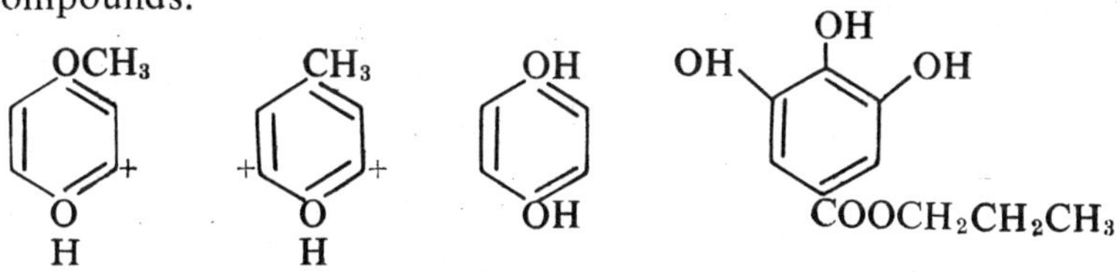

(BHA)   (BHT)   Hydroxyquinone   N-propylgalate

Mechanism of action of antioxidants:

The first step of action of antioxidant may be its giving up a H free radical which may then break the free radical chain mechanism of the fat.

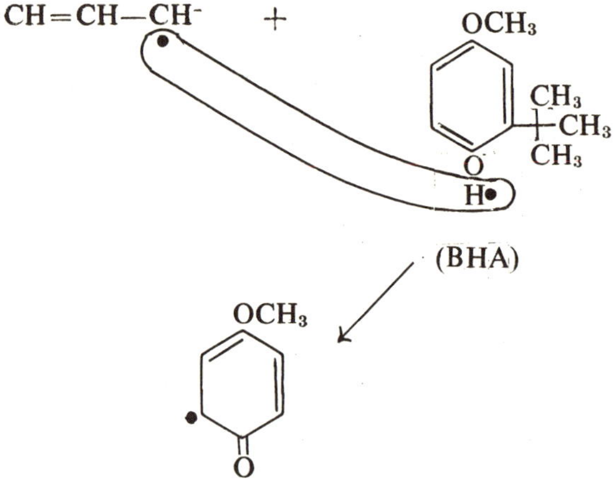

$$CH=CH-\underline{C}H^- \quad + \quad \text{(BHA structure with OCH}_3\text{, C(CH}_3\text{)}_3\text{, OH)}$$

$$\downarrow$$

(BHA)

$$\text{(OCH}_3\text{ cyclohexadienone radical)}$$

In this case, the free radical from the antioxidant reacts with the free radical of the fat, thus neutralizing it and breaking the oxidation chain mechanism. Vitamin E is a common natural antioxidant in foods. It is widely distributed in the plant kingdom, and becomes a major antioxidant in animals, as they derive it from plants. Vitamin E helps stabilize fats against oxidation. Vitamin E being a relatively weak antioxidant, its level needed for protection of foods may bring about off-flavors, providing the uneconomical aspect of its use as a commercial antioxidant. Nevertheless, natural vitamin E presence in foods renders it as an important antioxidant in this particular respect.

Synergists

In the absence of a primary antioxidant, a synergist is of no effect, but in presence of the antioxidant, it will enhance its action. Many synergists normally contain COOH or OH groups such as citric acid and phospholipids which are common synergists. The exact mechanism of action of synergists is not known.

SOURCES OF OXIDATION

The compound 2-4 decadienal produces the deep-fat-frying-oxidized flavor. In early stages in the oxidation of lard, this contributes a sweet flavor. As oxidation proceeds, lard acquires a pungent flavor. In milk, the phospholipid content is the source of the oxidized flavor. The phospholipid content of milk makes it very susceptible to oxidation, as their polyunsaturated fatty acids with their pentadiene system are very susceptible to oxidation. Like the early stages of milk oxidation, the reversion flavor in soybean oil is difficult to prevent through use of antioxidants. Linolenic acid content of soybean oil is thought to be responsible for its reversion flavor, where oxidation at the double bonds of the acid gives rise to the-enal type of carbonyls. Enals usually cause a cardboard flavor, in contrast to the tallowy flavor produced by oxidation of butter. In the case of butter, after an initial phospholipid oxidation of the monoenoic system, a later and typical hydrolytic type of oxidation occurs. After the peroxides have formed, in this case, the actual rancidity occurs when oxidation of the mono-unsaturated fatty acid has taken place. On the other hand, reversion occurring in the highly unsaturated FA takes place at an early stage. Cu^+, sunlight, and lipoxidase enzymes are catalysts which enhance the onset of oxidation in milk, especially lipoxidase, which is present in many animal and plant tissues and which causes oxidation of foods sometimes even at freezing storage temperature. Milk's being more susceptible to oxidation at 4°C than at 20°C may be attributed to the enzymes' presence and characteristics. Heat treatment of milk may protect it against oxidation as the sulfhydryl groups caused by the heat effect on the lactoglobulin of the S-containing amino acids may act as antioxidant. Light may induce free radical formation and production of off-flavor in milk and other food products. When milk is exposed to light, two distinct off-flavors are produced. A "burnt" flavor is produced by exposure to sunlight, which is due mainly to reaction on methionine by B_2 vitamin and the subsequent production of methional. A "cardboard" flavor is produced when lipids of milk get oxidized. Copper and iron are catalysts for the development of the oxidized flavor in the following order:

$$Cu^{++} > Fe^{++} > Fe^{+++}$$

There is thought to be a close association between the ionization of copper and the destruction of ascorbic acid in milk. Inhibition or retardation of the ionization of copper will reduce the loss of vitamin C and, subsequently, retard oxidation.

LIPID METABOLISM

Unlike carbohydrates and proteins, fat is highly reduced and its lack of water renders it an excellent energy storage source. When triglycerides from a fatty meal reach the small intestines, they are in large, 500–1000-mu, spherical particles. Meanwhile, the bile salts which contain some cholesterol and phospholipids act as emulsifying agents for these neutral lipids in the intestines. The enzyme lipase hydrolyzes the triglycerides into β-monoglycerides, fatty acids (FA), fat-soluble vitamins, and cholesterol. The β-monoglyceride is synthesized into depot-fat (triglycerides); the FA are activated with Co-ASH and ATP, and undergo either β-oxidation or fat synthesis.

β-Oxidation of Fatty Acids (FA)

This is the process by which fatty acids are oxidized or taken apart, two carbons at a time, inside the mitochondria.

The first step in this process is to get the FA inside the mitochondria. The enzyme thiokinase and ATP activate the FA into the acyl-adenylate form which, upon condensation with Co-A-SH, is converted to the acyl-Co-ASH form that penetrates the mitochondria through the aid of carnitine. Acyl-carnitine crosses the mitochondrial membrane leaving the isolated FA inside the mitochondria to undergo β-oxidation.

$$CH_3 - (CH_2 - CH_2)_{N-1} - COOH + ATP + Co\text{-}A \xrightarrow{\text{thiokinase}}$$

fatty acid

$$CH_3(CH_2 - CH_2)_{N-1} - \overset{O}{\underset{\|}{C}} - S - CoA + AMP + PPi$$

acyl-Co-A

$$CH_3 - (CH_2 - CH_2)_{N-1} - \overset{O}{\underset{\|}{C}} - S - CoA + FAD \xrightarrow{\text{acyl-Co-A dehydrogenase}}$$

acyl-Co-A

$$CH_3 - (CH_2 - CH_2)_{N-1} - \overset{H}{\underset{}{C}} = \overset{}{\underset{H}{C}} \text{-} \overset{O}{\underset{\|}{C}}\text{-S-CoA} + FADH_2$$

$+ H_2O$

Enoyl
Hydrolase
↓

$$CH_3(CH_2 - CH_2)_{N-1} - \overset{OH}{\underset{H}{\overset{|}{C}}}\text{-}\overset{}{\underset{H}{\overset{|}{CH}}}\text{-}\overset{O}{\underset{\|}{C}} - S - Co\text{-}A$$

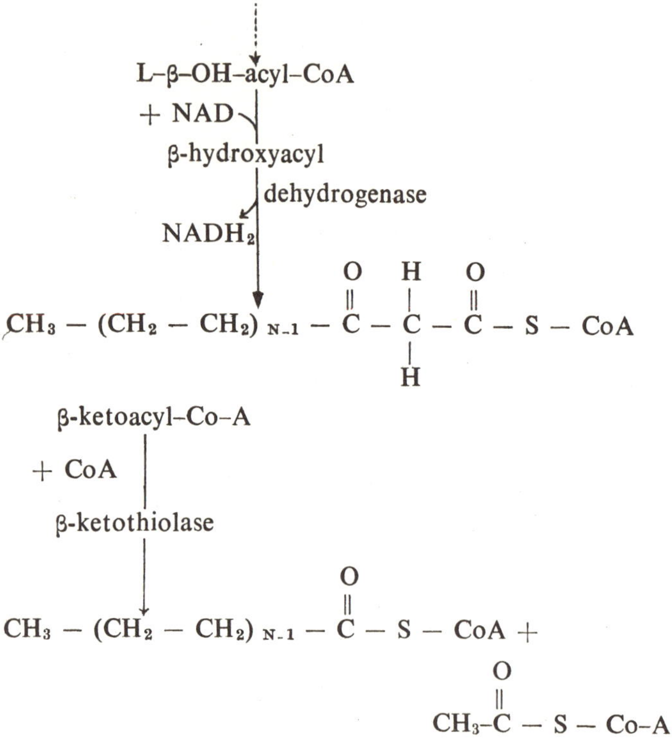

$$\text{L-}\beta\text{-OH-acyl-CoA} + \text{NAD} \xrightarrow[\text{NADH}_2]{\beta\text{-hydroxyacyl dehydrogenase}}$$

$$CH_3-(CH_2-CH_2)_{N-1}-\overset{O}{\underset{}{\overset{\|}{C}}}-\underset{H}{\overset{H}{\underset{|}{\overset{|}{C}}}}-\overset{O}{\overset{\|}{C}}-S-CoA$$

$$\beta\text{-ketoacyl-Co-A} + \text{CoA} \xrightarrow{\beta\text{-ketothiolase}}$$

$$CH_3-(CH_2-CH_2)_{N-1}-\overset{O}{\overset{\|}{C}}-S-CoA + CH_3-\overset{O}{\overset{\|}{C}}-S-Co\text{-}A$$

Role of Carnitine

$$R-COOH + \text{thiokinase} + ATP \rightleftharpoons (R-\overset{O}{\overset{\|}{C}}-AMP)$$
$$\text{acyladenylate} + PPi$$
$$+ \text{Co-A-SH}$$

$$\underset{\underset{CH_3}{|}}{\overset{CH_3}{\overset{|}{CH_3-\overset{+}{N}}}}-CH_2-\underset{OH}{\overset{|}{CH}}-CH_2-COO^- + $$

Carnitine

$$CH_3-\overset{O}{\overset{\|}{C}}-S-Co-ASH$$
acyl-Co-A

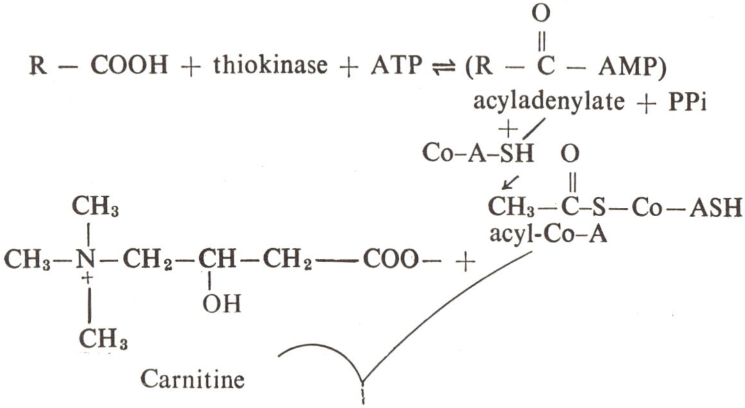

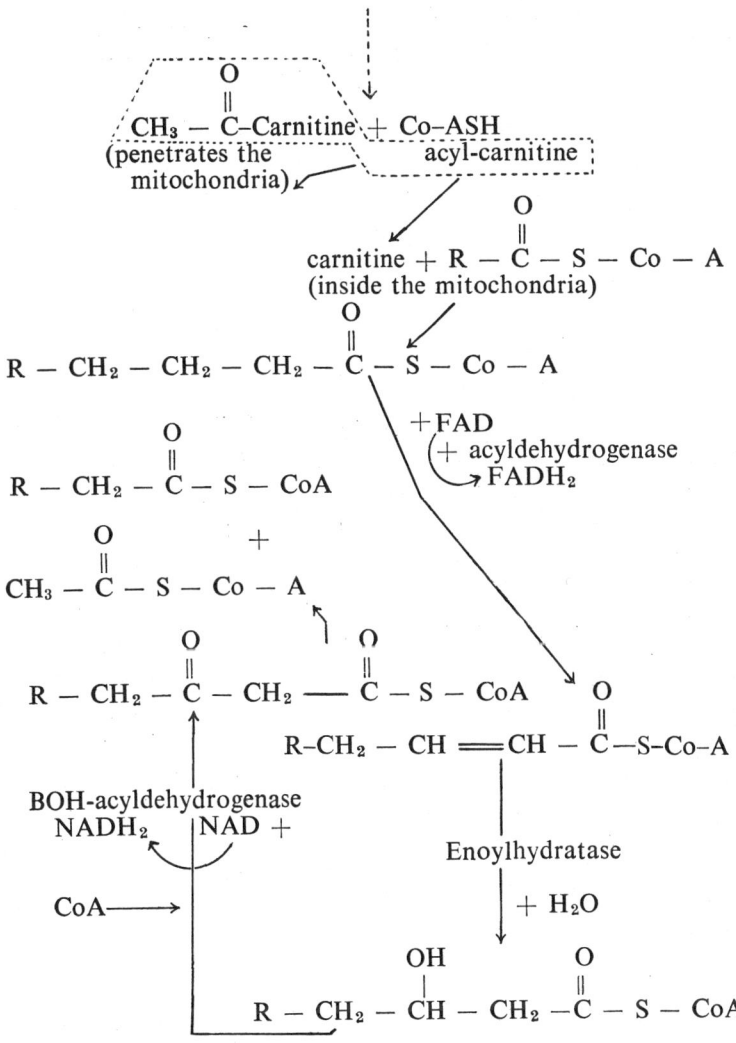

In this process, $FADH_2$ and $NADH_2$ are produced, and water is used each time acetyl-Co-A is made from an even-numbered carbon fatty acid. Thus, an enzyme takes the naturally occurring cis-FA, converts its cis-double bond into the trans form, and the two-carbon degradation process goes on.

If the 16-carbon palmitic acid is β-oxidized, there will be produced 7FADH$_2$, 7NADH$_2$ and 8 acetyl-Co-A which would mean the production of 131 ATPs.

The acetyl-Co-A which is the final product of β-oxidation of fatty acids may end up as trapped energy in ATPs, as indicated, or it may again go into stored fat, or it may go into the formation of ketone bodies (i.e., acetoacetic acid, β-OH-butyric acid and acetone), or it may go into synthesis of cholesterol. For the end product of β-oxidation of fat to go into ATP usable energy production (through the TCA cycle), it requires the carbohydrate sparker, that is, oxaloacetic acid by which, by condensation with acetyl-Co-A, entrance to the TCA cycle is made. In cases where carbohydrates cannot be used, such as in diabetes, starvation, or very high fat diet, the fire of carbohydrate in which only fat may burn is here lacking. If, thus, the carbohydrate sparker (oxaloacetic) is not present, two of the acetyl-Co-A's produced in β-oxidation of fat combine, producing acetoacetyl-Co-A which, in turn, condenses with a third acetyl-Co-A in the liver forming β-methyl-β-OH-glutaryl-Co-A, a compound that may leave the mitochondria and enter into the biosynthesis of cholesterol, or it may lose one acetyl-Co-A, thus, reverting to acetoacetate which, in this case, may be decarboxylated to acetone or reduced to β-OH-butyric acid and leave the mitochondria.

Mostly, fatty acids occur in nature in even-numbered carbons that build up and break down two carbons at a time. Odd-numbered fatty acids, such as propionate, end up with an odd-numbered carbon as they break down by β-oxidation.

Fatty Acids Synthesis

Since Co-A or acetyl-Co-A do not move freely across the mitochondrial membrane, and since fatty acid synthesis occurs outside of the mitochondria, a smaller compound than Co-A is, thus, needed that would transfer the acetyl-Co-A outside the mitochondria. This small compound is carnitine which, when activated to the carnitine-Co-A form, moves freely across the mitochondrial membrane.

$$(CH_3)_3 - N^+ - CH_2 - \underset{\underset{H}{|}}{\overset{\overset{OH}{|}}{C}}-CH_2 - \overset{\overset{O}{\|}}{C} - O^- + CH_3 \text{------}$$

Carnitine

$$\text{-------} \overset{\overset{O}{\|}}{C} - S - Co - A \longrightarrow$$

acetyl-Co-A $\rightleftharpoons$ ----

------ $\rightleftharpoons$ acetyl-carnitine + Co-A.

So, for fat synthesis, acetyl-Co-A must be available outside the mitochondria by either reacting with carnitine as indicated above, or through getting citrate outside the mitochondria and its subsequent cleavage to acetyl-Co-A and oxaloacetic acid. Hence, citrate accumulates and moves outside the mitochondria where the enzyme citrate lyase cleaves it (outside the mitochondria) into oxaloacetic acid and acetyl-Co-A. Acetyl-Co-A (now outside the mitochondria) gets carboxylated to malonyl-Co-A as follows:

$$CH_3 - \overset{\overset{O}{\|}}{C} - S - Co - A + ATP + CO_2 \underset{\underset{\text{(Biotin, Mn}^{++}, CO_2)}{\text{carboxylase}}}{\overset{\text{acetyl-Co-A}}{\longleftrightarrow}}$$

acetyl-Co-A

$$\underset{\underset{H}{|}}{\overset{\overset{H}{|}}{H}}C - \underset{\underset{H}{|}}{\overset{\overset{COOH}{|}}{C}} -- \overset{\overset{O}{\|}}{C} - S - Co - A + ADP + Pi$$

Malonyl-Co-A

This step is the first and is the rate-limiting reaction step. Polycarboxylic acids as citric and α-ketoglutaric, (from the TCA cycle) stimulate this rate-limiting reaction, while excess acetyl-Co-A will inhibit it. High ADP presence will stimulate the isocitric dehydrogenase rate-limiting reaction governing the TCA cycle. Thus, with ADP and citrate availability, both the TCA cycle and fatty acid synthesis are activated.

Synthesis of Triglycerides

Dihydroxy acetone phosphate coming from glycolysis gets reduced with $NADH_2$ into α-glycerol-phosphate (or glycerol-phospho-kinase may phosphorylate glycerol with ATP producing α-glycerol phosphate). The two proper acyl-Co-A's then condense in ester form with α-glycerol phosphate, giving rise to the important phosphatidic acid. A phosphatase enzyme splits off the phosphate group from phosphatidic acid, thus creating a diglyceride, with which another and third acyl-Co-A condenses, creating, this time, a triglyceride as shown below:

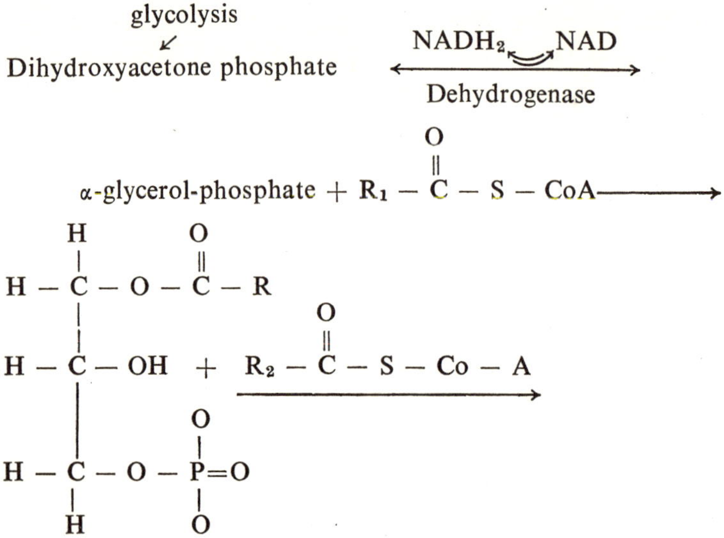

$$\text{Dihydroxyacetone phosphate} \xrightarrow[\text{Dehydrogenase}]{NADH_2 \rightleftarrows NAD} $$

(from glycolysis)

$$\alpha\text{-glycerol-phosphate} + R_1 - \overset{O}{\overset{\|}{C}} - S - CoA \longrightarrow$$

$$\begin{array}{l} H \quad\quad\quad O \\ | \quad\quad\quad\quad \| \\ H - C - O - C - R \\ | \\ H - C - OH \quad + \quad R_2 - \overset{O}{\overset{\|}{C}} - S - Co - A \longrightarrow \\ | \\ H - C - O - \overset{O}{\underset{O}{\overset{|}{P}}} = O \\ | \\ H \end{array}$$

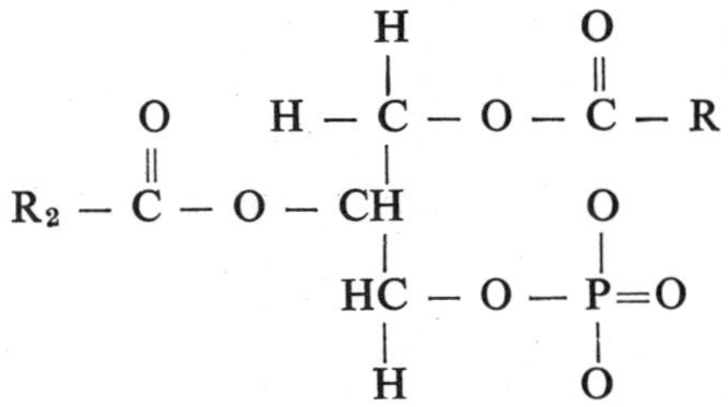

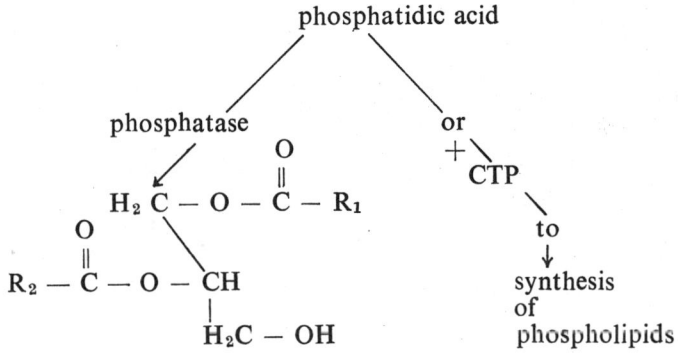

phosphatidic acid

Diglyceride

Triglyceride

Phosphatidic acid is a key intermediate in the biosynthesis of lipids. Animals cannot, however, synthesize linoleic, linolenic, or arachidonic acids. These are, thus, the essential fatty acids that must be provided in the diet.

Biosynthesis of Cholesterol

All carbon atoms of cholesterol are derived from acetate.

Acetate→→→squalene⟶lanosterol⟶cholesterol.

Or:

2 acetyl-Co-A ⟶ acetoacetyl-Co-A $\xrightarrow{+ \text{ acetyl Co-A}}$ ------

------⟶ β-OH-β-methyl-glutaryl-Co-A

$$\begin{array}{c} CO_2 \\ | \\ CH_2 \\ | \\ HO-C-CH_3 \\ | \\ CH_2 \\ | \\ C=O \\ | \\ S-Co-A \end{array} \xrightarrow[\text{Reductase}]{NADPH \quad NAD} \xrightarrow[\text{Reductase}]{NADPH \quad NADP} \begin{array}{c} CO_2 \\ | \\ CH_2 \\ | \\ HOC-CH_3 \\ | \\ CH_2 \\ | \\ CH_2OH \end{array}$$

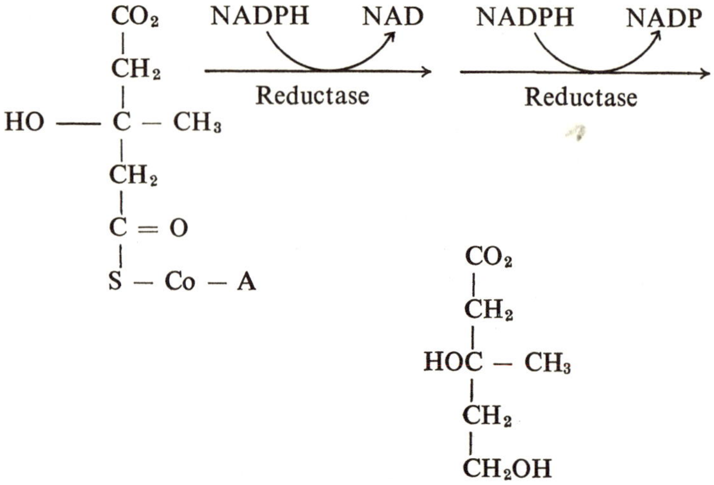

Mevalonic acid

High cholesterol intake inhibits the above reductase enzyme, causing a negative feedback control on endogeneous cholesterol biosynthesis.

REFERENCES

Bailey. A. E. 1951 *Industrual oil and fat products*. New York: Interscience, Publishers.

Bonner, J. 1950 *Plant biochemistry*. New York: Academic Press.

Chipault, J. R.; Mizuno, G. R.; and Landberg W. O. 1955. Antioxidant properties of spices in oil-in-water emulsions. *Food Research* 20;443.

Deuel, H. J. 1951. *The lipids*. New York: Interscience Publishers.

Devine, J., and Williams, P. N. 1961. *The chemistry and technology of edible oils and fats*. New York: Pergamon Press.

Feuge, R. O. 1960. *Effect of Processing on composition of edible oils in nutritional evaluation of food processing*. New York: John Wiley and Sons.

Hilditch, T. P. 1947. *The chemical composition of natural fats*. 2nd. ed. London: Chapman and Hall.

Holman, R. T. 1954. *Progress in chemistry of fats and other lipids*. 2 vols. London: Pergamon Press.

Lehman, A. J.; Fitzhugh, O. G.; Nelson, A. A.; and Woodward, G. 1951. The pharmacological evaluation of antioxidants. *Advances in Food Research* 3:197.

Lovern, J. A. 1955. *The chemistry of lipids of biochemical significance*. London: Methuen and Co.

Meyer, L. H. 1960. *Food chemistry*. New York: Reinhold Publishing Co.

Trotman-Dickenson, A. F. 1959. *Free radicals*. London: Methuen and Co.

Chapter 3

CARBOHYDRATES

INTRODUCTION

Carbohydrates are made up of sugar molecules. They are polyhydroxy aldehydes or polyhydroxy ketones. The aldehyde derivatives are called "aldoses" where the sugar has an aldehyde group on Carbon-1; meanwhile, the ketoses have a ketone group instead. The simplest aldose sugar is glyceraldehyde. Carbohydrates' being the first product of photosynthesis, they make up half of all organic matter on earth.

MONOSACCHARIDES

Monosaccharides are usually the important carbohydrates in foods. The dominant hexoses and pentoses occurring free in plants are D-glucose, D-mannose, D-galactose, D-fructose, and L-sorbose.

The open chain formulas of these sugars are:

```
    H−C=O              H−C=O              H−C=O
    |                  |                  |
  HC − OH          HO − C − H          H − C − OH
    |                  |                  |
  HOC − H          HO − C − H         HO − C − H
    |                  |                  |
  HC − OH            HC − OH          HO − C − H
    |                  |                  |
  HC − OH            HC − OH          H − C − OH
    |                  |                  |
   CH₂OH              CH₂OH              CH₂OH
  D-glucose          D-mannose         D-galactose
```

```
     CH₂OH              CH₂OH
     |                  |
     C = O              C = O
     |                  |
  HO − C − H         HO − C − H
     |                  |
  H − C − OH         H − C − OH
     |                  |
  H − C − OH         HO − C − H
     |                  |
    CH₂OH              CH₂OH
   D-fructose         L-sorbose
```

In the above formulas, glucose and mannose differ only in the configuration of one carbon; thus, they are epimers of each others.

The L-form of a sugar means it rotates a polarized plane of light to the left; the D-form rotates it to the right. These D- and L-forms are derived from glyceraldehyde.

```
        H−C=O                    H−C=O
        |                        |
   H − C − OH              HO − C − H
        |  ―                ―    |
        D                   L
        |                        |
       CH₂OH                    CH₂OH
  D-form of glyceraldehyde      L-form
```

These sugars are like and are unlike aldehydes in their reactions. They are like aldehydes in reducing cupric sulfate ions to the cuprous form ($Cu^{++} SO_4 \longrightarrow Cu_2 \downarrow$) in Fehling solution; they also reduce silver nitrate in ammonia into the silver metal. But they are unlike aldehydes in that they do not form typical bisulfite complexes and they do not possess the typical aldehyde odors. These sugars tend to form a hemiacetal instead of the free aldehyde.

$$
\begin{array}{ccc}
\text{H—C=OH} & \text{H—C=O} & \text{HO—C—H} \\
\text{H—C—OH} \; \alpha & \text{H—C—OH} & \text{H—C—OH} \\
\text{HO—C—H} & \text{HO—C—H} & \text{HO—C—H} \\
\text{H—C—OH} & \text{H—C—OH} & \text{H—C—OH} \\
\text{H—C} & \text{H—C—OH} & \text{H—C} \\
\text{CH}_2\text{OH} & \text{CH}_2\text{OH} & \text{CH}_2\text{OH} \\
[\alpha]_D{}^{20} + 113° & [\alpha]_D{}^{20} + 52.5 & [\alpha]_D{}^{20} + 19
\end{array}
$$

α–D-glucopyranose

The above forms of α–D-glucose in solution are in equilibrium manifesting the mutarotation property of glucose.

The α- and β-forms differ only in the configuration of one carbon. Fructose is known as the fruit sugar. It is found in fruits and honey.

$$
\begin{array}{c}
\text{CH}_2\text{OH} \\
\text{C=O} \\
\text{HO—C} \\
\text{H—C—OH} \\
\text{H—C} \\
\text{CH}_2\text{OH}
\end{array}
$$

D-furanose

The possession of a free or potentially free aldehyde gives these sugars their reducing property.

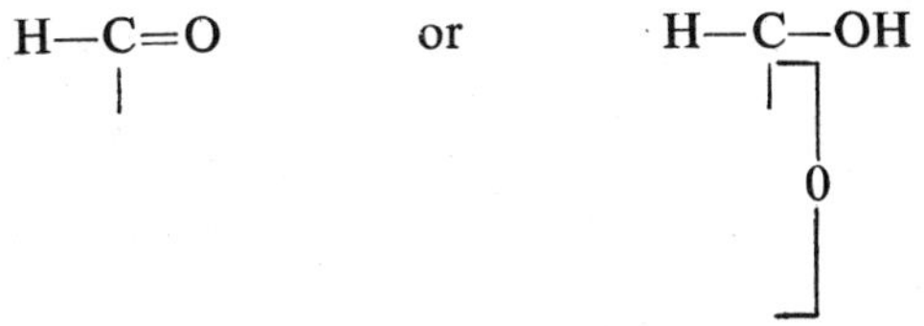

$$H-C=O \quad \text{or} \quad H-C-OH$$

In Fehling solution, they reduce the cupric ions into cuprous red precipitate

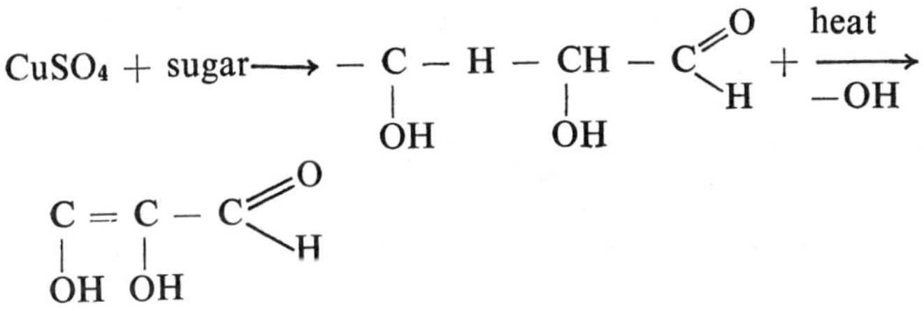

$$CuSO_4 + sugar \longrightarrow -\underset{OH}{C}-H-\underset{OH}{CH}-C\overset{O}{\underset{H}{\diagdown}} + \xrightarrow{\text{heat}} -OH$$

$$\underset{OH}{C}=\underset{OH}{C}-C\overset{O}{\underset{H}{\diagdown}}$$

Cu^{++} picks up an electron from the reductone, splitting it, and forming sugar acids.

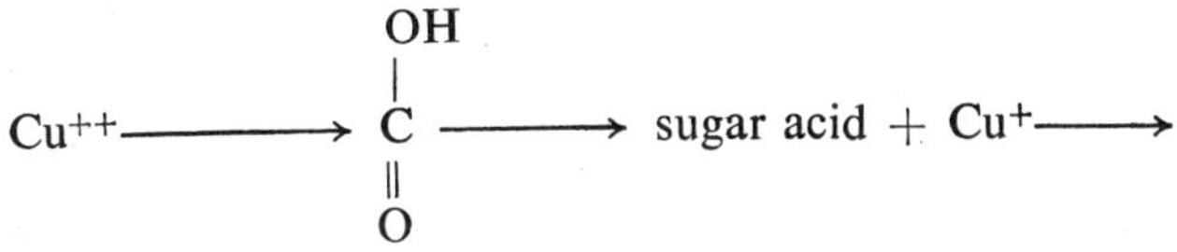

$$Cu^{++} \longrightarrow \underset{O}{\overset{OH}{\underset{\|}{C}}} \longrightarrow \text{sugar acid} + Cu^{+} \longrightarrow$$

$(Cu_2)O \downarrow$ (precipitate)

The five-carbon sugar pentoses are present in foods, mostly in the form of polymers rather than free. Pentosans polymers of these sugars are the most common in occurrence in foods.

Pentose sugars most likely to occur in foods are L-arabinose, D-ribose, D-xylose and L-rhamnose.

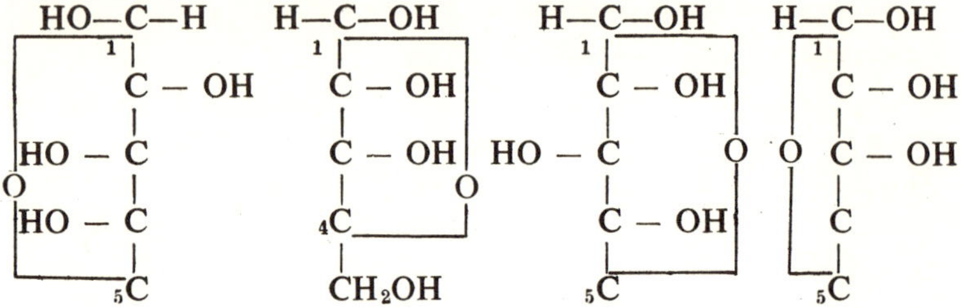

α–L-arabinose   α–D-ribose   α–D-xylose   α–L-rahmnose

If we treat glucose with alkali and mild heat, we get a mixture of 65% glucose, 31% fructose, and 25% mannose.

The most probable structure of a simple sugar, like glucose in solution, is in the ring form. Glucose undergoes the formation of hemiacetal, where one of its hydroxyl groups reacts with the aldehyde group. (In solution, normally glucose is 0.024% in the free aldehyde form.)

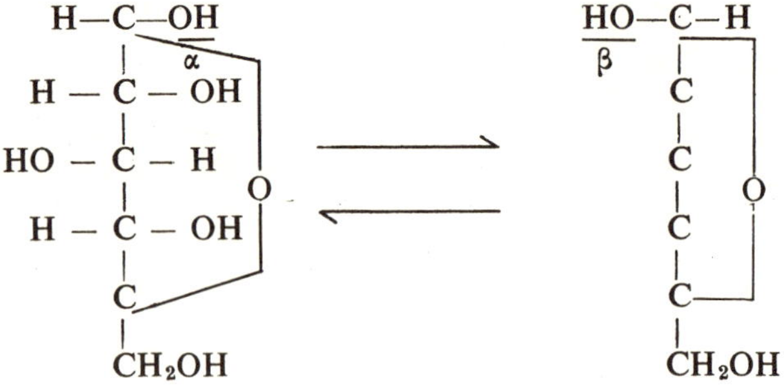

α-form   β-form

Fisher Formula of Glucose

Here the OH on C–5 reacts with the aldehyde group on C–1 forming a hemiacetal. The α- and β-forms shown above are interconvertible in aqueous solutions, where they usually exist in an equilibrium mixture of the two forms.

The Haworth formula refers to glucose in solution which is usually in the ring form.

```
         CH₂OH                              CH₂OH
       6 |                                    |
    H   C————O    H                      H    C————O    OH
     \ 5/         \                       \  /         |β
      C            C                       C           C
    4/   HO   H   /                       /   HO   H   |
     C    \   |  C                       C     \   |   C
    /      \  | / \OH                   / \     \  |  / \
   HO   C————C    α                    HO   C————C     H
      3|    2|                            |     |
       H    OH                            H    OH
```

Ring Form of Glucose

The OH group which is on the right hand side in Fisher's formula will be below the plane on the ring formula; the OH group on the left hand side in the Fisher formula will be above the plane of the ring. When the OH group on C–1 in the ring form is below the plane, the sugar is in the β-form. If it is above the plane, then it is in the α-form. α- and β-forms differ only in the arrangement at the carbonyl carbon. The keto sugars have ketone groups. Fructose, for example, has a ketone group on C–2, which may react with the OH group on C–6.

```
        CH₂OH                      CH₂OH
         |                          |  α
         C = O                      C————
         |                          |    |
    HO – C – H                      C    |
         |                          |    |
    H  – C – OH                     C    O
         |                          |    |
    H  – C – OH                     C    |
         |                          |    |
        CH₂OH                      H₂C————

        Fructose               α-form of fructose
```

In the ring form, α-fructose will be as follows:

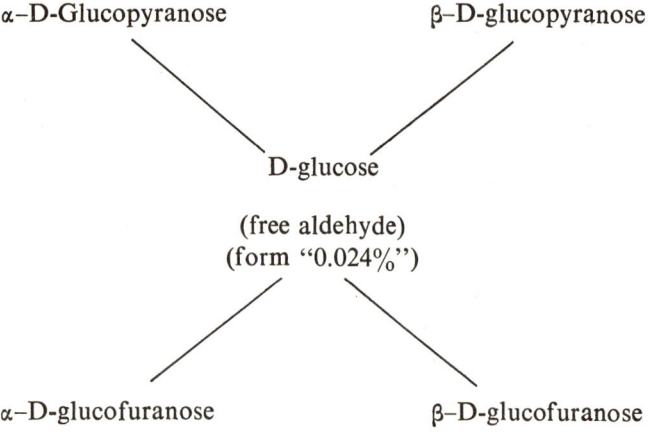

α-form   β-form

The α- and β-forms are spontaneously interconvertible in solution.

VARIOUS FORMS OF GLUCOSE IN SOLUTION

α–D-Glucopyranose   β–D-glucopyranose

D-glucose

(free aldehyde)
(form "0.024%")

α–D-glucofuranose   β–D-glucofuranose

Anomers differ only in the arrangement at the carbonyl carbon. When two sugars differ only in the arrangement at one of their carbons, they are called "epimers."

In the ring form, fructose forms a hemiacetal, where the OH group on its anomeric carbon becomes very reactive.

In a reducing sugar, the linkage between the individual units cannot involve the two anomeric carbons.

OLIGOSACCHARIDES

Oligosaccharides are made of 2-10 monosaccharide units. The most common oligosaccharides are the disaccharides which are composed of two monosaccharides. Maltose, for instance, is composed of two glucose units, where a molecule of water is split out from between the glycosidic linkage, making maltose the anhydride of the two glucoses.

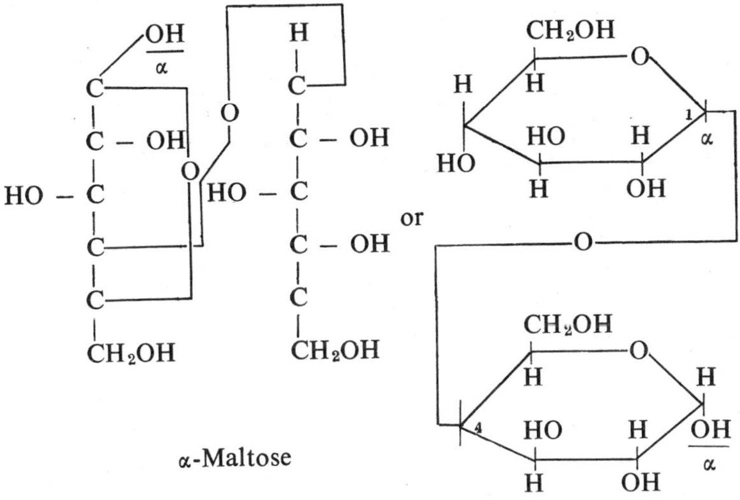

α-Maltose

α-maltose having a free glycosidic OH group renders it a reducing sugar. It, thus, can be oxidized, may undergo mutarotation, and goes in equilibrium between the α-and β-forms. Maltose is the sugar present in malt and cereals. It is a sugar highly digestible by both young and old. It is a product of the hydrolysis of starch, and is a reducing sugar.

Cellibiose

Cellibiose is a product of the hydrolysis of cellulose. It is a β-1-4 linkage between two glucoses.

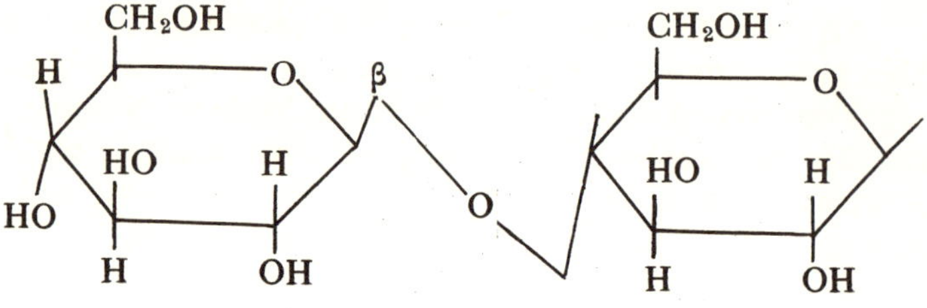

4-D-glucose-β-D-glucopyranoside

Cellibiose, like maltose, has a free glycosidic group which renders it a reducing sugar and which undergoes mutarotation in solutions.

Regardless of the fact that cellulose is the most abundant material on earth, man so greatly fears lack of food that he often goes into wars with his fellow man simply to take over his land and secure its food for himself. Unfortunately, nature has not provided him with the intestinal microorganisms that are capable of breaking the β-1-4 glucose-to-glucose linkage of cellibiose so he may subsequently utilize the glucose units. Thus, although cellulose is so abundant, man cannot utilize it beyond the cellibiose form. Ruminants, however, do have the microorganisms in their rumen that break the β-1-4 linkage of cellibiose, and they have no problem utilizing cellulose.

Lactose

Lactose is the milk sugar. It also has a free glycosidic OH group; thus, it is a reducing sugar. It is 5% in cow's milk and about 7% in human milk. If the level of lactose in solution reaches 12% at 0°C, it will crystallize out, largely in the form of a-lactose. The a-form's being relatively insoluble is the cause of ice cream grittiness. Lactose is one-fifth the sweetness of sucrose.

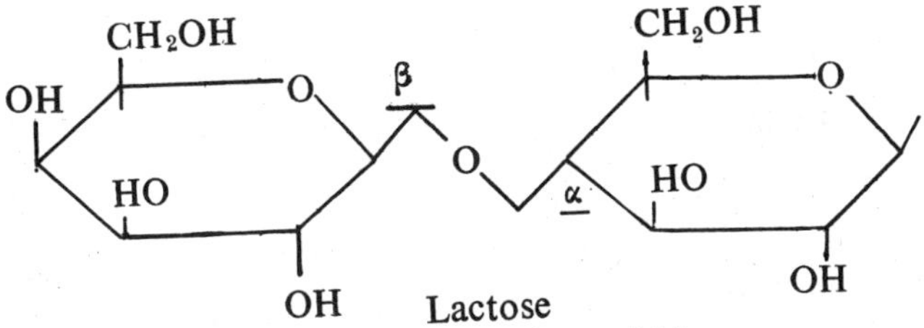

Lactose
(β-D-galactopyranoside)

Sucrose

Sucrose is the most widely distributed sugar in foods. It is composed of glucose and fructose. Sucrose is not a reducing sugar, as both of its glycosidic groups are tied up. Sucrose is easily hydrolyzed with a weak acid or with the enzymes sucrase or invertase, giving glucose and fructose. The fact that sucrose is hydrolyzable with these enzymes grants it the property of an "invert sugar." The specific rotation $[a]^{20}_D$ of sucrose, being + 65°, which, after inversion or splitting of the two sugars, drastically changes into − 20°.

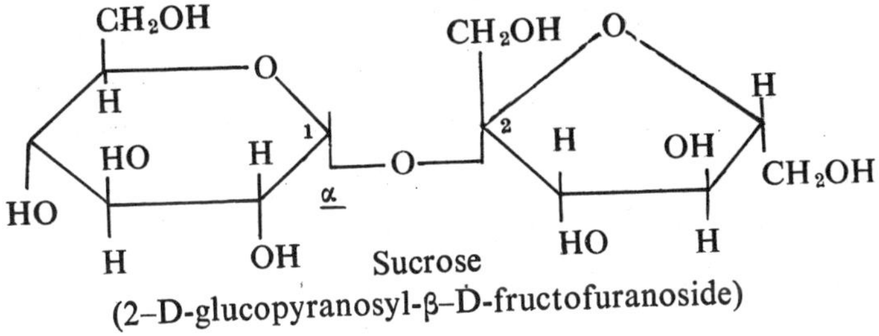

Sucrose
(2-D-glucopyranosyl-β-D-fructofuranoside)

Trehalose

Trehalose is also a nonreducing sugar of two glucose units in α-1-1 linkage. Its α-glucoses are tied up at both the glycosidic linkages. Trehalose is found in yeast, fungi, and in insect blood.

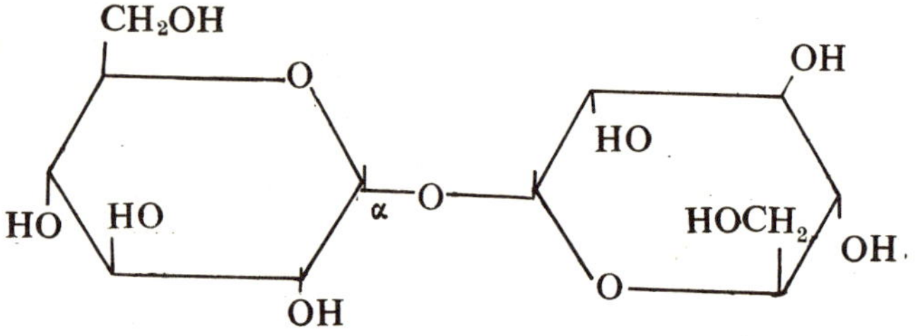

(1-α-D-glucopyranosyl-α-D-glucopyranoside)

Maltose

Glucose-glucose α-1-4

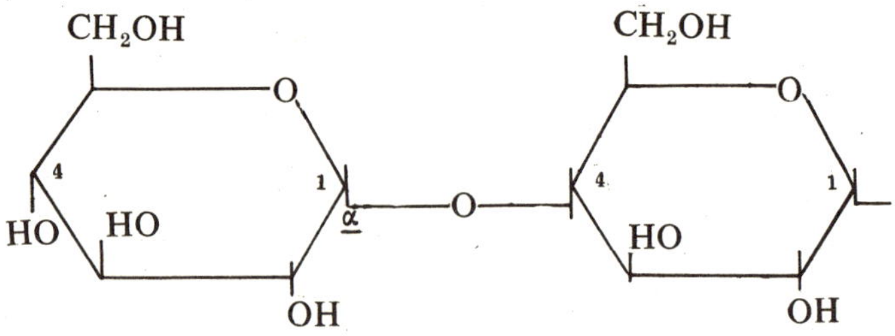

(4-D-α-glucose-α-D-glucopyranoside)

Isomaltose is a degradation product of amylopectin.

TRISACCHARIDES

The important trisaccharide in foods is raffinose, which is composed of galactose, glucose, and fructose.

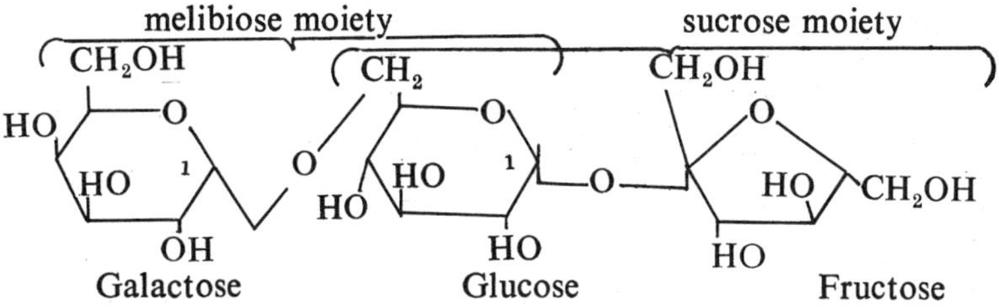

Raffinose

α–D-galactopyranosyl–(1–6)–α–D-glucopyranosyl–(1–2)–
β–D-fructofuranoside

On a weak acid treatment, raffinose yields mellibiose and fructose. Gentibiose is similar to raffinose except fructose is replaced with a glucose in a β-1-6 linkage.

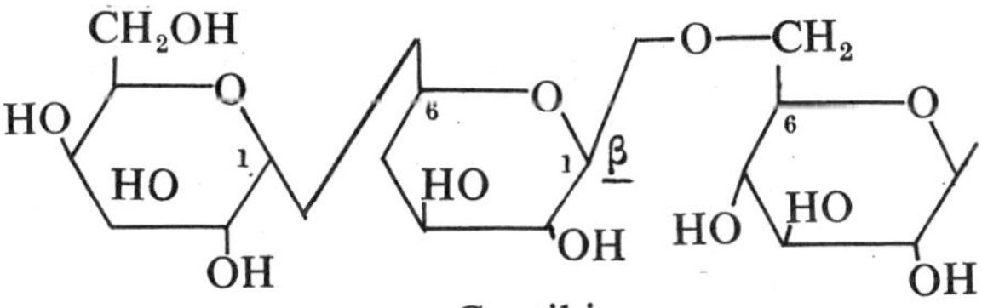

Gentibiose

ALDONIC ACID

Is a glucose with one carboxyl group at C-1

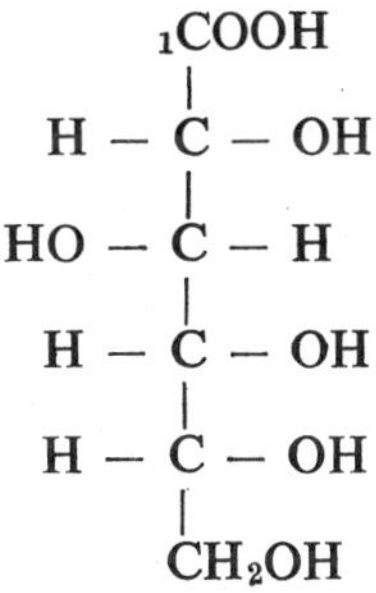

$$\begin{array}{c}
_1COOH \\
| \\
H-C-OH \\
| \\
HO-C-H \\
| \\
H-C-OH \\
| \\
H-C-OH \\
| \\
CH_2OH
\end{array}$$

(D-gluconic acid)

URONIC ACID

Has COOH group on C-6

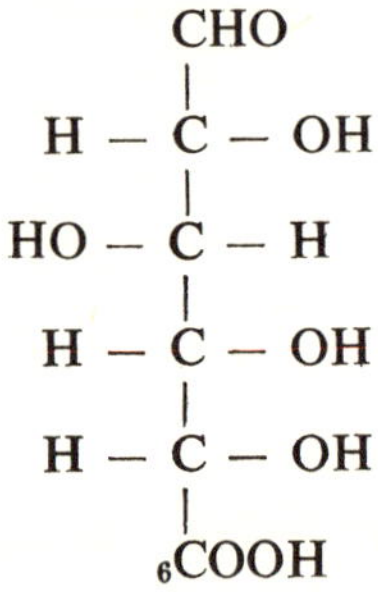

$$\begin{array}{c} CHO \\ | \\ H-C-OH \\ | \\ HO-C-H \\ | \\ H-C-OH \\ | \\ H-C-OH \\ | \\ _6COOH \end{array}$$
(D-glucuronic acid)

SACCHARIC OR ALDARIC ACID

Has two COOH groups, on C-1 and C-6

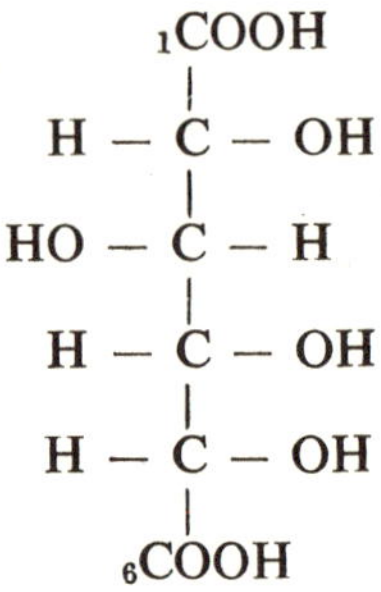

$$\begin{array}{c} _1COOH \\ | \\ H-C-OH \\ | \\ HO-C-H \\ | \\ H-C-OH \\ | \\ H-C-OH \\ | \\ _6COOH \end{array}$$
(D-saccharic acid)

When the carbonyl on C-1 of the sugar is reduced to C-OH, the corresponding alcohol of the specific sugar is produced. Examples:

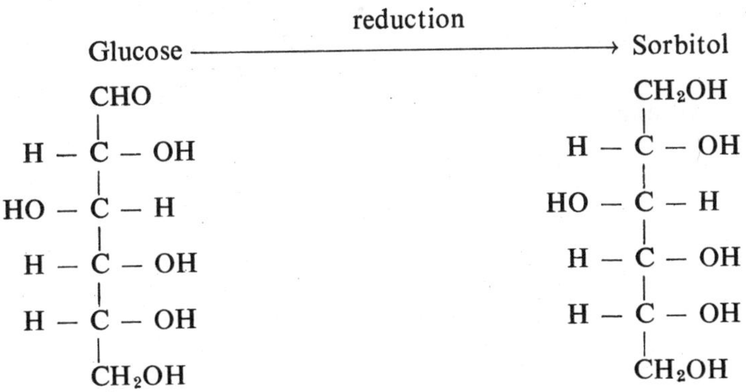

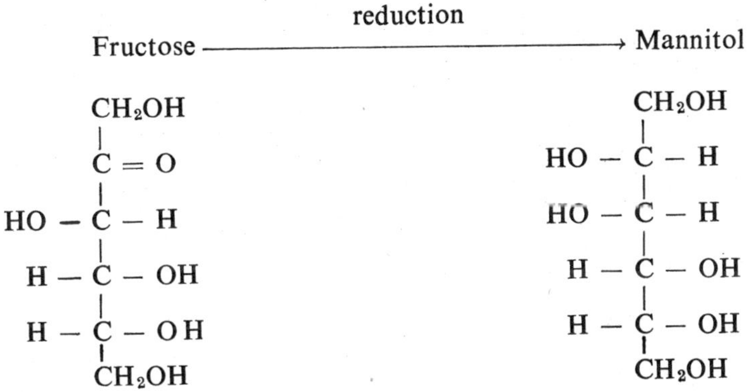

Glucose —reduction→ Sorbitol

```
    CHO              CH₂OH
H — C — OH       H — C — OH
HO — C — H       HO — C — H
H — C — OH       H — C — OH
H — C — OH       H — C — OH
    CH₂OH            CH₂OH
```

Fructose —reduction→ Mannitol

```
    CH₂OH            CH₂OH
    C = O        HO — C — H
HO — C — H       HO — C — H
H — C — OH       H — C — OH
H — C — OH       H — C — OH
    CH₂OH            CH₂OH
```

MUCOPOLYSACCHARIDES

Hyaluronic acid

glucose-N-acetylglucoseamine 1-3-β

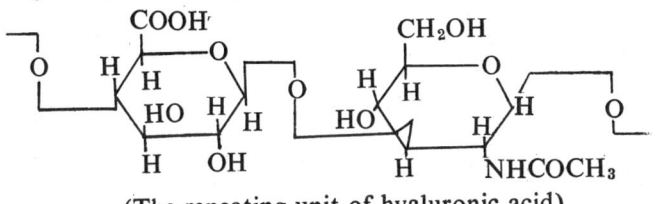

(The repeating unit of hyaluronic acid)

It is a heteropolysaccharide in skin and soft tissue, of linear chains or it may have only small branches of alternative units of 2-N-acetyl-β-glucosylamine and glucuronic acid in β-1-4 linkage.

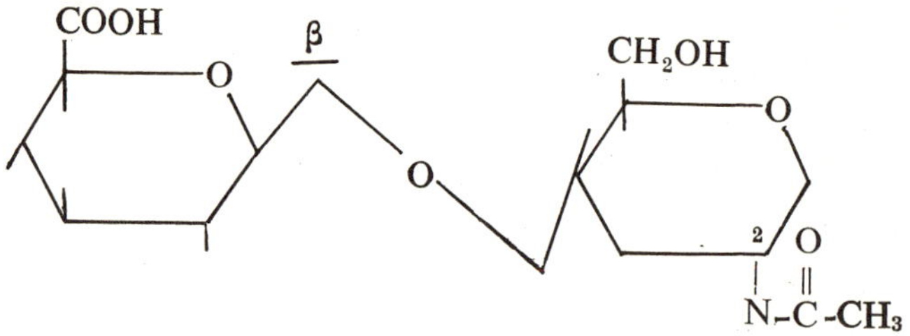

Their jelly-like property gives them softness and elasticity. They are usually present in vitrous and aqueous humor.

Chitin

A structural component in shrimps, lobsters, and insects, it is a homopolysaccharide of 2-N-acetyl glucoseamine in β-1-4 linkage.

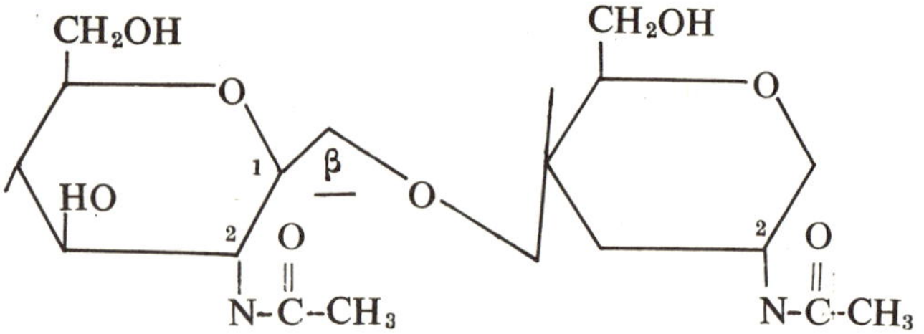

Chitin (linkage part)

Chondriotin sulfate

Chondriotin sulfates are polymers of 2-N-acetyl galactosylamines, glucuronic acid, and sulfuric acid in β-1-4 or β-1-3 linkage. It is a component of cartilage and tendon.

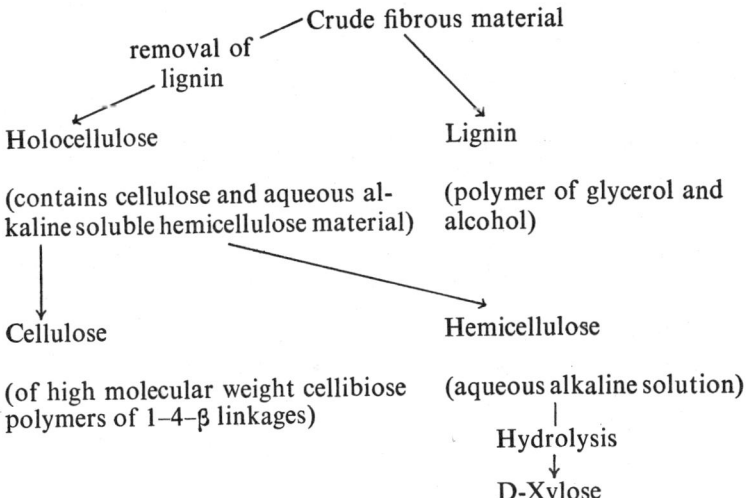

Hemicellulose

It is a crude fibrous material imbibed in a matrix of cemented material, and is obtained from the parent compound as follows:

```
                    ┌─ Crude fibrous material
         removal of │
           lignin   │
                    ▼                    ▼
       Holocellulose                  Lignin
```

Holocellulose
(contains cellulose and aqueous alkaline soluble hemicellulose material)

Lignin
(polymer of glycerol and alcohol)

Cellulose
(of high molecular weight cellibiose polymers of 1–4–β linkages)

Hemicellulose
(aqueous alkaline solution)
  │
  Hydrolysis
  ↓
  D-Xylose

POLYSACCHARIDES

Polysaccharides are usually polymers of pentoses and hexoses, and they form colloidal suspensions in solution. Cellulose gives rigidity to plants, and upon complete hydrolysis, 95 to 96% of cellulose is converted to its basic unit, glucose. The molecular weight of cellulose ranges from 200,000 to 2,000,000, depending on its source. The purest form is usually isolated from cotton and exists in crystalline and partially amorphous structure. Cellulose is usually accompanied by lignin, which is composed of polymers of aromatic alcohols. Another polymer associated with cellulose is xylan and which is a polymer of xylose in β-1-4 linkage between each twenty to forty units.

Most important of all the polysaccharides for man, however, is starch. Starch forms the reserve energy of almost every part of a plant. Thus, starch contributes, in human diet, more calories than any other substance. Starch consists of granules of various kinds and shapes. Amylose is a straight-chain starch of glucose units, while amylopectin is branched.

When reacted with iodine, amylose produces a blue color, while amylopectin gives a red color. The two colors are due to the optical properties of starch. Amylose generally composes 20-30% of the starch content, the balance being amylopectin. Starches differ in properties depending on amount of amylose versus amylopectin, on their degree of polymerization, on the homogeneity of their chain, and on the extent of branching in amylopectin.

The degree of polymerization in amylose is small compared to other polysaccharides, like cellulose.

The main part of amylopectin is the straight chain of α-1-4 glucose linkages that runs to between twenty and thirty units. The molecular weight of amylopectin ranges between 300,000 to 400,000.

The amylose chain, on the other hand, may be composed of 250 to 300 glucose units, with some that may get as high as 2000 units. There is some evidence of α-1-3 glucose-to-glucose linkage.

Glycogen is closely related to amylopectin, except that it is more highly branched, with branching occurring every 6-7 glucose units. Glycogen is the carbohydrate storage form in animals.

ENZYMATIC ACTION ON STARCH

The α-amylases act on α-1-4 linkages attacking large chunks and producing dextrin (glucose polymers with six units), thus, liquifying starch.

β-amylases split the chain from the nonreducing end to two sugar fragments, thus producing β-maltose units plus glucose. In other words, β-amylases split the starch chain up to the α-1-6 branching; ending up with β-limit dextrin.

Thus, if we have amylose type of a starch, the whole chain will be enzymatically hydrolyzed to glucose units, with the α-and β-amylases.

The β- and α-amylases together degrade the starch molecule to maltose, isomaltose, maltotriose, and glucose. The action of β-amylase enzyme on amylopectin and glycogen produces maltose and limit dextrin, while α-amylase action on amylopection and glycogen gives rise to maltose, isomaltose, and small oligosaccharides. Thus, the combined action of α- and β-amylases, in this respect, gives rise to maltose, glucose, isomaltose, and maltotriose.

Phosphorylase is another enzyme that acts on starch. It splits glycogen into glucose-1-phosphate, down to six units from the branch point. The maltase enzyme is widely used in brewing industry. It attacks maltose, producing glucose units.

Transglucosylase Action

This enzyme takes six glucose units from the nonreducing end of a polysaccharide and transfers them onto the branch point of the chain.

nonreducing end⟶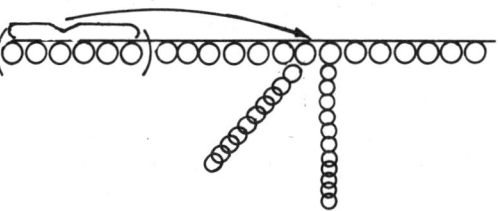

DEGREE OF SWEETNESS OF SUGARS
(MEASURED BY ORGANOLEPTIC ANALYSIS)

Sucrose:	100
Invert sugar:	130
Fructose:	173
Glucose:	74
Corn syrup:	30
Molasses:	74
Honey:	97
Sorghum:	69
Lactose:	16
Raffinose:	23
Xylose:	40
Artificial sweeteners	
Cyclanols:	3000
Saccharin:	30,000–50,000

A very slight difference in the stereo-chemical structure of the sugar may mean a big difference in its sweetness. For an example, D-glucose is sweet while the L-glucose is not. α-mannose is slightly sweet, while its β-form is bitter. A slight modification in the structure of a sugar may bring about a considerable difference in its sweetness. Mostly, however, it is the stereo-chemical structure that has to do with the sweetness of a sugar. The polysaccharides are less sweet when compared to the oligo- and mono-saccharides.

PECTIC SUBSTANCES

Pectic substances are complex colloidal derivatives of carbohydrates occurring in or prepared from plants. They contain large portions of anhydrogalacturonic acid units. These units may be esterified with methanol and partly or completely neutralized with base. Thus, they are polymers of galacturonic acids or the methyl esters of galacturonic acids, joined with

α-1-4 linkages between the galacturonic acids and the methyl esters.

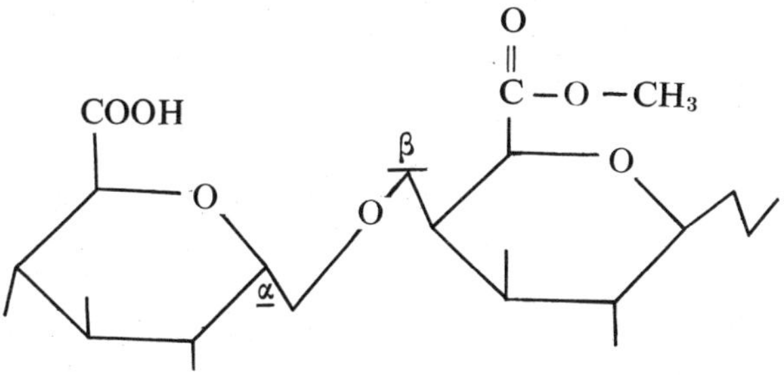

Pectic substances are polymers of large molecular weight varying with the different sources.

Protopectin, the precursor of pectin, occurs in plants and is water-insoluble. Upon restricted hydrolysis, it yields pectinic acid.

Pectinic acid is, thus, a polymer of galacturonic acid containing a more than negligible amount of galacturonic acids in methyl ester form. It forms a gel with sugar and acid. If pectinic acid is suitably low in the methoxy groups content, it can form a gel with metallic ions. The salt of pectinic acid is called "pectinate."

Pectin

Pectin is water-soluble pectinic acid of varying methyl ester content and extent of neutralization. It is capable of forming a gel with sugar and acid.

Pectic acid

Pectic acid is a polymer of galacturonic acids that are free from methyl esters. Its salt is called "pectate."

The pectic substance occurs in between the cells or the cell walls of plants. Its molecular weight may vary between 20,000 and 100,000.

```
G — COOH           G — CO — OCH₃
G — COOH           G — COOH
G — COOH           G — CO — O — CH₃
                   G — CO — OCH₃
```

Pectic Acid   Pectinic Substance

(units of methyl esters and galacturonic acids in α-1-4 links)

The pectic substance in plants lies in close contact between cellulose.

Protopectin is insoluble in water because it exists as calcium or magnesium salt. Thus, it is a macromolecule made of many pectinic acids joined together by Ca or Mg bridges.

```
        COO — Ca — OOC
         |           |
```

Protopectin is, thus, the cementing material in between cells that give hardness to plant material.

Pectin in fruits has been used for many years in making sugar-acid gels. A pectic substance that brings about a firm gel should have a high molecular weight, a high percentage of methyl esters, and a relatively low percentage of free carboxyl groups. The ratio of 1 : 1, methyl esters to galacturonic acid, is the optimum for gelling. One percent pectin is necessary to produce satisfactory gelling. Pectin with a low molecular weight produces weak gel. More gelling is produced when half of the carboxyl groups of the galacturonic acids is methylated. 65% sugar is the optimum for maximum gelling. Above this, crystallization may occur.

A pH of 3.5 or less is necessary for gelling. Rapid onset of gelling usually occurs at 88°C, and it is a slow gelling at 54°C, this being influenced by the amount of salt present and by the pH. Adjustment to a low pH is done by use of citric or lactic acid. To raise the pH, sodium citrate or sodium tartarate is used. If methyl esters are less than seven percent, a divalent ion such as Ca^{++} is needed for gel-setting.

If the level of pectin becomes high in wines, turbidity or cloudiness develops. This is usually removed by adding positively charged colloidal polymers which precipitate out the negatively charged hydrocolloids, thus removing the pectic material, or the enzyme pectin esterase might instead be added, which splits off the methyl groups leaving behind pectic acid which is more soluble than pectinic acid. The enzyme polygalacturonase may also be used in removing the pectic material which causes cloudiness in wines, as it acts through splitting the galacturonic acid chains. In certain foods, such as fruit juices, it might become a desirable end instead to stabilize the cloud. In this case, the pectic material is added so it may help cause the suspension of the insoluble material in the juice.

GUMS AND MUCILAGES

Gums and mucilages are used in foods that contain finely divided particles, liquid, or gasses for even distribution and stabilization of these particles, as well as for the smoothness, juiciness, and desirability of the product. They help rid the food product of excessive water, change its electrical charge, increase its viscosity, and influence its interface tension.

Many gums are polysaccharides. Plant gums are usually polymers of pentoses, hexoses, and uronic acids. Gum arabic is derived from the acacia tree and serves as a thickener and emulsifying agent in confections, as it aids in the formation of a very strong gel.

Gum ghatti comes also from plant sources. Its properties generally are similar to gum arabic, except that its solutions are more viscous and less adhesive. It is mostly used for oil-water emulsion formation.

Gum karaya is derived from several East Indian trees, its general use being as a stabilizer in whipped products.

TYPES AND SOURCES OF CARBOHYDRATES IN THE AMERICAN DIETARY[a]

Carbohydrates	Approx. percentage total Carbohydrate intake[b]	Chief food sources	Endproducts of digestion	Remarks
Polysaccharides				
Indigestible	3			
Celluloses and hemicelluloses		Stalks and leaves of vegetables; outer covering of seeds		May be partially split to glucose by bacterial action in large bowel
Pectins		Fruits		Chemical hydrolysis yields galactose and arabinose
Partially digestible	2			
Inulin		Jerusalem, artichokes, onions, garlic	Fructose	
Galactogens		Snails	Galactose	Digestion uncomplete: further splitting by bacteria may occur in large bowel
Mannosans		Legumes	Mannose	
Raffinose		Sugar beets	Glucose, fructose and galactose	
Pentosans		Fruits and gums	Pentoses	
Digestible				
Starch and dextrins	50	Grains, vegetables (especially tubers and legumes)	Glucose	The most important group quantitatively. Usually accompanied by some maltose
Glycogen	Negligible	Meat products and sea food	Glucose	
Disaccharides				
Sucrose	25	Cane and beet sugars, molasses maple sirup	Glucose and fructose	

[a] From Soskin, S. and Levine, R. 1947. *Carbohydrate metabolism*, Chicago; Press, University of Chicago. Calculated from the average dietary of the middle-income group in the United States.

[b] Ibid.

Carbohydrates	Approx. percentage total Carbohydrate intake[b]	Chief food sources	Endproducts of digestion	Remarks
Lactose	10	Milk and milk products	Glucose and galactose	
Maltose	Negligible	Malt products	Glucose	
Monosaccharides				
Hexoses				
Glucose	5	Fruits, honey, corn sirup	Glucose	In fruits and vegetables the contents of glucose and fructose depend on species, ripeness, and state of preservation
Fructose	5	Fruits, honey	Fructose	
Galactose			Galactose	These monosaccharides do not occur in free form in foods; see under lactose and mannosans
Mannose			Mannose	
Pentoses				
Ribose			Ribose	These monosaccharides do not occur in free form in foods. They are derived from pentosans of fruits and from the nucleic acids of meat products and sea food
Xylose			Xylose	
Arabinose			Arabinose	
Carbohydrate derivatives				
Ethyl alcohol	variable	Fermented liquors		These substances are the products of natural or induced carbohydrate break down
Lactic acid	negligible	Milk and milk products	Absorbed as	
Malic acid	negligible	Fruits		
Citric acid	negligible	Fruits		

67

Gum tragacanth is derived from Asiatic or East European species of the Astragalus herb. It is used as a stabilizer and thickening agent in salad dressing, citrus oil emulsions, ice creams, and candies.

Seaweed gums are produced by extraction from the various seaweeds with boiling water or alkaline solution. Agar, which is a galactan sulfate, is produced by boiling red seaweed in water. It is used in icing baked goods, ice creams, fruits, and vegetables, etc.

Irish moss (carageenan) is extracted with hot water and is used in chocolate milk. It prevents sedimentation from forming in the chocolate.

Giant kelp gum (alginic acid plus sodium alginate) is extracted with a sodium carbonate solution after harvesting kelp from beds of the Pacific Ocean off the coast of California. It is used as a stabilizer in ice creams and water ices, as a thickening and suspending agent in fruit juice, and as an emulsifier.

Some gums from seed sources are locust bean gums and guar gums. The former is obtained from the seeds of carob tree. It is 70–80% mannose and the rest is galactose, and is used as a stabilizer in ice creams, salad dressing, bakery products, etc.; the latter is obtained from seeds of guar plants, is a galactomannans polymer, and is used in stabilization of ice creams and cheeses, for giving texture to doughs, and for binding and lubrication purposes in sausages.

ANALYTICAL PROCEDURES FOR DETECTION OF CARBOHYDRATES

Enzymes have been used, of course, in characterization of carbohydrates. The enzymes' specificity is valuable in distinguishing between the α and β, and between the L- and D-forms of sugars.

The first thing to be done before identification of carbohydrates would be the removal of contaminants, proteins, and lipids. Solvent-extraction or ion-exchange methods may be applied here. The next step should be the breaking up of the long polysaccharide chain into the monosaccharides. This is usually done by acid hydrolysis. Treatment with water and mineral acid cleaves the glycosidic linkage between the sugars.

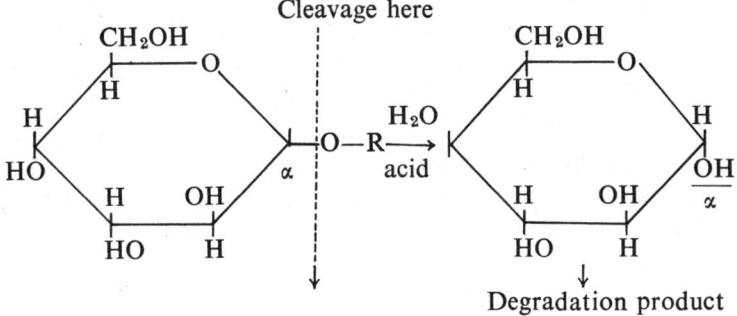

Degradation product

Use of a base should be avoided as it gives rise to fragmentation, rearrangement, and side reactions.

If starch is hydrolyzed in 1N H_2SO_4 under heat at 100°C for five hours, a 99% conversion of polysaccharides to glucose takes place. Hydrochloric acid should substitute for H_2SO_4 if amino sugars are present. In this case, 3N HCl and digestion for 15–16 hours should apply, in order to avoid the brown polymer formation that usually develops.

The time for the previously mentioned treatment with H_2SO_4 may be drastically shortened by the use of 35N H_2SO_4 under 100°C heating for 10–15 minutes.

The acid treatment of hexoses produces hydroxymethylfurfural, while that of pentoses produces furfural. These compounds being colorless and of low absorptivity, when reacted with orsinol they produce a complex which shifts their absorption from the ultraviolet to the visible region as a red color develops, the intensity of which is measured, and subsequent spectrophotometrical quantitation of the sugar is made. An advantage of this method, of course, is that it distinguishes between hexoses and ketoses. A disadvantage of it, however, is that it cannot distinguish between polysaccharides and monosaccharides.

Paper chromatographic methods are used in separation and identification of sugars; however, the disadvantage, in this case, is that the soluble carbohydrates may wash off, and their successful quantitation may prove hard to achieve.

Enzymatic procedures for detecting sugars, as mentioned earlier, are of special importance, as their specificity makes it possible to distinguish between the various isomers of sugars. For example, glucose oxidase is an oxidizing enzyme that catalyzes the oxidation of β-D-glucopyranose to gluconolactone. The color produced in shifting from oxidation-reduction states makes it possible to measure it in the visible region of the spectrum. An advantage of this test is its specificity for β-glucose.

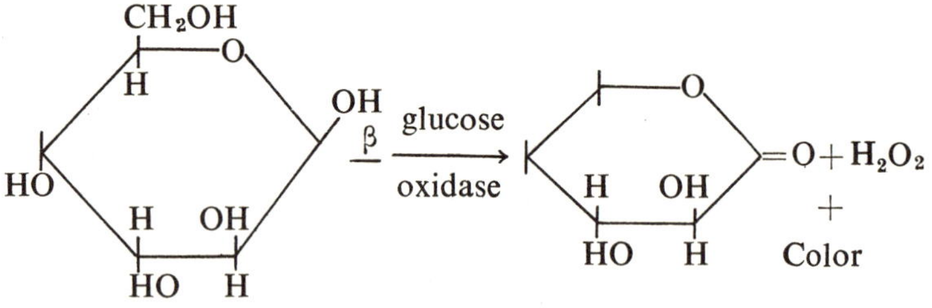

Glucose $\xrightarrow{\text{glucose oxidase}}$ Gluconolactone $= O + H_2O_2 +$ Color

In 1960, gas chromatography came into play in the identification and quantitation of sugars as it became possible to make derivatives of the nonvolatile sugars which become volatilized upon injection into the gas chromatogram. In this case, glucose or other sugars are dissolved in pyridine and esterified into a trimethyl-silo-ether, which becomes effectively volatile in passing through the gas chromatogram.

ENZYMATIC BROWNING

Enzymatic browning was first observed as a reaction involving enzymes by Lindet in 1895. It was not until 1920, however, when it was discovered that the O-dihydroxyphenol compounds were responsible for the color production. These compounds, along with other related phenolic compounds of generally similar structure, such as tannins, are common in nature and probably are formed from the anthocyanin flavanoids. The dehydroxyphenolic compounds associated with enzymatic browning may include the following:

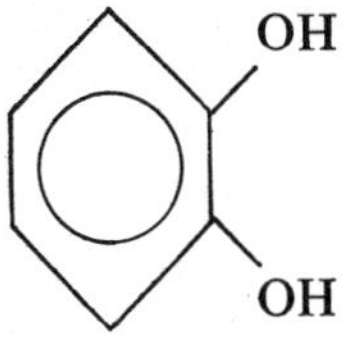

Catechol

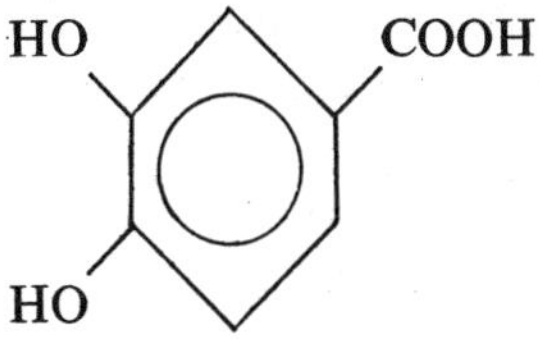

Protocateahuic acid

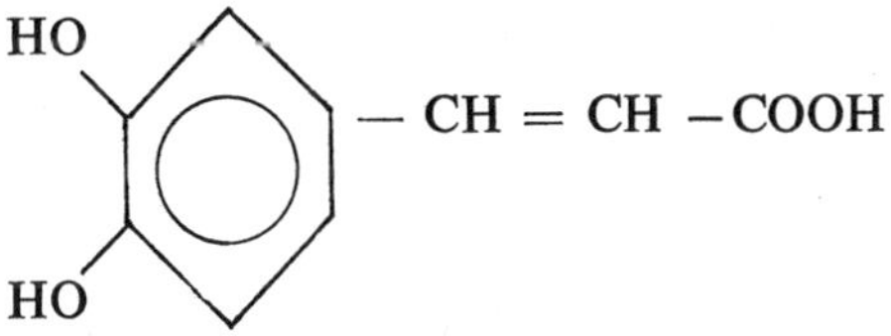

Caffeic acid

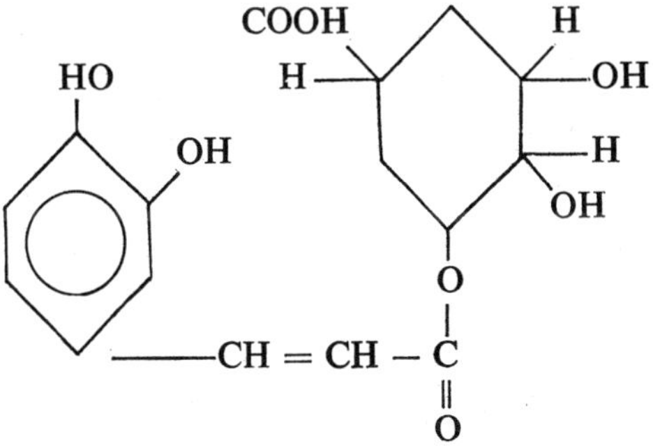

Chlorogenic acid

The phenolases are the major parent enzymes that catalyze the enzymatic production of the brown coloration. Tyrosinases are a subgroup of these phenolases. They are monophenol oxidases and require Cu^{++} ions for activation. Tyrosinase catalyzes the conversion of the amino acid tyrosine to O-quinone. Foods that contain the dihydroxyphenol compounds and the oxygenases enzymes (phenolases, polyphenolases, polyphenol oxidases) will undergo enzymatic browning. Such foods may include the following: apples, peaches, apricots, cherries, figs, strawberries, pears, potatoes, and beets. On the other hand, foods that contain the peroxidases or catalases enzymes do not undergo enzymatic browning; such foods are: citrus products, pineapples, melons, carrots, most berries, and tomatoes.

The tyrosinase enzyme is a specific phenolase. It catalyzes the reaction of tyrosine conversion to a quinone. The quinones themselves are not responsible for the brown color. Instead, their polymerization products produce the brown coloration.

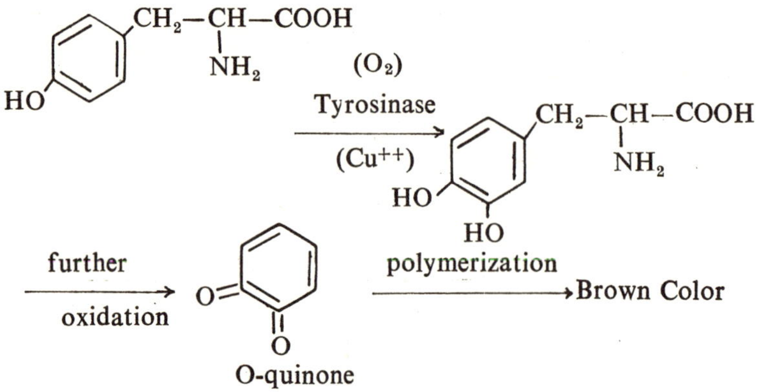

Catecholase, which is a polyphenolase enzyme, converts dihydroxyphenol compounds to quinones, which upon polymerization produce brown color.

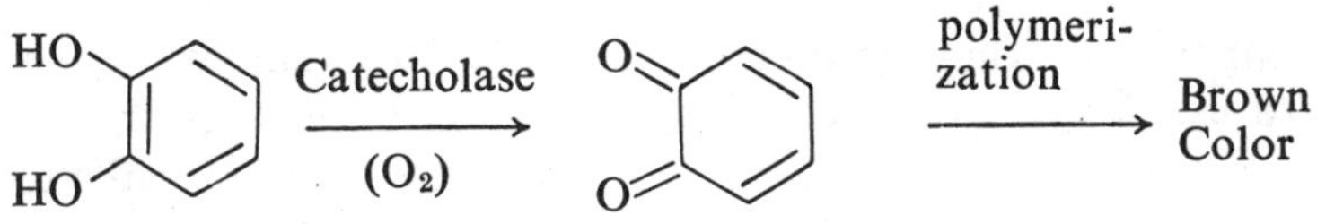

$$\text{Hydroxy-quinone} \xrightarrow[\text{(O}_2\text{)}]{\text{Catecholase}} \text{Quinone} \xrightarrow{\text{polymerization}} \text{Brown Color}$$

Lacol, which is a material from latex, is converted by the polyphenol oxidase enzyme *laccase* to a compound that is converted upon oxidation into melanoidin, which is a black pigment.

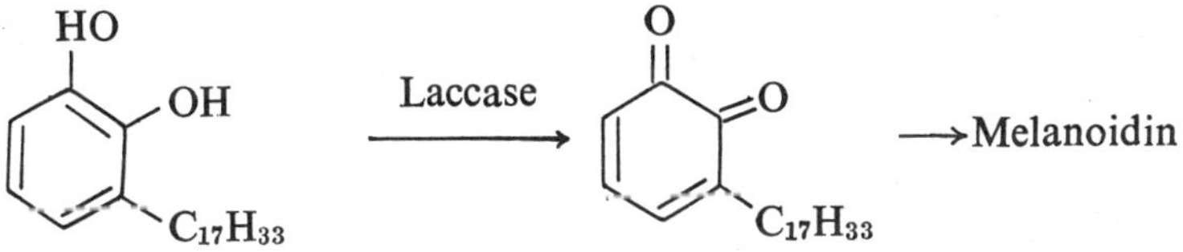

$$\text{Lacol} \xrightarrow{\text{Laccase}} \text{o-quinone derivative} \longrightarrow \text{Melanoidin}$$

The enzyme ascorbinase is responsible for the reversible oxidation-reduction of ascorbic-dehydroascorbic acid.

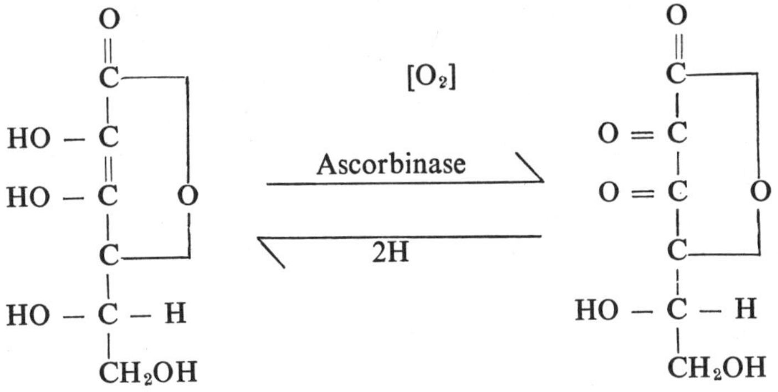

$$\text{Ascorbic Acid} \underset{2H}{\overset{[O_2] \; \text{Ascorbinase}}{\rightleftarrows}} \text{Dehydroascorbic Acid}$$

Properties of the Phenolase Enzymes

Unlike peroxidases, the phenolases are very sensitive to temperature. They are thus, unstable even at rather low temperatures. For an example, at temperatures of 85-95°C, the peroxidases would require five minutes for inactivation, while the phenolases are immediately inactivated. Practically all phenolases are inactivated by heavy metal ions, halogens, cyanides, sulfites, and ultrasonic and radiowaves of high frequency.

Optimum Temperature Range of Activity of Phenolases

Enzyme	Optimum Temperature°C
Apple phenolase	40.0
Olive oxidase	31.5
Grape peroxidase	36.3
Guava peroxidase	50.0

To inactivate peroxidases, however, dipping a food product such as potato slices in salt brine before they are dehydrated will prevent enzymatic browning. Treatment of cuts of fruits and vegetables with free SO_2 gas immediately after slicing but before dehydration will prevent the occurrence of browning in these products.

Enzymatic browning is inhibited in acid pH. The optimum pH for apple phenolase is 4.0; its activity diminishes at pH 3.7, and stops at pH 2.5. For plum phenolases, the optimum pH is 7.0, and they are completely inactivated at pH 2.5.

Usually after peaches and apricots are peeled in lye, they are then dipped in acid (pH 4 or below) to prevent browning. 0.03% of ascorbic acid is added, which acts as a reducing agent or as an antioxidant.

Possible mechanism of enzyme inhibition

$SH_2 + E \longleftrightarrow E \cdot SH_2$ or the formation of an enzyme-substrate complex.

$E \cdot SH_2 + O_2 \longleftrightarrow SH_2 \cdot E \cdot O_2$, where the enzyme-substrate complex reacts with oxygen.

$SH_2 \cdot E \cdot O_2 \longleftrightarrow S + H^+ + HO_2 - E \cdot$, where the enzyme-substrate-oxygen complex breaks up into oxygen, substrate, proton, and the production of a free radical.

$HO_2 - E \cdot + SH_2 \longleftrightarrow HO_2 - E \cdot SH_2$, where the free radical formed complexes with the substrate. $HO_2 E \cdot SH_2 \longrightarrow E + S + HO_2 + OH^-$, where the new complex formed breaks up to its components.

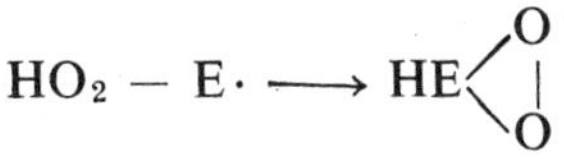

$$HO_2 - E \cdot \longrightarrow HE\!\!\begin{array}{c}\diagup O \\ | \\ \diagdown O\end{array}$$ where free radicals poison the enzyme.

Possible mechanism of melanoidin formation

Initial hydroxylation

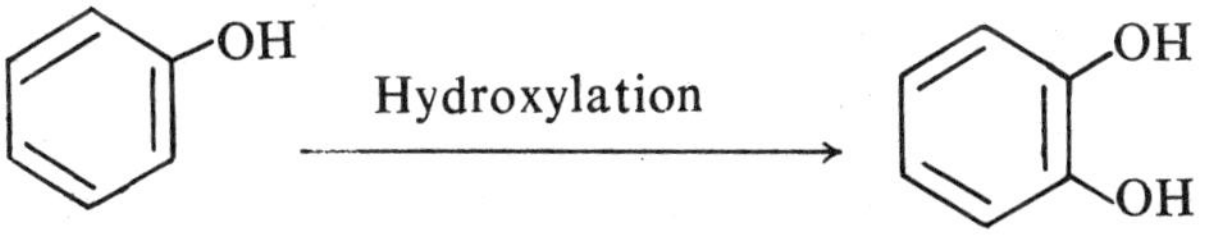

[phenol] —Hydroxylation→ [catechol (1,2-dihydroxybenzene)]

Oxidation

[catechol] —+O₂, Oxidation→ [ortho-benzoquinone]

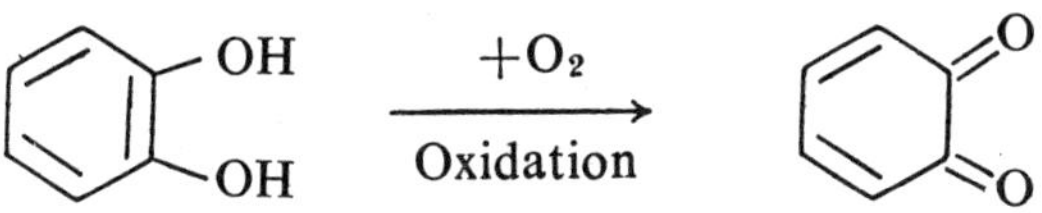

Secondary hydroxylation

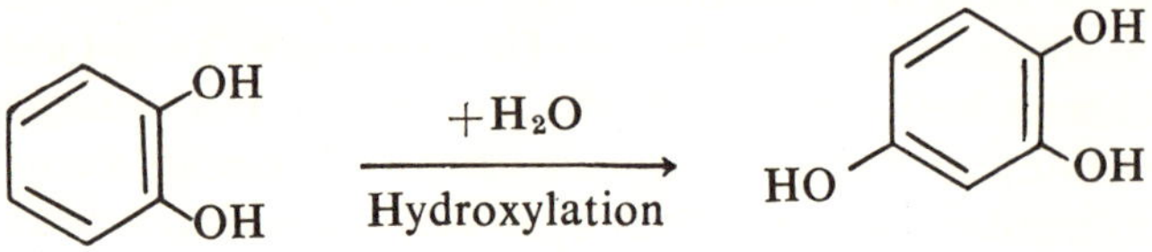

[Reaction: catechol + H₂O → 1,2,4-trihydroxybenzene via hydroxylation]

Intramolecular rearrangements

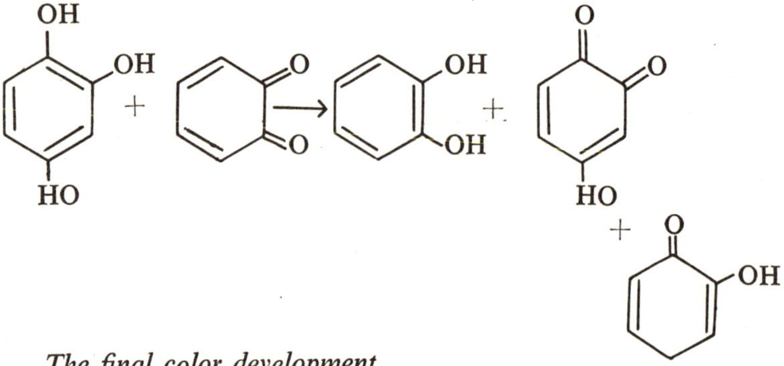

[Reaction scheme showing rearrangements of hydroxylated benzene derivatives to quinone forms]

The final color development

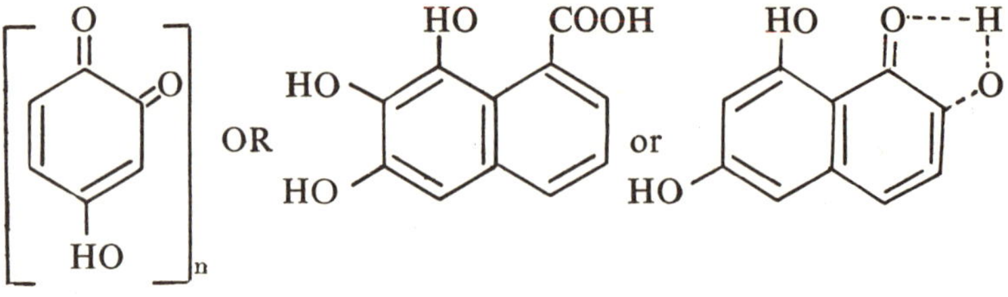

[Structures showing polymeric quinone, and two naphthalene-derived polyhydroxy structures (one with COOH, one with intramolecular H-bond)]

(From a review by Joslyn and Ponting)

NONENZYMATIC BROWNING

Nonenzymatic browning may be beneficial in that it develops a desirable flavor, and it may be nonbeneficial when it develops an undesirable flavor.

Nonenzymatic browning contributes more to the flavor of processed foods than any other factor, such as beneficial contributions to the flavor of coffee, maple syrup, date goods, pastry, beer, and toast. On the other hand, adverse contributions to flavor by nonenzymatic browning occurs in dry milk, dry eggs, concentrated and dehydrated fruit juices, starches, and protein hydrolases. In such foods, an off-flavor develops by the nonenzymatic-browning reactions that occur during storage. There are three mechanisms proposed for the nonenzymatic-browning process. These are:

1. Maillard reaction
2. Ascorbic acid mechanism (in citrus fruits)
3. Active aldehyde mechanism (carmelization in absence of amines)

In the Maillard-type reaction, the initial stage is colorless, and there is no U.V. absorption at this stage. This initial reaction is due either to a sugar-amine condensation or to an Amadori rearrangement. For the former type of a reaction to occur, a free aldehyde group should be available. For the latter type, dehydration, i.e., loss of water, should occur. Even though no color change occurs in the initial stage, yet, flavor changes may develop. This early stage of the reaction requires high temperatures, but once the Amadori rearrangement occurs, browning will proceed even at considerably lower temperatures.

The intermediate stage can be either colorless or pale yellow, and there is U.V. absorption at this stage at the 270–280 mu wavelength. This intermediate type of reaction may be the function of the following:

1. Sugar dehydration
2. Sugar fragmentation
3. Amino acid degradation
4. Strecker degradation

This reaction does develop a color, which becomes intense at the final reaction.

$$\begin{array}{c} H-C=O \\ | \\ C-OH \\ | \\ HO-C \\ | \\ C-OH \\ | \\ C-OH \\ | \\ CH_2OH \end{array} \quad \xrightarrow[1]{+RNH_2} \quad \begin{array}{c} H-N-R \\ | \\ C-OH \\ | \\ C-OH \\ | \\ HO-C \\ | \\ C-OH \\ | \\ C-OH \\ | \\ CH_2OH \end{array}$$

Aldehyde Sugar

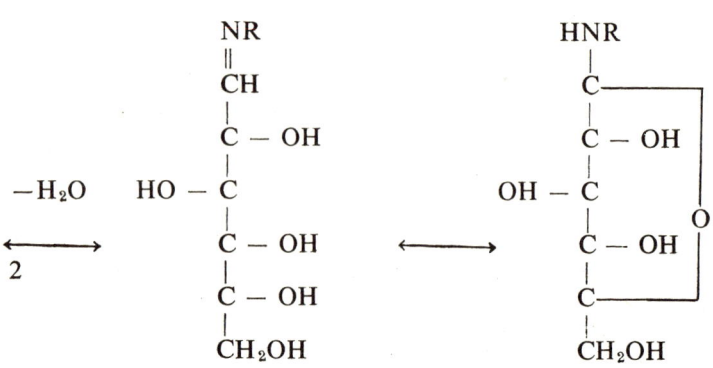

$$\xrightarrow[2]{-H_2O} \quad \begin{array}{c} NR \\ \parallel \\ CH \\ | \\ C-OH \\ | \\ HO-C \\ | \\ C-OH \\ | \\ C-OH \\ | \\ CH_2OH \end{array} \quad \longleftrightarrow \quad \begin{array}{c} HNR \\ | \\ C \\ | \\ C-OH \\ | \\ OH-C \\ | \\ C-OH \\ | \\ C \\ | \\ CH_2OH \end{array}$$

Schiff base   N-glucosylamine

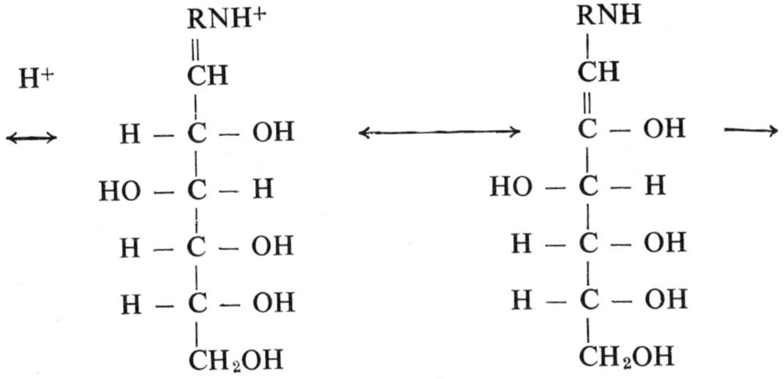

$$\xrightarrow{H^+} \quad \begin{array}{c} RNH^+ \\ \parallel \\ CH \\ | \\ H-C-OH \\ | \\ HO-C-H \\ | \\ H-C-OH \\ | \\ H-C-OH \\ | \\ CH_2OH \end{array} \longleftrightarrow \begin{array}{c} RNH \\ | \\ CH \\ \parallel \\ C-OH \\ | \\ HO-C-H \\ | \\ H-C-OH \\ | \\ H-C-OH \\ | \\ CH_2OH \end{array} \longrightarrow$$

Enol form

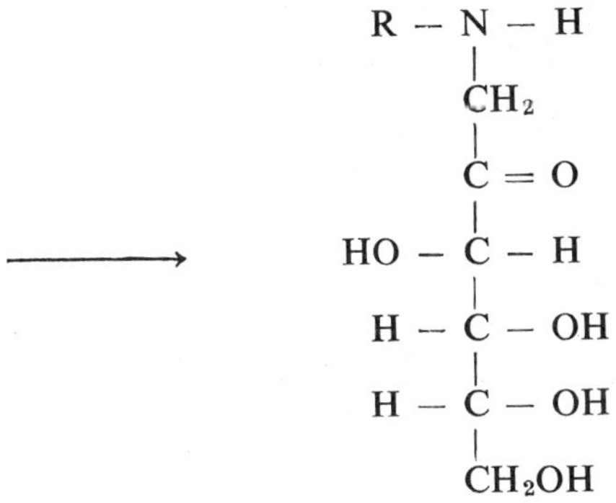

$$\longrightarrow \quad \begin{array}{c} R-N-H \\ | \\ CH_2 \\ | \\ C=O \\ | \\ HO-C-H \\ | \\ H-C-OH \\ | \\ H-C-OH \\ | \\ CH_2OH \end{array}$$

The Amadori Rearrangement

The Amadori product then loses two moles of water, forming a reductone which easily loses hydrogen to form ketoses.

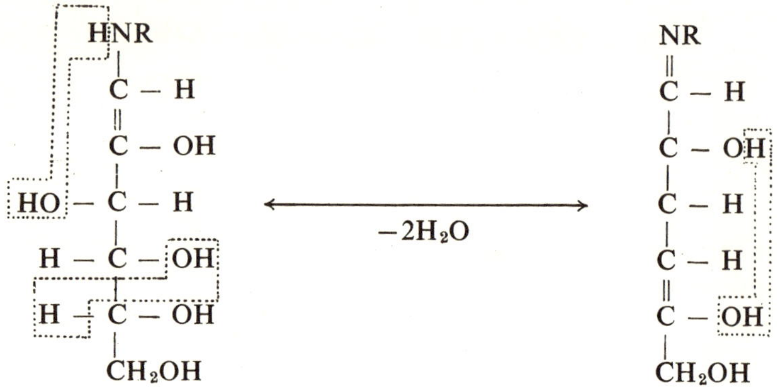

$$\begin{array}{c}\text{H}\text{NR}\\|\\\text{C}-\text{H}\\||\\\text{C}-\text{OH}\\|\\\text{HO}-\text{C}-\text{H}\\|\\\text{H}-\text{C}-\text{OH}\\|\\\text{H}-\text{C}-\text{OH}\\|\\\text{CH}_2\text{OH}\end{array} \quad \xleftrightarrow{-2\text{H}_2\text{O}} \quad \begin{array}{c}\text{NR}\\||\\\text{C}-\text{H}\\|\\\text{C}-\text{OH}\\|\\\text{C}-\text{H}\\|\\\text{C}-\text{H}\\||\\\text{C}-\text{OH}\\|\\\text{CH}_2\text{OH}\end{array}$$

Reductone

The partial dehydration of the sugar may be written in the following form as well:

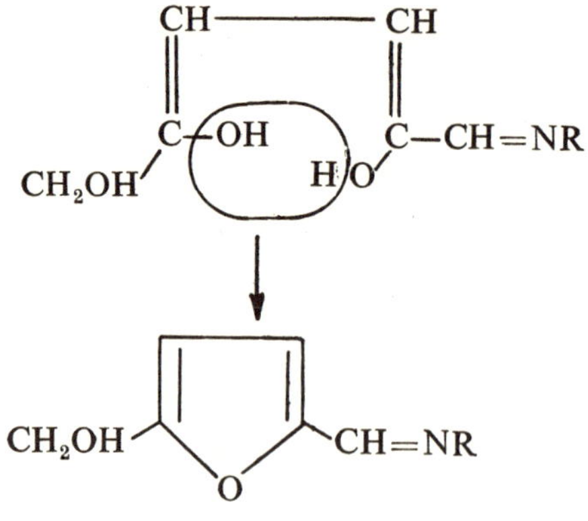

[Structural diagram showing cyclization with loss of H_2O to form OH-methyl furfural with CH_2OH and $CH=NR$ substituents]

OH-methyl furfural

Measuring the amount of OH-methyl furfural serves as an estimate of the extent of browning and the establishment of the shelflife of a product. The OH-methyl furfural will also undergo polymerization (aldehyde-amine polymerization) or aldol condensation reactions.

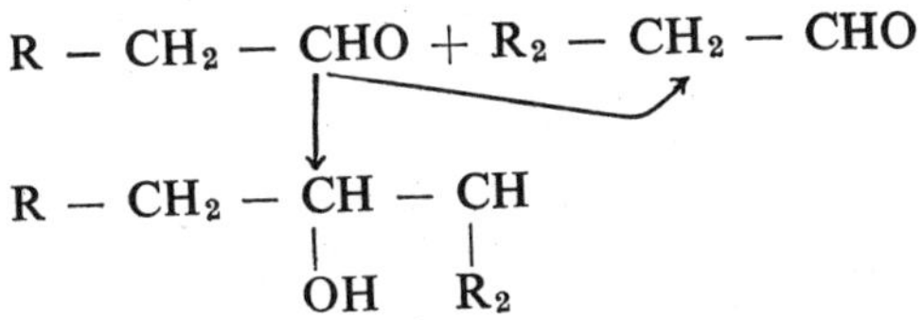

$$R-CH_2-CHO + R_2-CH_2-CHO$$

$$R-CH_2-\underset{OH}{\overset{}{CH}}-\underset{R_2}{\overset{}{CH}}$$

At the presence of an amine, the C–C bond a to carbonyl is weakened after enolization, i.e., the occurrence of dealdolization.

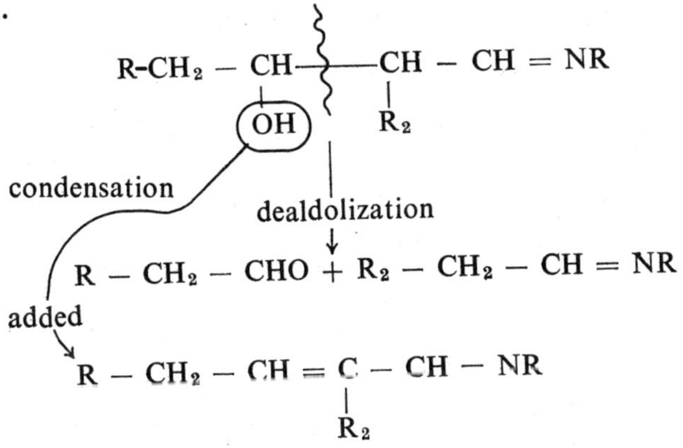

$$R-CH_2-\underset{\text{(OH)}}{CH}\!\!\not\!\!-CH-CH=NR$$
$$\underset{R_2}{|}$$

condensation / dealdolization

$$R-CH_2-CHO + R_2-CH_2-CH=NR$$

added

$$R-CH_2-CH=\underset{R_2}{\overset{}{C}}-CH-NR$$

α–β-unsaturated corbonyl

The dealdolization reaction is usually catalized by the presence of an amino group.

The Strecker degradation reaction is where an amino acid is degraded to an aldehyde of one carbon less. Here, the reaction between an amino acid and a dicarbonyl gives rise to the aldehyde with one carbon less than the original amino acid. 90 to 100% of the CO_2 formed in the browning reaction is due to this Strecker reaction.

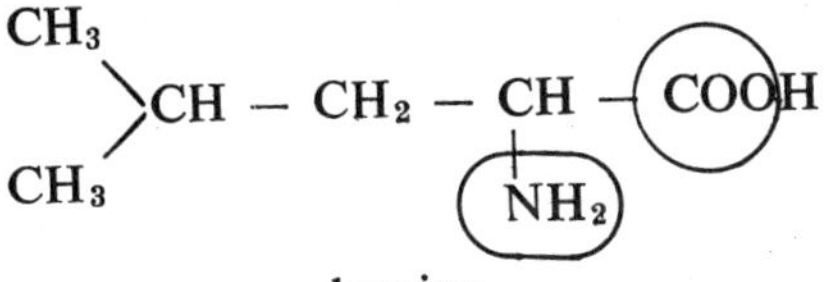

$$\underset{CH_3}{\overset{CH_3}{\diagdown}}CH-CH_2-\underset{(NH_2)}{\overset{}{CH}}-(COOH)$$

leucine

$$-\!\!\mid\!\!- CH_3 - \overset{O}{\underset{\|}{C}} - \overset{O}{\underset{\|}{C}} - OH$$

dicarbonyl

$$\longrightarrow \begin{array}{c} CH_3 \\ \diagdown \\ \diagup \\ CH_3 \end{array}\!\!CH - CH_2 - CHO + CH_3 - \underset{\underset{NH_2}{|}}{CH} - COOH + CO_2 \downarrow \quad (90\text{--}100\%)$$

Aldol condensation

This reaction is catalized by amino compounds.

Simple aldehydes and amines react at low temperatures forming highly colored polymers. Once the initial stage occurs at high temperatures, low moisture content, and a pH between 6 and 10, the reaction progresses even if the temperature is drastically lowered.

$$R - CHO + RNH_2 \longrightarrow RCH = NR$$

The unsaturated carbonyl then reacts with the amine.

$$R - CH = CH - CHO + RNH_2 \longrightarrow$$

$$R - \underset{\underset{RNH}{|}}{CH} - CH = CHOH \xrightarrow{+RNH_2}$$

$$R - \underset{\underset{RNH}{|}}{CH} - CH = CH - \underset{\underset{R}{|}}{NH}$$

Browning in citrus fruits is brought about by the dehydration of ascorbic acid at pH 6-10. This becomes more pronounced at lower pH.

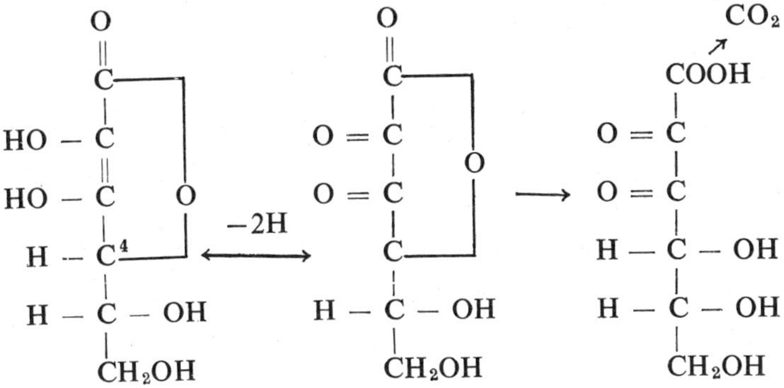

$$\text{Ascorbic Acid} \xrightarrow{-2H} \text{Dehydro-ascorbic Acid} \longrightarrow \begin{array}{c} CO_2 \nearrow \\ COOH \\ | \\ O=C \\ | \\ O=C \\ | \\ H-C-OH \\ | \\ H-C-OH \\ | \\ CH_2OH \end{array}$$

Ascorbic Acid — Dehydro-ascorbic Acid

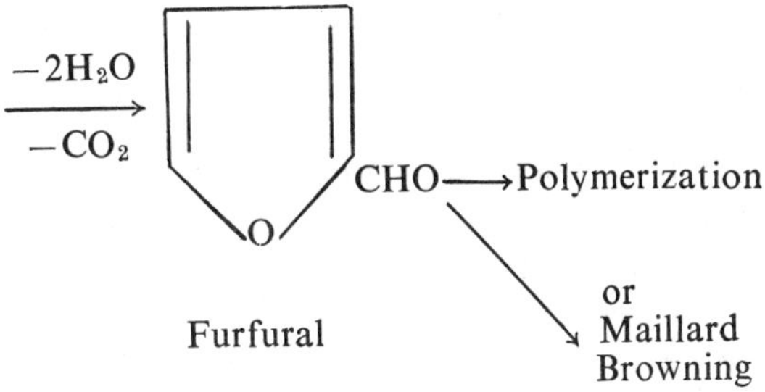

$$\xrightarrow[-CO_2]{-2H_2O} \text{Furfural (CHO)} \longrightarrow \text{Polymerization}$$
$$\searrow \text{or Maillard Browning}$$

Carmilization

This type is the active aldehyde reaction where dehydration of the sugar in an acid media occurs even in the absence of amines.

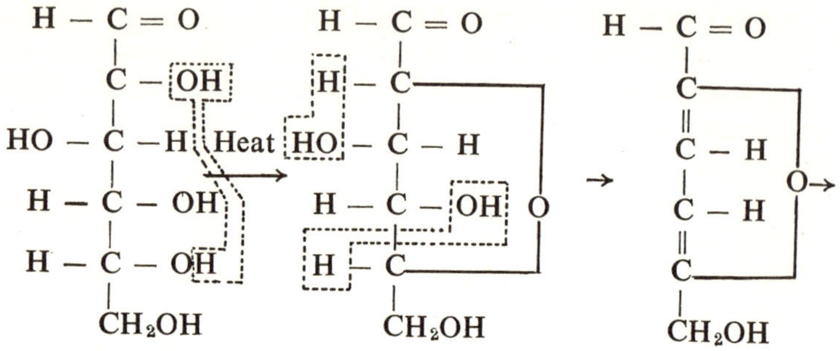

$$\begin{array}{c} H-C=O \\ | \\ C-OH \\ | \\ HO-C-H \\ | \\ H-C-OH \\ | \\ H-C-OH \\ | \\ CH_2OH \end{array} \xrightarrow{Heat} \begin{array}{c} H-C=O \\ | \\ H-C \\ \| \\ HO-C-H \\ | \\ H-C-OH \\ | \\ H-C \\ | \\ CH_2OH \end{array} O \rightarrow \begin{array}{c} H-C=O \\ | \\ C \\ \| \\ C-H \\ | \\ C-H \\ \| \\ C \\ | \\ CH_2OH \end{array} O \rightarrow$$

Sugar ⟶ Hydroxymethyl furfural.

PREVENTION OF MAILLARD BROWNING

1. By avoiding use of high temperature.
2. By addition of SO_2 which ties up the carbonyls.
3. Use of a low pH (optimum for Maillard is 6-10).
4. By reducing the concentration of sugar and amines in media.
5. By replacement of reducing sugars with sucrose (which does not undergo Maillard reaction).
6. By using fructose (does not undergo Maillard reaction).
7. Removal of reducing sugars from media through fermentation reactions before drying or by enzymatic reactions (as products low in soluble sugars and soluble proteins are usually more resistant to browning reactions).
8. Protein may be removed by coagulation or by absorption on resins.

CARBOHYDRATE METABOLISM

Metabolism is the total of all chemical processes that are performed by an organism. There are two types of metabolic reactions:

1. *Catabolism,* which is the breaking-down or degradation of large molecules to small ones. Energy is derived from such exergonic reactions, and they have negative δG. Energy produced is usually trapped as ATP.

2. *Anabolism.* Here build-up or synthesis occurs in an endergonic reaction that has a positive δG.

Usually, not the same enzymes are chosen for catabolism and anabolism. For example, it takes twelve enzymes to degrade glycogen to lactic acid, while it takes only nine enzymes for the synthesis of glycogen from lactic acid. Fatty acid oxidation occurs through a set of enzymatic reactions in the mitochondria while fatty acid synthesis takes place by another set af enzymes outside the mitochondria. The two processes, thus, occur in different compartments within the same cell.

The glycolytic system sets the stage for carbohydrate metabolism where glucose is used as source of energy. Here energy is derived, stepwise, from glucose, and is trapped as adenosine triphosphate (ATP), due to the conversion of hexoses to 3-carbon fragments that get oxidized and used. This is the primitive pathway that dominated before the production of O_2 on earth. The first reaction in glycolysis is a phosphorylation reaction where a hexokinase enzyme transfers a phosphate group from ATP into glucose at number 6 carbon. This reaction is thermodynamically irreversible. The addition of a phosphate group to glucose gives it a polar handle that helps it break the membrane barrier and enter the cell for its subsequent metabolism. Hexokinase is not so specific; it may, as well, phosphorylate other sugars, like mannose and galactose, and it is activated with insulin. Glucose-6-phosphate gets converted inside the cell into fructose-6-phosphate by an isomerase enzyme. There is no importance energetically to this reaction. The enzyme phosphofructokinase (PFK) transfers a phosphate group from ATP to carbon-1 of fructose-6-phosphate, forming fructose-1, 6-diphosphate and, thus far, two ATPs have been used, which is a loss energy-wise. The fructose-1, 6-diphosphate is cleaved by an aldolase enzyme to 3-carbon-unit, simple phosphorylated keto- and aldo-sugars (dihydroxyacetone phosphate and 3-phosphoglyceraldehyde). Here, aldolase catalyzes cleavage of the C–C bond between C-3 and C-4, producing dihydroxyacetone phosphate from the top carbons, and 3-phosphoglyceraldehyde from the bottom three carbons.

There is 96% dihydroxyacetone phosphate produced, compared to 4% of the 3-phosphoglyceraldehyde; however, there is equilibrium between the two catalyzed by an isomerase enzyme, and eventually all the dihydroxyacetone phosphate gets converted to the 3-phosphoglyceraldehyde (the active fragment in glycolysis). This isomerase works on the bottom carbon of dihydroxyacetone phosphate and, thus, converts it to 3-phosphoglyceraldehyde.

There are two enzymes, glyceraldehyde-3-ph-dehydrogenase and phosphoglycerokinase, involved in the next step, the production of ATP and the conversion of NAD to $NADH_2$. Here oxidation of glyceraldehyde to glyceric acid by NAD, forming a mixed anhydride and energy conservation, takes place. This step is inhibited by iodoacetate, arsenate, and maybe by fluoride ions, as well. This inhibition presumably is done through the oxidation of the sulfhydryl group (SH) of the dehydrogenase enzyme that is necessary for its activity.

This step is the important, first oxidation-reaction where energy is derived through a direct anaerobic dehydrogenation reaction. NAD causes the top carbon on the phosphoglyceraldehyde to be oxidized to the acid form, and the production of $NADH_2$, which means the generation of 3ATPs. The next step is the addition of an inorganic phosphate, and thus the production of 1, 3-diphosphoglyceric acid. To get energy out of 1, 3-diphosphoglyceric acid, the high-energy phosphate is transferred to ADP, thus, first trapping of energy into

ATPs occurs, and 3-phosphoglyceric acid is subsequently produced. Next is the transfer of the phosphate group from the C-3 to the C-2 position by a mutase enzyme, this reaction not being important energy-wise, and 2-phosphoglyceric acid is produced. Next is an important key enzyme reaction involving an enolase enzyme which cleaves out H_2O from 2-phosphoglyceric acid, thus dehydrating it to phosphoenol pyruvate. This dehydration generates another high-energy reaction. The enolase enzyme is inhibited by fluoride ions which replace Mg^{++} (the trace-metal cofactor activating the enzyme). Phosphoenolpyruvic acid is produced from the enolase reaction. Thermodynamically, the most important reaction which drives glycolysis is the one-way reaction that catalyzes the conversion of phosphoenol pyruvate to pyruvate by the enzyme pyruvic kinase.

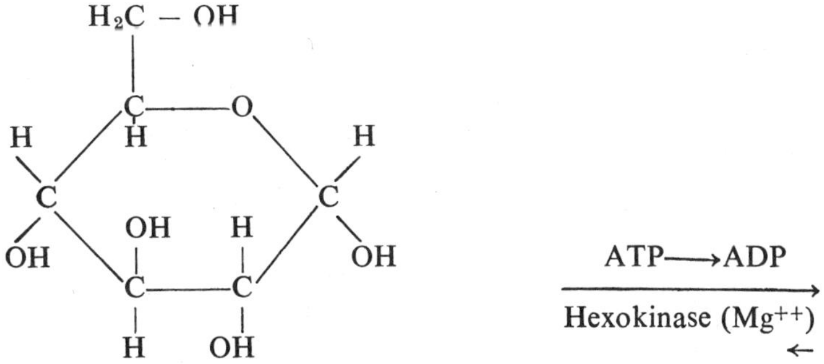

Glucose

$$\xrightarrow[\text{Hexokinase } (Mg^{++})]{ATP \longrightarrow ADP}$$

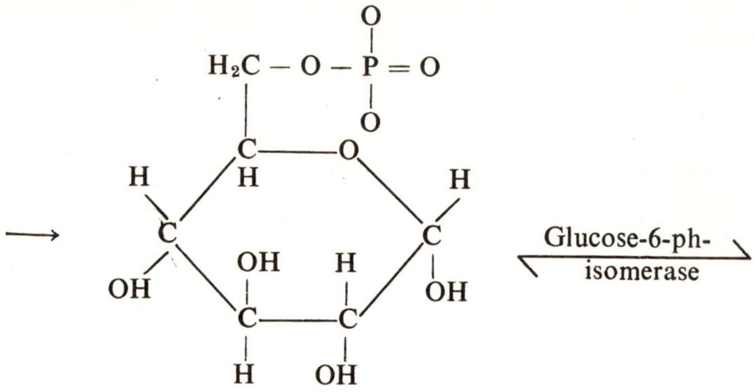

→

Glucose-6-ph.

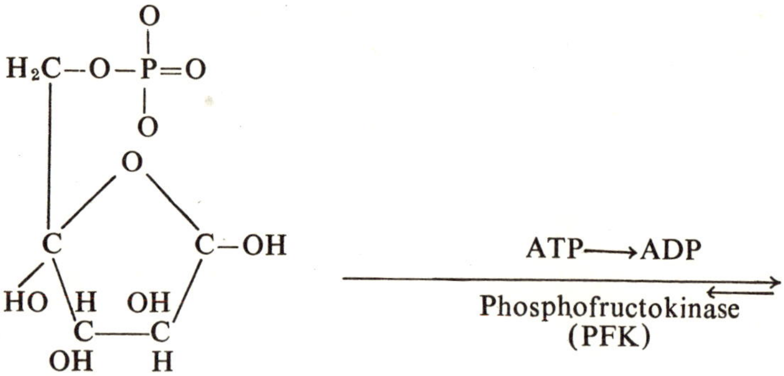

$\xrightarrow{\text{Glucose-6-ph-isomerase}}$

Fructose-6-ph.

$\xrightarrow[\text{Phosphofructokinase (PFK)}]{\text{ATP} \longrightarrow \text{ADP}}$

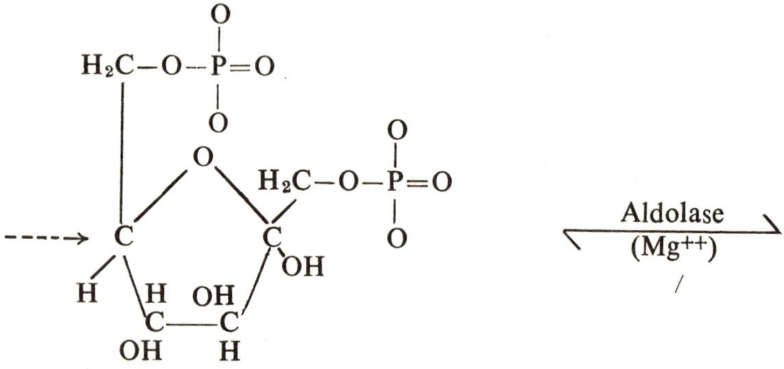

⇢ Fructose 1, 6 diph.

$\xrightarrow{\text{Aldolase } (Mg^{++})}$

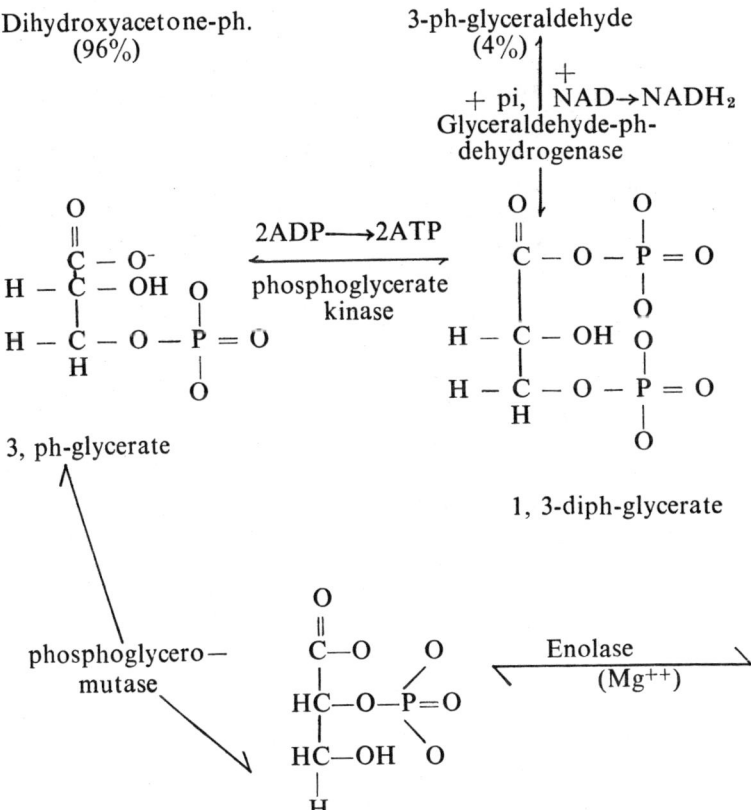

$$\underset{\text{ph-enol-pyruvate}}{\begin{array}{c}\text{O}\\\|\\\text{C—O}\quad\text{O}\\|\quad\quad|\\\text{C—O—P=O}\\\|\quad\quad|\\\text{H—C}\quad\text{O}\\|\\\text{H}\end{array}} \xrightarrow[\text{Pyruvate kinase}]{\text{ADP}\longrightarrow\text{ATP}} \underset{\text{Pyruvate}}{\begin{array}{c}\text{O}\\\|\\\text{C—O}^-\\|\\\text{C=O}\\|\\\text{H—C—H}\\|\\\text{H}\end{array}}$$

Control Mechanisms

1. Control points in metabolism may be accomplished through induction or repression of certain enzymes, or through the inhibition or activation of enzymes that are already there.
2. Positive or negative feedback by the end products of metabolism, like the accumulation of these end products, affects the equilibrium of reactions.
3. Pacemaker enzymes, which are thermodynamically irreversible.
4. Anaplerotic control points, which occur at the side and not on the main pathway.
5. Small-size products may affect the conformation of enzyme protein through the "induced fit" reaction.
6. Phosphofructokinase is an important rate-limiting enzyme in glycolysis. High levels of readily available energy like ATP, or high-potential energy that will soon be available, such as citrate, starch, glycogen, and inorganic phosphate, will all decrease phosphofructokinase activity, and, consequently, inhibit glycolysis. On the other hand, ADP and AMP will increase phosphofructokinase activity and activate glycolysis.

Since glycolysis is anaerobic, where energy is obtained without O_2, it, thus, requires an abundance of sugar, contrary to the aerobic metabolism where O_2 presence requires much less sugar and produces abundant energy in the meantime. It, therefore, becomes evident that O_2 will inhibit glycolysis (Pasteur's effect), while lack of O_2 activates it (Crabtree's effect).

THE TRICARBOXYLIC ACID (TCA) CYCLE

The first step in the tricarboxylic acid (TCA) cycle is a nonreversible, one-way reaction where decarboxylation followed by oxidation occurs, utilizing thiamine pyrophosphate coenzyme (TPP), biotin, and Co-A. The pyruvate dehydrogenase enzyme complex mediates this reaction and NAD, FAD, lipoate, and dehydrolipoic dehydrogenase (DLD) are all involved in the reaction. The enzyme citrate synthase activates the condensation of acetyl-Co-A with oxaloacetic acid, giving rise to citrate. The TCA cycle actually starts with this condensation of the active 2-carbon unit, acetyl-Co-A, with oxaloacetate. Citrate requires three points of attachment to the enzyme, and carbons that are added last come from oxaloacetic and will be last to be oxidized. The enzyme aconitase splits out water from citrate and, thus, produces cis-aconitate. Another aconitase adds water to cis-aconitate, giving rise to isocitrate. Isocitrate undergoes a similar reaction to that catalyzing the conversion of pyruvate to acetyl-Co-A, where an isocitrate dehydrogenase enzyme complex converts it to α-ketoglutarate. Here, as well, NAD, FAD, Co-A, biotin, TPP, lipoate, and DLD are involved in this oxidative decarboxylation reaction where the first CO_2 is produced. α-ketoglutarate is converted to succinyl-Co-A by the α-ketoglutaric dehydrogenase enzyme complex that requires the same cofactors involved in the preceeding reaction, and a second CO_2 is produced. Succinyl-Co-A is converted to succinate by a thiokinase enzyme and use of GDP. Here GTP and Co-A are produced. Succinate is converted to fumarate by succinic dehydrogenase, which requires FAD for a cofactor that is covalently bound to it, and catalyzes the removal of 2Hs from the central carbon of succinate, giving rise to fumarate.

Fumarate loses water through a fumarase enzyme and is converted to malate. Fumarase is a good example of a non-heme iron enzyme, and is inhibited by malonate. Malate is converted to oxaloacetate through malic dehydrogenase and NAD. The two Co-A's produced at the isocitrate and α-ketoglutarate dehydrogenase reactions come originally from the oxaloacetic and not from the acetyl-Co-A part.

Isocitric dehydrogenase catalyzes the rate-limiting reaction of the TCA cycle. It is a good example of an allosteric enzyme that is activated with ADP and AMP, and is inhibited by ATP.

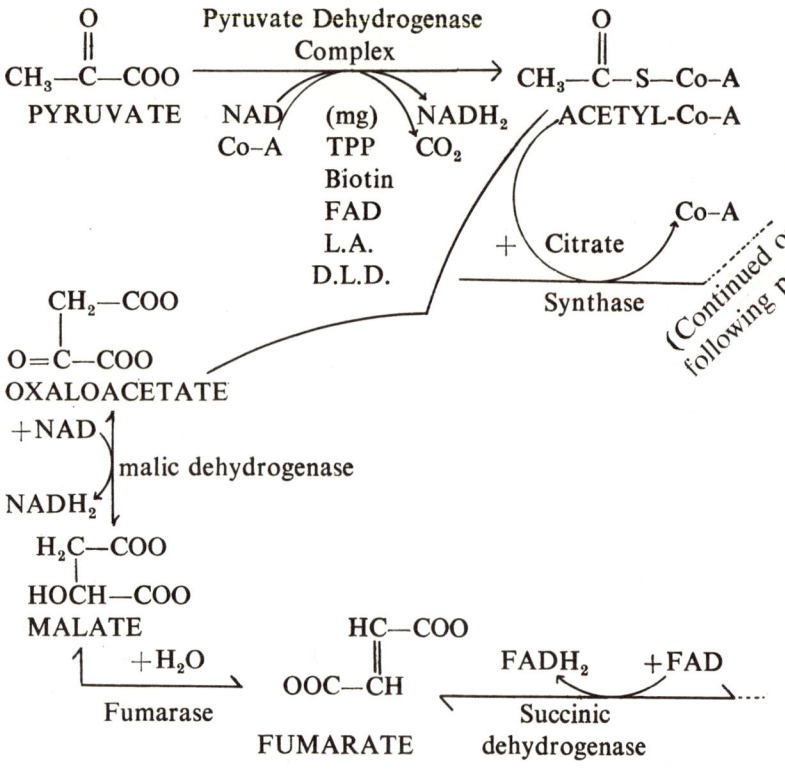

TCA CYCLE

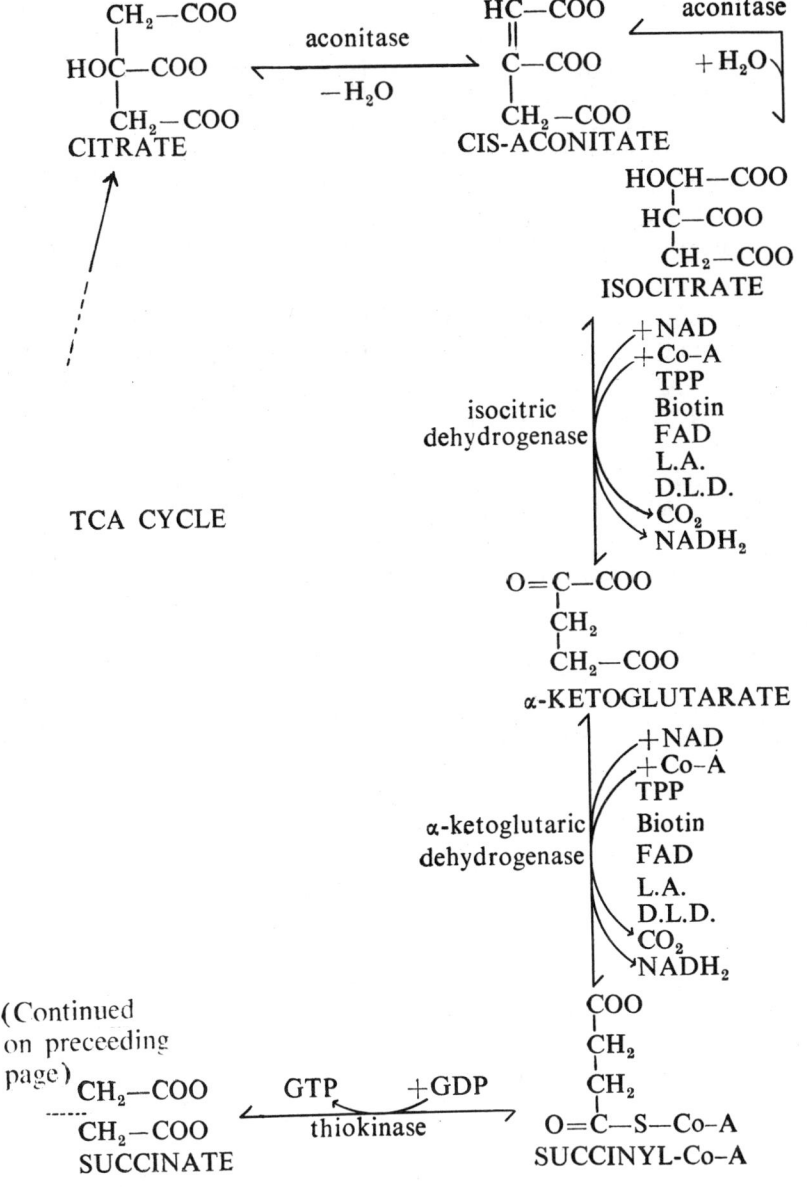

Energy Production

Reaction	Coenzyme	ATPs Produced
Isocitric dehydrogenase	$NADH_2$	3
α-ketoglutaric dehydrogenase	$NADH_2$	3
Succinyl-Co-A	GTP	1
Succinic dehydrogenase	$FADH_2$	2
Malic dehydrogenase	$NADH_2$	3

Thus, starting with one acetyl-Co-A, we obtain twelve ATPs. However, since two acetyl-Co-As are produced from glucose, therefore, there are 24 ATPs produced from one glucose in the TCA cycle. There are eight ATPs produced from the anaerobic glycolytic metabolism of glucose just before getting into the TCA cycle. The complete oxidation of one glucose through the pathway of glycolysis followed by the TCA cycle to CO_2 and water, however, produces 38 ATPs, which represents a 45% efficiency.

While the discussed pathway is a major one for glucose metabolism in most tissues, an alternative pathway of carbohydrate metabolism in some tissues exists through the phosphogluconate oxidative pathway, or the pentose shunt, through which up to 30% of glucose metabolism may take place in tissues such as liver, kidney, mammary gland, and adipose (skeletal muscle does not use this cycle).

THE PENTOSE SHUNT

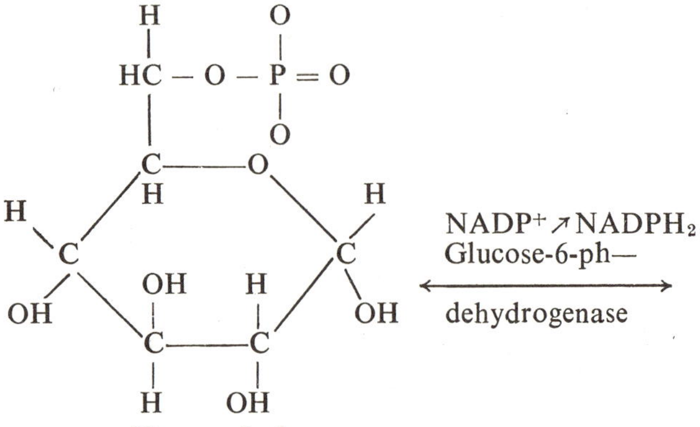

Glucose-6-ph.

$NADP^+ \nearrow NADPH_2$
Glucose-6-ph—
⟵⟶ dehydrogenase

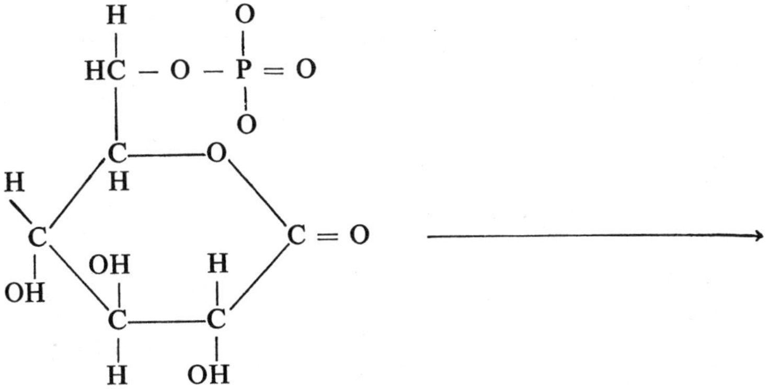

Gluconolactone-6-ph

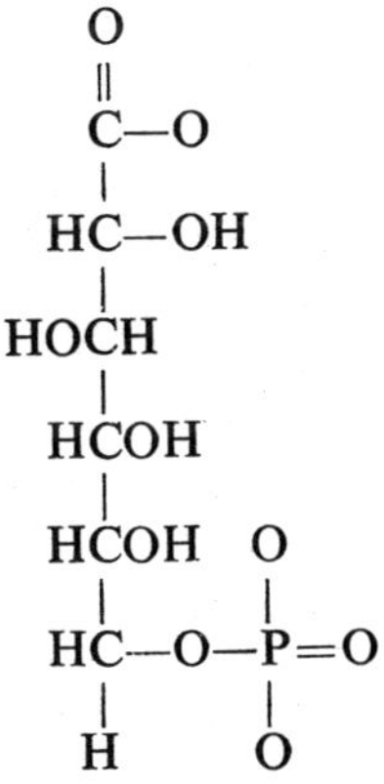

$\xrightarrow{}$

$$\begin{array}{c} O \\ \parallel \\ C-O \\ | \\ HC-OH \\ | \\ HOCH \\ | \\ HCOH \\ | \\ HCOH \quad O \\ | \quad\quad | \\ HC-O-P=O \\ | \quad\quad | \\ H \quad\quad O \end{array}$$

6-phosphogluconate

$\xrightleftharpoons[\text{dehydrogenase}]{\text{NADP}^+ \longrightarrow \text{NADPH}_2 \\ \text{6-ph-gluconic}}$

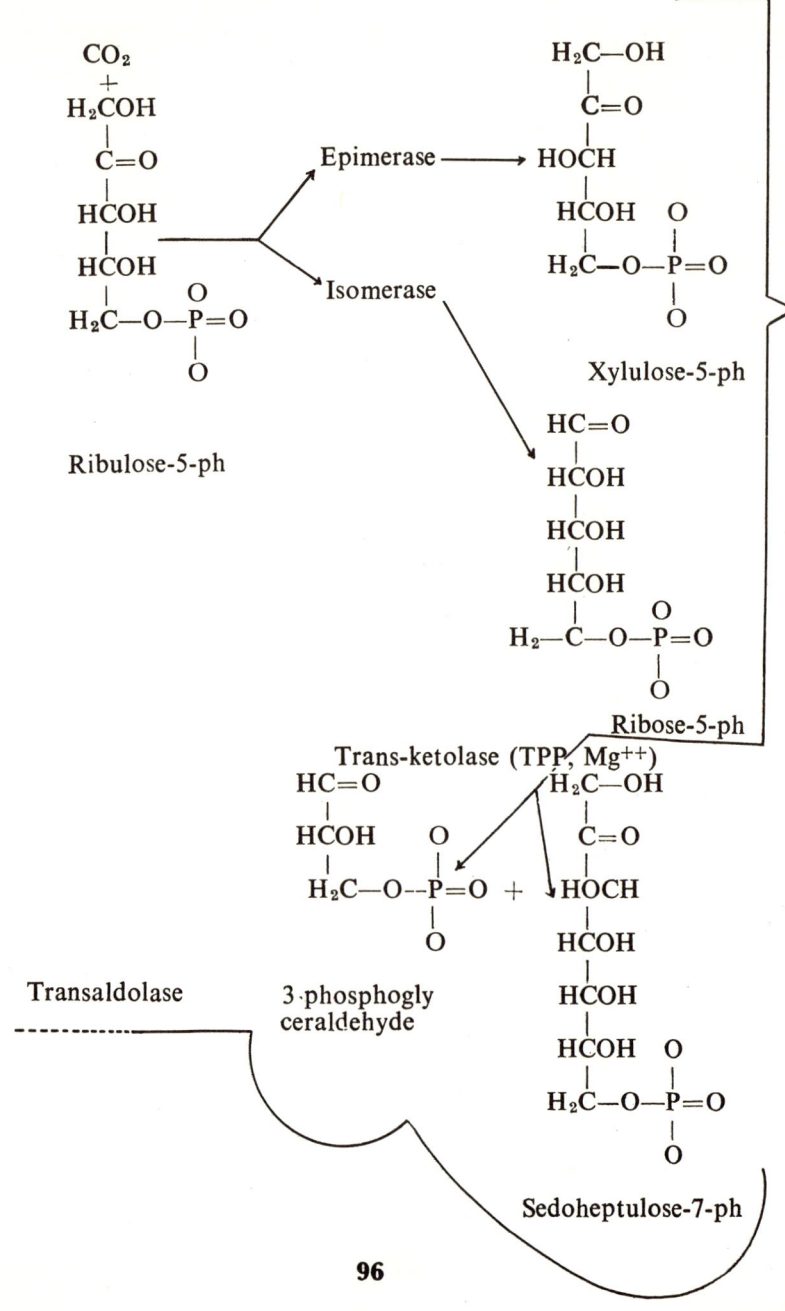

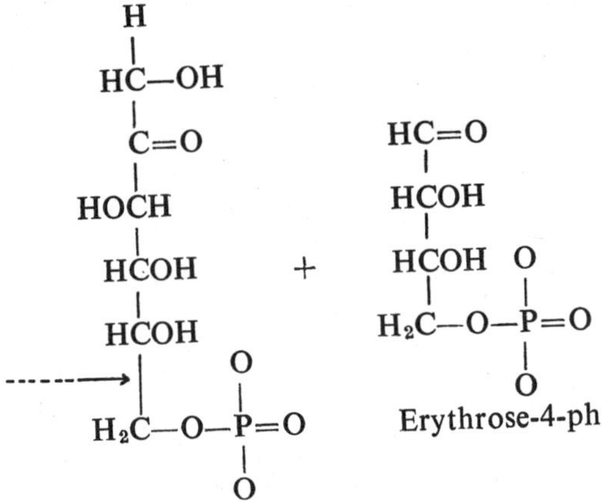

$$\begin{array}{c}\text{H}\\|\\\text{HC-OH}\\|\\\text{C=O}\\|\\\text{HOCH}\\|\\\text{HCOH}\\|\\\text{HCOH}\\|\\\text{H}_2\text{C-O-P(=O)(O)O}\end{array} \quad + \quad \begin{array}{c}\text{HC=O}\\|\\\text{HCOH}\\|\\\text{HCOH}\\|\\\text{H}_2\text{C-O-P(=O)(O)O}\end{array}$$

Fructose-6-ph   Erythrose-4-ph

In summary of the pentose sugar shunt:

1. 6-glucose-6-ph + 12 NADP$^+$ $\longrightarrow$ 5 glucose-6-ph + 6CO$_2$ + 12 NADPH + 12H$^+$

Carbon-1 of glucose is oxidized to CO_2, and we end up with 5 glucose-6-ph and 12 NADPH$_2$ which, when multiplied by three, will mean 36 ATPs have been producd from the oxidation of one mole of glucose through this pathway, which allows specific tissues to bypass glucose-6-phosphate.

2. NADPH$_2$ produced here in cytosol is used for fat and steroids synthesis.

3. CO_2 production goes into photosynthesis.

4. The 5-carbon sugars go into nucleic acid synthesis.

5. Various other sugars produced in this pathway are 3, 4, 5, 6, and 7-carbon sugars, which are important in metabolism in allowing for a considerable interconversion of sugars.

6. The 36 ATPs produced from one mole of glucose in this pathway make it almost as efficient as the glycolytic-TCA cycle pathway which was discussed earlier.

Since from phosphoenol pyruvate to pyruvate and from pyruvate to acetyl-Co-A are both one-way reactions, there is no way in getting the carbons back to a carbohydrate and, thus, acetyl-Co-A needs to be bypassed. To bypass acetyl-Co-A, anaplerotic reactions which add back intermediates, particularly oxaloacetate, to the cycle come into play. Here, for example, pyruvate is directly converted to oxaloacetate or directly to malate; or phosphoenol pyruvate is directly converted to oxaloacetate. These bypass reactions become important in a high-fat diet where there is an **abundancy** of acetyl-Co-A, but in the meantime, it cannot revert to pyruvate.

Anaplerotic Reactions That Generate Oxaloacetic

1. $H_3C-\overset{O}{\overset{\|}{C}}-COO^- + CO_2 + ATP \xrightarrow[\text{(Biotin}\;Mn^{++},\;CO_2)]{\text{Pyruvic-Carboxylase}} \begin{array}{c} COO^- \\ | \\ CH_2 \\ | \\ C=O \\ | \\ COO^- \end{array}$

   Pyruvate

   oxaloacetic + ADP + Pi

2. $H_3C-\overset{O}{\overset{\|}{C}}-COO^- + CO_2 + NADPH \xrightarrow{\text{malic dehydrogenase}} \begin{array}{c} COO \\ | \\ CH_2 \\ | \\ HC-OH \\ | \\ COO \end{array} + NADP$

   Pyruvate

   Malate

3. $H_2C=C-COO$
   $\quad\quad |$
   $\quad\quad O$
   $\quad\quad \diagdown\!\!\diagup$
   $\quad O-P=O$
   $\quad\quad |$
   $\quad\quad O$

   Phosphoenol pyruvate

   $+\;\;CO_2 + GDP \xrightarrow{\text{ph-enol-pyruvic carboxylase}} \begin{array}{c} COO \\ | \\ CH_2 \\ | \\ C=O \\ | \\ COO \end{array} + GTP$

   Oxaloacetate

If, in the meantime, blood glucose is low (due to high insulin secretion or low epinephrine or glucagon secretion) and the animal does need glucose in his blood, he may obtain it from his glycogen storage in the liver (animals do not have large storage of carbohydrates; usually 4% carbohydrate storage in the form of glycogen is the maximum). Release of glucose into blood from liver glycogen is a hormonal-mediated process where epinephrine or glucagon play the role of the master activator. Here, epinephrine or glucagon activates the large enzyme adenyl cyclase which sits on a specific receptor on the cell membrane and requires Mg^{++} for a cofactor. Upon activation of adenyl cyclase by the hormone, it makes cyclic-AMP (c-AMP) from ATP. c-AMP, the second messenger, in its turn, activates the inactive phosphorylase-b by further phosphorylating it into the "a" form (this requires 4ATPs and B_6 vitamin. The phosphorylase-a is composed of four polypeptide chains, each with a phosphorylated serine residue). This active phosphorylase-a acts at the nonreducing end of the glycogen chain, attacking the sugar bond and phosphorylating the sugar in the meantime, thus producing glucose-1-phosphate which, in turn, is converted to glucose-6-phosphate by the enzyme phosphoglucomutase. The enzyme phosphatase then splits the phosphate group at carbon-6, and glucose is consequently released into blood. This latter process, however, cannot occur in muscle, as phosphatase is not present in muscle (present in endoplasmic reticulum, liver, kidney, intestinal mucosa). In the case of muscle, glucose-6-phosphate goes through glycolysis to lactic acid and finally to glycogen in liver:

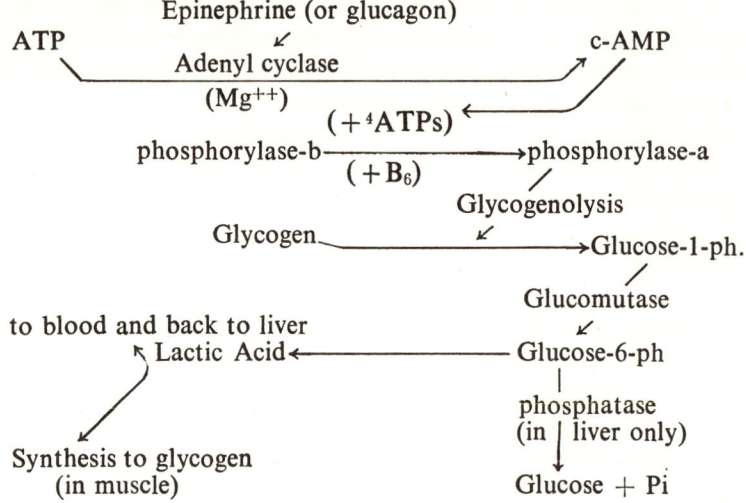

Metabolism of Galactose

Galactose is an important sugar as it is the only carbohydrate present in milk. Infants depending on milk for survival may have a severe problem if they lack the necessary enzymes required for galactose metabolism, namely, UDP-galactose pyrophosphorylase and uridyl transferase.

For its metabolism, galactose has to be converted to glucose through the aid of a uridine derivative.

$$\text{Galactose-1-ph} + \text{UTP} \xrightarrow{\text{UDP-galactose pyrophosphorylase}} \text{UDP-galactose} + \text{PPi}$$

or

$$\text{Galactose-1-ph} + \text{UDP-glucose} \xrightarrow{\text{uridyl transferase}} \text{UDP-galactose} + \text{glucose-1-ph}$$

Once galactose is activated to UDP-galactose, it gets easily converted by an epimerase enzyme to the UDP-glucose form and hence metabolized: UDP-galactose Epimerase, UDP-glucose. When uridyl transferase enzyme is missing (hereditary effect in some infants), interconversion of galactose to glucose is inhibited and galactosemia ensues.

REFERENCES

General

Braverman, J. B. 1949. *Citrus Products, chemical composition and chemical technology.* New York: Interscience.

Davis, P. R., and Prince, R. N. 1955. Liquid sugar in the food industry, in use of sugars and other carbohydrates in the food industry. *Advances in Chem., Ser. No. 12.*

Deuel, H., and Stutz, E. 1958. Pectic substances and pectic enzymes. *Advances in Enzymology 20:341.*

Haynes, W. 1953. *Cellulose: the chemical that grows.* New York: Doubleday and Co.

Kertesz, Z. I. 1951. *The pectic substances.* New York: Interscience.

Percival, E. G. 1950. *Structural carbohydrate chemistry.* London: F. Muller.

Pigman, W. W. 1957. *The carbohydrates, chemistry, biochemistry, physiology.* New york: Academic Press.

———, and Goepp, R. M. 1958. *Chemistry of carbohydrates.* New York: Academic Press.

Radley, J. A. 1943. *Starch and its derivatives.* London: Chapman and Hall.

Solms, J. 1960. Some structural aspects of the gel formation of pectins and related polysaccharides, in physical functions of hydrocolloids. *Advances in Chem., Ser. No. 25.*

Tauber, H. 1949. *The chemistry and technology of enzymes.* New York: J. Wiley and Sons.

Torr, L. W. 1923. A study of the factors affecting the jellying of fruits. *Univ. Delaware Agr. Exptl. Sta. Bull. 133:14.*

Enzymatic Browning

Biale, J. B. 1950. Post-harvest physiology and biochemistry of fruits. *Plant Physiol. Ann. Rev. 1:183.*

Dixon, M. 1949. *Multi-enzyme systems.* Cambridge: The University Press.

Griffiths, L. A. 1959. Detection and identification of the polyphenoloxidases substrate of the banana. *Nature* 184:58.

Joslyn, M. A., and Ponting, J. D. 1951. Enzyme-catalyzed oxidative browning of fruit products. *Advances in Food Research* 3:1.

Kertesz, Z. I. 1933. The oxidase system of non-browning yellow peach. *N.Y. Agr. Exptl. Sta. Tech. Bull.* 216.

Lu Valle, J. E., and Goddard, D. R. 1948. The mechanism of enzymatic oxidations and reductions. *Quart. Rev. Biol.* 23:197.

Onslow, M. W. 1920. Oxidizing enzymes III. The Oxidizing enzymes of some common fruits. *Biochem. J.* 14:541.

Szent-Gyorgyi. 1939. *Studies on biological oxidation and some of its catalysts.* Baltimore: Williams and wilkins Co.

Nonenzymatic Browning

Berk, Z, and Braverman, J. B. 1958. Some observations on nonenzymatic browning of citrus concentrates. *Symposium of Fruit Juice Concentrates,* Bristol: Fruit Juice Union.

Braverman, J. B. 1953. The mechanism of the interaction of SO_2 with certain sugars. *J. Sci. Food Agr.* 4:540.

Danehy, J. P., and Pigman, W. W. 1951. Reactions between sugars and nitrogenous compounds and their relationship to certain food problems. *Advances in Food Research* 3:241.

Ellis, G. P. 1959. The Maillard reaction. *Advances Carbohydrate Chem.* 14:63.

Hodge, J. E. 1953. The chemistry of browning reactions. *J. Agr. Food Chem.* 1:92.

Isbell, H. S., and Frush, H. L. 1957. Reaction of the glycosylamines. U.S. Atomic Energy Comm. Rept. No. NBS-5352, National Bureau of Standards, Washington.

Joslyn, M. A., and Braverman, J. B. 1954. The chemistry and technology of the pre-treatment and preservation of fruit and vegetable products with SO_2. *Advances in Food Research* 5:97.

Katchalsky, A. 1941. Interaction of aldoses with α-amino acids or peptides. *Biochem. J.* 35:1024.

Lamden, M. P., and Harris, R. S. 1950. Browning of ascorbic acid in pure solutions. *Food Research* 15:79.

Maillard, L. C. 1912. Action of amino acids on sugars. *Compt. rend* 154:66.

Moore, E. L.; Errelen, W. B.; and Fellers, C. R. 1942. Factors responsible for the darkening of packaged orange juice. *Fruit Products J.* 22:100.

Proctor, B. E., and Goldblith, S. A. 1951. Electromagnetic radiation fundamentals and their application in food technology. *Advances in Food Research* 3:119.

Stadtman, E. R. 1948. Non-enzymatic browning in fruit products. *Advances in Food Research* 1:325.

Wolfrom, M. L.; Schultz, R.D.; and Calvalieri, L. E. 1948. *J. Amer. Chem. Soc.* 70:514.

Chapter 4

PROTEINS

INTRODUCTION

Proteins, more than any other ingredient, characterize foods. They are not only important components in man's diet, but, indeed, they are essential for his maintenance and growth. Proteins are synthesized by plants, upon which man depends for his protein supply and requires practically one gram of protein a day for each kilogram of his body weight. This requirement increases in pregnancy, rapid growth, lactation, and during some emergency situations.

The biological value of proteins varies with their amino acid content. Any essential amino acid deficiency will limit the value of protein. In this regard, protein of eggs, milk, and meat is superior to other proteins. Proteins from plant sources may lack or are low in certain essential amino acids, such as maize corn which is deficient in tryptophane and low in lysine, and wheat which is low in lysine.

AMINO ACIDS IN FOODS

Valine

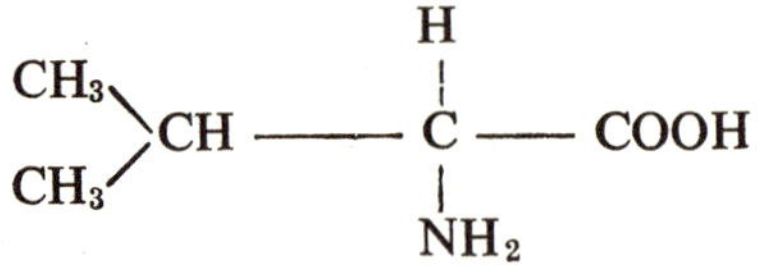

$$(CH_3)_2CH - \underset{\underset{NH_2}{|}}{\overset{\overset{H}{|}}{C}} - COOH$$

Valine is present in all foods. It is high in milk and eggs (8-10%) while in meat, it is 5%. In gelatin it is 3%, and in cereals, wheat, and corn, its content is rather low.

Leucine

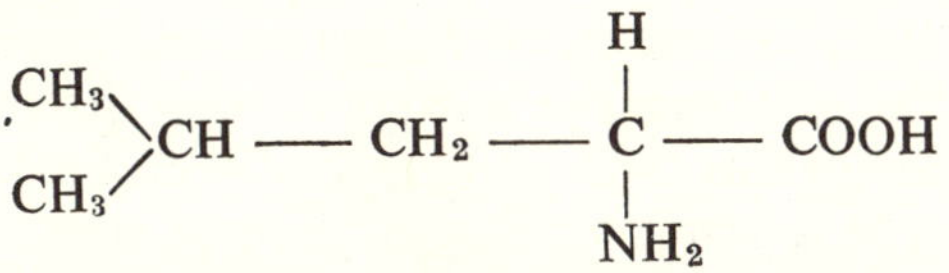

$$\begin{array}{c}CH_3\\ \diagdown\\ CH\end{array}\!\!-\!\!CH_2\!-\!\underset{\underset{NH_2}{|}}{\overset{\overset{H}{|}}{C}}\!-\!COOH$$

Leucine composes 6-15% of most food proteins. Only gelatin is low in leucine content (less than 3%).

Isoleucine

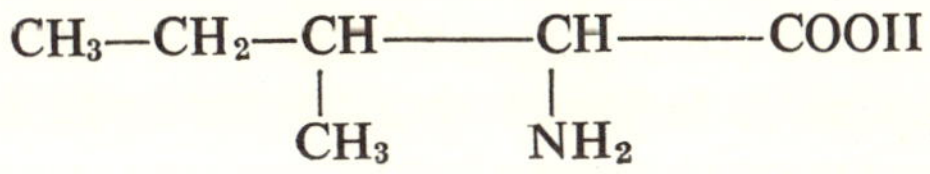

$$CH_3\!-\!CH_2\!-\!\underset{\underset{CH_3}{|}}{CH}\!-\!\underset{\underset{NH_2}{|}}{CH}\!-\!COOH$$

The content of isoleucine is highest in meat, amounting to 5-6%, rather less in milk and eggs, and is considerably less in cereals.

Threonine

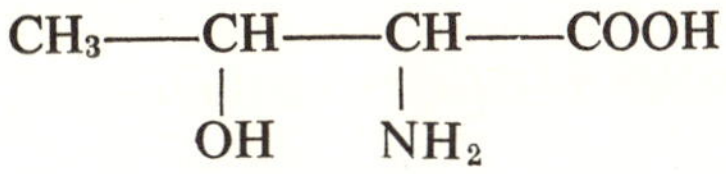

$$CH_3\!-\!\underset{\underset{OH}{|}}{CH}\!-\!\underset{\underset{NH_2}{|}}{CH}\!-\!COOH$$

Threonine is a hydroxy amino acid and is found in all food proteins. Its content in milk, meat, and eggs is about the highest, ranging from 4.5 to 5.0%. In cereals, it is 2.5 to 4.5%.

Lysine

$$CH\text{—}CH_2\text{–}CH_2\text{–}CH_2\text{–}CH\text{—}COOH$$
$$\phantom{CH\text{—}}|\phantom{CH_2\text{–}CH_2\text{–}CH_2\text{–}}|$$
$$\phantom{CH\text{—}}NH_2\phantom{CH_2\text{–}CH_2\text{–}CH_2\text{–}}NH_2$$

Lysine is a basic amino acid. Its terminal amino group is of specific importance. Its content is high in eggs, milk, and meat (7-9%). It is, however, low in grains, and some cereals contain no lysine, although the germ portion of the grain may contain levels as high as those of eggs and meat. Soybean and leafy vegetables have levels similar to those of meat.

Methionine

$$CH_3\text{–}S\text{–}CH_2\text{–}CH_2\text{–}CH\text{–}COOH$$
$$\phantom{CH_3\text{–}S\text{–}CH_2\text{–}CH_2\text{–}}|$$
$$\phantom{CH_3\text{–}S\text{–}CH_2\text{–}CH_2\text{–}}NH_2$$

Methionine is a sulfur-containing amino acid. Its content in eggs, milk, and meat ranges from 3.5 to 4.0%. In cereals, the level may be as low as 1 to 1.5% with the exception of rice which contains levels similar to eggs and milk. Soybeans level may be over 2%.

Phenylalanine

$$\text{C}_6\text{H}_5\text{–}CH_2\text{–}CH\text{–}COOH$$
$$\phantom{\text{C}_6\text{H}_5\text{–}CH_2\text{–}}|$$
$$\phantom{\text{C}_6\text{H}_5\text{–}CH_2\text{–}}NH_2$$

Phenylalanine may be abundant in most foods. In eggs, it is 6%. Most other foods may contain 4-5%.

Tryptophane

$$\text{(indole)}\text{–}CH_2\text{–}CH\text{–}COOH$$
$$\phantom{\text{(indole)}\text{–}CH_2\text{–}}|$$
$$\phantom{\text{(indole)}\text{–}CH_2\text{–}}NH_2$$

Tryptophane occurs in small amounts in all foods. Its content in eggs, milk, and meat ranges from 1.5 to 2.0%. Corn protein is low in tryptophane.

Arginine

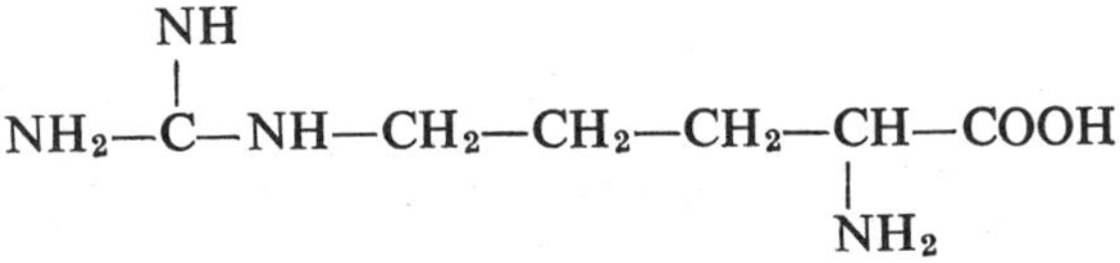

$$NH_2-\underset{\underset{NH}{\|}}{C}-NH-CH_2-CH_2-CH_2-\underset{\underset{NH_2}{|}}{CH}-COOH$$

Arginine is an essential amino acid for rat, chick, and human infants, but is not essential for the grownup man. It contains a guanidyl group and is abundant in protamines (protamine of salmon has 20% arginine). Rat can synthesize arginine, but not in the optimum amount required for its need. Arginine is a required amino acid for the growth of chicks. Eggs, milk, and meat have 5 to 8%, while wheat and rye have 4%.

Histidine

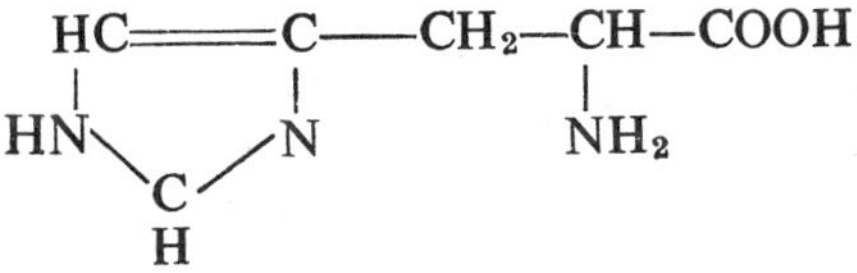

$$\underset{\underset{H}{\underset{|}{C}}}{\underset{HN\diagdown \quad \diagup N}{HC=C}}-CH_2-\underset{\underset{NH_2}{|}}{CH}-COOH$$

Histidine is present in protein of most foods ranging in 1 to 3% in content. Histidine is not an essential amino acid for the adult man, but it is essential for infants.

Cystine

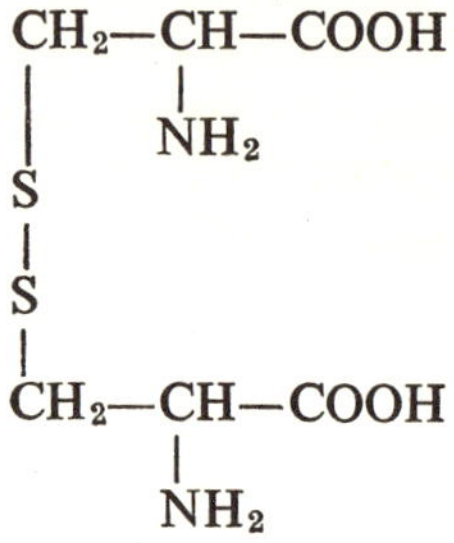

$$\begin{array}{c} CH_2-CH-COOH \\ | \quad\quad | \\ S \quad\quad NH_2 \\ | \\ S \\ | \\ CH_2-CH-COOH \\ | \\ NH_2 \end{array}$$

Cystine is a sulfur-containing amino acid that is usually, in one form or another, is present in spices in considerably high quantities. It is also appreciably present in eggs.

Glycine

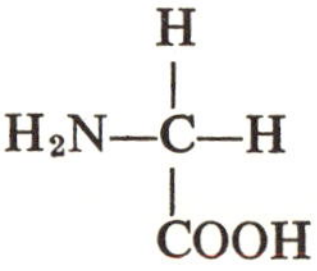

$$\begin{array}{c} H \\ | \\ H_2N-C-H \\ | \\ COOH \end{array}$$

Glycine content of gelatin is 30%. Unlike other amino acids, it has no asymmetric carbon.

Proline

(HO) → Hydroxyproline

$$C\diagdown_{\substack{N \\ H}}\diagup C-COOH$$

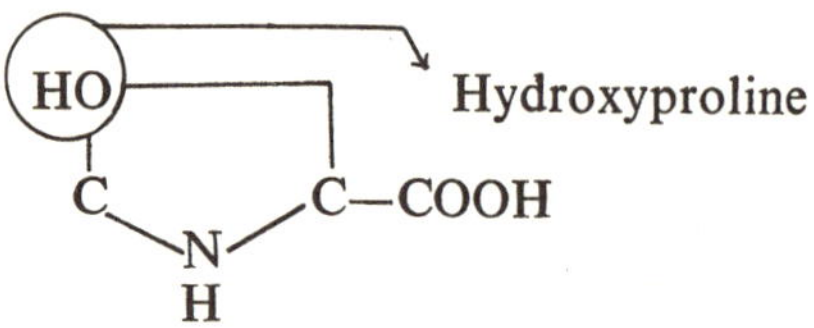

Proline is an imino acid. It is an important imino acid in collagen, where it is present in large quantity. Proline gets hydroxylated into hydroxyproline after it is incorporated into collagen.

FORMATION OF PEPTIDE LINKAGES BETWEEN AMINO ACIDS

The splitting out of a molecule of water between the amino acids of protein results in the formation of a peptide linkage between the amino acids.

$$R-CH(NH_2)-C(=O)-OH + NH_2-CH(R)-COOH$$

$$\downarrow$$

$$R-CH(NH_2)-C(=O)-NH-CH(R)-COOH$$

$$\downarrow$$

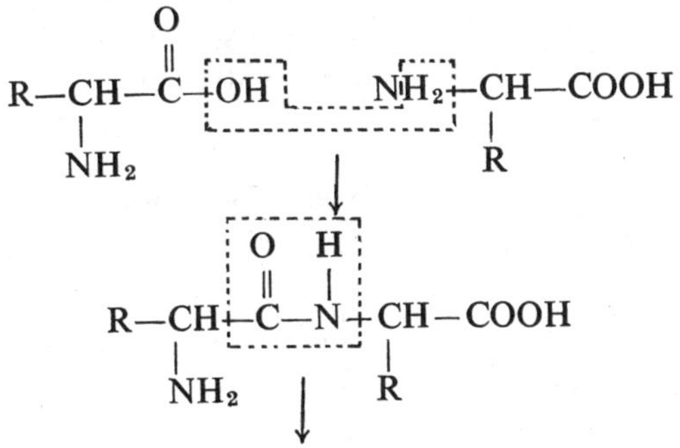

Peptide Linkage

ACID-BASE CHARACTERS OF AMINO ACIDS

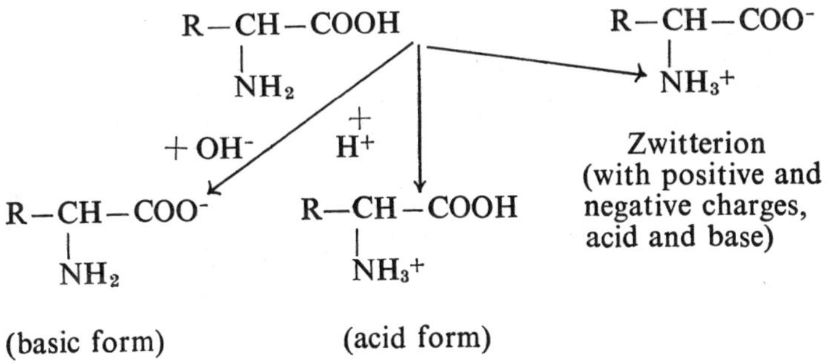

$R-CH(NH_2)-COOH$

$+OH^-$ ↙   ↓ H^+ ↘

$R-CH(NH_2)-COO^-$ $R-CH(NH_3^+)-COOH$ $R-CH(NH_3^+)-COO^-$

(basic form)   (acid form)   Zwitterion (with positive and negative charges, acid and base)

If the amino acid is basic with respect to its isoelectric point, it will have a negative charge (almost all natural proteins contain negative charges). If it is acid with respect to the isoelectric point, it has positive charge and is in the acid form.

AMINO ACIDS IN FOODS
Calculated to 16.0 G Nitrogen

	Cow's Milk	Human Milk	Egg	Whole Wheat	(Maize) Whole Maize	White Rice	Whole Rye	Rolled Oats	Cotton Seed	Soy Bean	Peas	Beef Muscle
Argenine	4.3	4.3	6.4	4.2	4.8	7.2	4.3	6.0	7.4	7.1	8.9	7.7
Histidine	2.6	2.8	2.1	2.1	2.2	1.5	1.7	2.2	2.6	2.3	1.2	2.9
Lysine	7.5	7.2	7.2	2.7	2.0	3.2	4.2	3.3	2.7	5.8	5.0	8.1
Tyrosine	5.3	5.2	4.5	4.4	5.5	5.6	...	4.6	3.2	4.1	...	3.4
Tryptophan	1.6	1.9	1.5	1.2	0.8	1.3	1.3	1.2	1.3	1.2	0.7	1.3
Phenylalanine	5.7	5.6	6.3	5.7	5.0	6.7	5.6	6.6	6.8	5.7	4.8	4.9
Cystine	1.0	3.4	2.4	1.8	1.5	1.4	...	1.8	2.0	1.9	1.2	1.3
Methionine	3.4	2.2	4.1	2.5	3.1	3.4	1.3	2.4	2.1	2.0	1.0	3.3
Threonine	4.5	4.6	4.9	3.3	3.7	4.1	3.0	3.5	3.0	4.0	3.9	4.6
Leucine	11.3	9.8	9.2	6.8	22.0	9.0	6.2	8.3	5.0	6.6	6.4	7.7
Valine	8.4	8.8	7.3	4.5	5.0	6.3	5.0	6.3	3.7	4.2	4.0	5.8
Glycine	2.3	...	2.2	...	...	10.3	...	...	5.3	High	...	5.0
Isoleucine	8.5	7.5	8.2	3.6	4.0	5.3	4.0	5.6	3.4	4.7	4.1	6.3

If it has both negative and positive charges, it is, then, both an acid and a base, i.e., it is Zwitterion.

CLASSIFICATION OF PROTEINS

In the human body, there probably are over 100,000 various kinds of proteins. Their classification generally is made according to their different solubilities and reactions in various solutions.

Simple Proteins

Simple proteins are the naturally occurring ones and which, upon hydrolysis yield only a-amino acids or their derivatives. This type of protein may fall into the following groups:

Albumins

These are proteins that are soluble in water, and dilute salt solution and coagulate with heat. Examples of this kind of protein are egg white, serum albumin, and lactalbumin of milk.

Globulins

These are insoluble in water, but are soluble in dilute neutral salt solutions. They coagulate by heat. Examples are blood serum globulins, muscle myosin, and egg-white protein.

Glutelins

These are insoluble in all neutral solvents, but are readily soluble in very dilute acid and alkali solutions. Examples are glutenin from wheat and oryzenin from rice.

Prolamines

These are alcohol-soluble proteins, are insoluble in water, absolute alcohol, or neutral solvents, but are soluble in 70–80% alcohol. Examples are zein of corn and gliadin of wheat.

Albuminoids

These are insoluble in all neutral solvents, dilute acid or base or salt solutions. Examples are keratin of horns, hides, hoofs, hair, and feathers.

Histones

Characterized by their high content of basic amino acids and soluble in H_2O, they are insoluble in dilute NH_3 solution, are not coagulated by heat, and are easily soluble in dilute acid solutions. Upon hydrolysis, histones yield predominantly basic amino acids such as lysine, argenine, and histidine. They are usually present in the cell nuclei as part of the nucleoprotein.

Protamines

These are composed of large polypeptides of 14–20 peptide linkages. They are highly basic, soluble in H_2O and NH_3, and form a true salt with mineral acids. They are not coagulated by heat, and they may precipitate other proteins.

Conjugated Proteins

These are simple proteins that are conjugated with some other nonprotein groups. Their various forms may include the following:

Nucleoproteins

These are combinations of one or more proteins, such as histones and protamines with nucleic acids. They are soluble in dilute sodium chloride solution.

Glycoproteins

These are conjugated forms of proteins with carbohydrates, such as mucin of saliva.

Phosphoproteins

These are proteins with phosphoric acid esters and with hydroxy amino acids as prosthetic groups. Examples are casein of milk and vitellin of eggs.

Chromoproteins

These are proteins with heme or other pigments, such as hemoglobulin; cytochromes, and flavoproteins.

Lipoproteins

These are conjugates of lipid and protein.

Derived Proteins

1. Denatured and coagulated.
2. Peptides, such as proteoses, peptones, and polypeptides.

SHAPE AND STRUCTURE OF PROTEINS:

Fibrous

This type takes the form of elongated filaments, joined together laterally through cross linkages and forming fairly stable and insoluble structures, such as keratin, myosin, and collagen.

Globular

This type is oval or eleptical in shape with an abundance of foldings, such as antigens and enzymes.

MILK PROTEINS

Milk protein is not a single one but a combination of several proteins. Total protein content of milk is 3.5%.

When acidifying to pH 4.6, casein, which composes 3.0% of milk protein, is precipitated. The soluble proteins after this treatment constitute the whey proteins or milk serum proteins and amount to 0.55% of the total milk proteins. When heat is applied to this soluble part, it is separated to the proteose-peptone fraction and the heat-labile milk serum proteins. The proteose-peptone fraction is soluble at pH 4.6, while the serum proteins are precipitated with temperature as high as 100°C and pH 4.6. This becomes important in the manufacture of "high heat" and "low heat" nonfat dry milk. It may also have an influence on yield and quality of cheeses.

Fractionation of Milk Proteins

If 100% skim milk protein is heated to 100°C and acidified to pH 4.6, a filtrate is separated that contains proteose-peptone on top of a precipitate. This filtrate is whey protein. If the above skim milk is acidified to pH 4.6 without heat application, casein (S, 76 to 86%) is precipitated. On the other hand, if the whey protein is saturated with magnesium sulfate or half-saturated with ammonium sulfate along with some pH adjustment, it becomes separated into a filtrate and precipitate fractions. The filtrate is composed of lactalbumins (9–18%) and proteose-peptones (2–6%), while the precipitate composes the immune globulins, i.e., the lactoglobulins (1.4–3.1%). The filtrate, i.e., the lactalbumins, proteoses, and peptones, are further separable to β-lactoglobulin (7–12%), α-lactalbumin (2–5%), and blood serum albumin (0.7–1.3%). Meanwhile, the precipitate of lactoglobulins is further separable into euglobulins, which is water-soluble, and pseudoglobulin, which is water-insoluble.

Casein Fractionation

As mentioned earlier, casein is precipitated at pH 4.6 (76–86%). This precipitate is nonsoluble when treated with 1.7 Molar (M) urea. It is, however, soluble in 6.6 M urea giving rise to α-complexes (45–63%), and it is soluble in 4.6 M urea giving rise to β-casein (19–28%); and is also soluble in 3.3 M urea when ammonium sulfate is added, giving rise, in this case, to γ-casein (3–7%).

The β-casein fraction (19–28%) mentioned above is further separable to beta$_a$-casein, beta$_b$-casein and beta$_c$-casein. The alpha complex (45–63%) also mentioned earlier is further separable to kappa-casein and alpha$_s$-casein. The kappa-casein is soluble in 0.4 M calcium chloride, pH 7 and 0–4°C. It is involved in the primary phase of rennet action forming the insoluble para-kappa-casein and glyco-macro-peptide. The alpha$_s$-casein meanwhile, is insoluble in 0.4 M calcium chloride, pH 7.0 and 0–4°C; thus, it is calcium sensitive. The alpha$_s$-casein is further separable to alpha$_s$-A, which prevents the firm-curd formation in cheese-making that is highly undesirable; besides its being very heat labile, it is converted to alpha$_s$-B fraction (found only in Ayrshire and shorthorn breeds), and to the Alpha$_s$-C fraction.

Role of Casein in Cheese-Making

Casein plays the role of clotting in cheese-making, as it precipitates out during the initial stages of cheese-making. Kappa-casein is attacked by the enzyme rennin (attacks kappa-casein 1000 times as rapidly as it attacks other proteins). Upon this attack, the stable colloidal form of the micelle is disturbed, and thus, the kappa-form loses its protective property over the α$_s$-casein form, and in the presence of Ca^{2+}, it will precipitate out. For a stable colloidal form, the ratio of α : β : kappa forms of casein should be 50 : 35 : 15 respectively, the ratio which becomes upset when rennin attacks kappa-casein.

Kappa-casein is believed to be a glycomacropeptide of small molecular weight, containing galactose, glucoseamine, and neuraminic acid. The fact that kappa-casein is a highly charged molecule gives it the ability to distribute itself around the outside part of the micelle, and prevents the small aggregates from coming together to form giant aggregates, and thus, separate them out.

Meanwhile, calcium-phosphate-type linkages, through serine, or lysine, link the various α, β, and γ protein coils together into calcium-caseinate-phosphate micelle (calcium phosphate makes up 2% of the micelle). The sites of attachment are the calcium appetites. When the charges that kappa-casein brought about around the external part of the colloidal micelle aggregate together, clotting, thus, occurs.

EGG PROTEINS

Eggs are produced by nature for the primary function of nourishing and sheltering the young bird, and providing its food. The white of the egg may serve as a protective barrier before its serving as a source of protein, while the yolk's function seems to be mainly as a source of food.

Egg Composition

The shell

The egg's shell is composed mainly of calcium carbonates (94%), 4% organic matter, about 1% magnesium carboantes, and about 1% calcium phosphate.

The outer shell membrane

This is an exterior layer of keratinized protein and another two mucin layers.

Inner shell membrane

This membrane is composed of two layers of mucin and two layers of keratinized protein.

A thick and a thin albumen layer

The vitelline membrane

This membrane is composed of a middle layer of keratinized protein and two outer mucin layers.

Egg yolk

The egg yolk is composed of 48% water, 32.5% fat, 17.5% protein, and 2% minerals.

Egg White Proteins

The avian egg white is so unique in properties because it is composed almost entirely of proteins of highly important functions. The superiority of the egg-white proteins that stems from their possession of all the essential amino acids, perhaps in the best possible ratios, their ready availability, and the low cost of their attainment puts egg-white proteins in first place when compared to other protein sources. Egg-white proteins may be divided into the following types:

Ovalbumin

Ovalbumin is present in the largest amount of all other egg-white proteins. It composes nearly 50% of the total net egg white. Ovalbumin contains four sulfhydryl groups that are unreactive to the conventional oxidizing reagents. Nevertheless, three of these SH groups may react with p-chloromercuric benzoate in the native form of protein, while the fourth SH group becomes reactive in the denatured protein form. Upon denaturation, however, all four SH groups become reactive. Ovalbumin is subgrouped into A_1, A_2, and A_3 subfractions, which are separable through electrophoretical methods and are thought to vary only in the number of phosphate groups present. The ovalbumin has a molecular weight of 45,000. Its isoelectric point occurs at pH 4.6, it is easily denatured with heat, its nitrogen content is 15.8%, and it contains about 2% polysaccharides, which consist of two glucosamines, four mannoses, plus a nitrogen group.

Conalbumin

Conalbumin is the iron-binding protein of the egg white. It is also thought to contain an antibacterial agent. Conalbumin is a homologue of serum transferrin. Using the starch gel electrophoresis technique, two forms, conalbumin 80% and conalbumin 20%, were reported as separated. The fact that conalbumin binds metals gives it a profound biological importance. Conlbumin has the molecular weight of 80,000. Its isoelectric point occurs at pH 6.6 and it binds two atoms of the iron metal per one mole. If it binds ferric (Fe^{+++}), the color is pink that absorbs at 470 mu. If it binds cupric ions (Cu^{++}) the color is yellow and maximum absorption is at 440 mu. If it binds zinc (Zn^{++}) or aluminum (Al^{+++}), there is no color produced.

A complex of protein with Fe or Al is stronger than with Cu, and is much stronger with Zn.

Lysozyme

This part of egg-white protein contains enzymes. It, at least, contains the carbohydrase type of enzyme that is thought to be a muramidase. Lysozyme is a basic protein and is noncoagulable with heat at acid pH, but becomes very heat sensitive in an alkaline media. Its molecular weight is 14,000. Its isoelectric point occurs at pH 10.7, and its nitrogen content is 18.6%. Using column chromatographic separation techniques, three forms of lysozymes are detectable. These are fractions A, B, and C. Dialysis causes the loss of fractions B and C, while interaction with lipids causes the inactivation of the whole activity of lysozyme. Components B and C are thought to be more reactive than component A.

Ovomucin

This type of egg-white protein is a carbohydrate protein or a mucoprotein that forms solutions that are very viscous in alkaline media. Fibrous material in egg white is composed mainly of ovomucin, thus giving it the property of contributing the thickness of egg white.

Upon dilution with water, or upon dialysis, ovomucin precipitates out. Ovomucin inhibits influenza virus from agglutinating red blood cells. This action is achieved through the action of the enzyme neuraminadase that hydrolytically cleaves sialic acid from the red blood cell surfaces. The inhibition mechanism is competitive and is due to sialic acid's presence in the mucoproteins.

The molecular weight of ovomucin is 7×10^6, its isoelectric point occurs at acid pH, its nitrogen content is 12.5%, and, generally, it is nonsoluble.

Ovomucoid

This is the trypsin inhibitor in egg white. It has a high content of carbohydrates and is highly heat stable. Its molecular weight is 28,000, its isoelectric point occurs at pH 3.9–4.3, and it is stable at considerably high temperatures. The carbohydrate content of ovomucoid is about 20%, of which hexoseamine is about 15%, a hexose 7%, and sialic acid 0.2 to 1.4%. Ovomucoid is a highly heterogeneous compound. Its nitrogen content is around 13%.

Ovoinhibitor

The ovoinhibitor in egg white which inhibits trypsin, bacterial, and fungal proteinase will, as well, inhibit chymotrypsin. This inhibitor is a species-specific. The ovoinhibitor comprises about 0.1% of the total egg-white protein. Its molecular weight is unknown, and its isoelectric point is 5.2.

Flavoprotein

There is a recognized and descrete protein that contains all the riboflavins in the egg white. Moreover, there is a distinct specificity in that the avian flavoprotein contains specifically riboflavin as the flavin moiety and not any other. This riboflavin becomes depleted when chickens are placed on riboflavin-low diets.

Avidin

Avidin is a trace protein in egg white. However, its ability to bind the B vitamin, biotin, makes it of special interest and importance.

Minor Proteins

A peptidase and a nucleic acid–containing constituent are suspected to be present in egg white. A component "18," globulins A_1 and A_2, have been reported as being present in egg white through starch gel electrophoresis studies.

Proteins in the Egg Yolk

Lipoproteins, lipovitellin, and lipovitellenin are believed to be present in egg yolk, along with the water-soluble fraction "levetin." However, due to the fact that the yolk contains large amounts of bound and unbound lipids, the above lipoprotein fractions are hard to study and characterize.

Moreover, it is believed that, with the exception of phosvitin, most proteins of the yolk have no enzymatic or biochemically reactive properties.

CEREAL PROTEINS

Generally, cereal proteins are:

Glutelins

These are the mixed proteins of the endosperm. They are the typical cereal proteins and are present in practically all cereals.

Prolamines

Although these are associated with glutelins, they vary widely in the different cereals.

Albumins

These are a constituent of the original protoplasm and constitute a minor protein of the endosperm.

Globulins

These are present in large quantity in the bran and germ but are present in considerably smaller amounts in the endosperm.

Proteoses

These are present in minute quantities in all cereals. They probably are derivatives from other parent proteins, or are fragmentations of them.

Amino acids

These are extremely minor components.

WHEAT PROTEINS

Composition of hard wheat

COMPONENT	%
Moisture	10
Protein	13.2
Fiber	2.6
Fat	1.9
Ash	1.8
Nitrogen free extracts	69.0

Morphological Composition of Wheat Proteins

COMPONENT	%
Endosperm	72
Aleurone	15
Scutellum	4.5
Pericarp	4.0
Embryo-germ	3.5

The bran is about 15% protein, half of which is prolamine. The albumin and globulin content are not well known. The germ content is from 2 to 3% of the kernel weight with 25% protein, mainly albumins and globulins.

The endosperm composes 85% of the weight of the kernel, of which 5 to 16% is protein (represents about 72% of the total grain proteins). These proteins may be subgrouped into the following protein classes:

Gluten protein

This type of protein is obtained by removal of the soluble starch. This is usually done by mixing the flour with about 65% water, and after about half an hour the soluble starch is washed away.

Gluten is composed of 75 to 85% protein, 5 to 10% lipids, and the remainder is probably occluded carbohydrates. Gluten may be further subdivided into its important components of gliadin and glutenin, which are fractionated using electrophoresis techniques. Modification of glutenin may be achieved through the reduction of the disulfide bonds that are present.

Soluble Proteins

Albumins: These are fractions of α, β, and γ albumins. In wheat flour of bread, however, the β fraction predominates and also contains from 9 to 11% carbohydrates.

Globulins: These are composed only of two fractions, α and γ. The γ fraction seems to be the predominant.

VEGETABLE PROTEINS

Proteins of plants are usually classified according to their function. Enzymes are, for the greatest part, present in the vegetation portion of the plant and they occur to a much lesser extent in the seeds. Unlike animals, plants have no albuminoids or protamines, and the characteristics of plant albumin and globulin are different from those of animals. Plant globulins are generally classified on the basis of their insolubility in water. Prolamines and glutelins are characteristic only of plants, as they do not occur in animals.

Upon hydrolysis, the prolamines give rise to ammonia, glutamic acid, and proline. Generally, both prolamines and glutelins contain high concentrations of glutamic acid; also, they contain small amounts of the basic amino acids, arginine and histidine.

Glutelins are extractable with dilute acids and dilute bases.

Globulin from plant seeds is unlike that from animal sources in that it is incompletely coagulated by heating in slightly acid solution. Many plant globulins are not precipitated with half-saturated $(NH_4)_2 SO_4$ or fully saturated $MgSO_4$ solutions (the classical method for separation of animal globulins). This, however, may not be the case with the plant seed globulin.

Most plant globulins may be recovered in a crystalline form by dialysis. Examples of plant globulins are legumin from pea and lentil, and tuberin from potato. Seed globulins usually have properties that are similar to proteoses.

Plant albumins, as well, differ from animal albumins. Some plant albumins are precipitated using saturated NaCl solution, while others are precipitated with less than half-saturated $(NH_4)_2 SO_4$ solution.

Some of the vegetable albumins are:

Leucosin—in wheat, barley and rye
Legumelin—in peas, lentil, and soybean
Phaselin—in kidney beans
Ricin—in castor beans

Legumelin, after precipitating out gets denatured.

Spermatophytes (in flowering plants) are usually obtained in a denatured form, as in freeing them the SH-groups get broken. Their solubility in water is very low. Spermatophytes of leaves are insoluble in water and in mildly alkaline buffers. They are, however, precipitated or coagulated with acid and heat (50°C).

MEATS

Meats are the properly dressed flesh that is derived from cattle, sheep, goats, or from swine that are in good health and are mature enough at the time of slaughter. This meat is usually restricted to the skeletal muscle type, and may refer to diaphram, tongue, heart, and esophagus, but should not include lip, ear, or snout meat.

Meat is also obtained from fish, fowl, mollusk, crustacean, or other animals that are used as source of food.

Types of Muscle

1. Smooth muscle: non-striated, involuntary near the intestines and other organs.
2. Cardiac muscle: cellularily type, multinucleated, involuntary.
3. Skeletal muscle: striated type.

Composition of Skeletal Muscle

Skeletal muscle is composed of 75% water, 24% protein, 1% ash (P, Na, K), 3.0% lipids, less than 5 parts per million (ppm) Ca, and 0.0 to 0.3% carbohydrates. If more than 5 ppm Ca is present in muscle tissues, contraction may be impaired, and if the amount of carbohydrates is greater than 0.3%, shelf life of meat will be shortened. As shown above, the ratio of moisture to protein is 3.5 to 1, or it may go as high as 4.1 to 1.

The origin of these muscles is the mesoderm that becomes spindle-shaped, develops muclei, and forms myoblasts. These continue to grow, become multinucleated, and develop into muscles.

The total number of muscle fibers is determined by the age of three months of the embryo. After this age, muscles get larger in diameter and increase in length, but the number of muscle fibers stays the same. Collagen and elastin are formed towards the middle and latter stages of embryonic development. In the newborn animal, the amount of connective tissue is greater than in any other stage of its life, and this does not adversely affect tenderness at that stage.

Protein Composition of Skeletal Muscle

Myosin	38%	} Myofibrillar
Actin	13%	
Myogen	31%	
Myoglobin	2%	
Hemoglobin		
Collagen	11% —	
Elastin		
Reticulin		

Types of Protein of Skeletal Muscle

1. Those soluble in water and in dilute salt solution (sarcoplasmic): (a) like myoglobin and myogen; and (b) enzymes of the glycolytic cycle.
2. Those soluble in concentrated salt solutions (myofibrillar type protein).
3. Those insoluble in concentrated salt solutions (like connective tissue protein).

Connective Tissue Proteins

Collagen

This type of protein is the most widely occurring group in the animal body. It is high in proline, and upon boiling it gets converted to gelatin.

Reticulin

This type is thought to be a precursor or a degradation product of collagen.

Elastin

This kind is the elastic type of ligaments of the connective tissues.

Keratins

These serve protective and structural functions. They are usually present in hair, hoofs, horns, and wool.

Connective Tissue Membranes of Skeletal Muscle

Endomysium

This is the interior membrane surrounding the entire muscle. It is a smooth, uniform membrane that completely surrounds the muscle fiber.

Perimysium

This is the intermediate connective tissue membrane. It is not uniform and may have thickenings at certain points and it surrounds large bundles of muscle fiber.

Epimysium

This is the exterior membrane around the entire muscle. It is a connective tissue sheath that binds muscle bundles forming single muscles.

The muscle fibers are believed to be short chains of protein held together by polar charges. Thus, the myofibrils and myofilaments are being held together through electrostatic charges that are stronger in the horizontal than in the vertical direction. There are two types of muscle fibers: the slow-contracting red muscle fiber, such as in legs, which contains large amounts of the oxidative type of enzymes; and the fast-contracting white muscle fibers, such as in the eyes, which contain large amounts of mitochondria. The membrane surrounding the muscle fiber is a continuous sheath, "the sarcolema," and within it lies the sarcoplasm containing everything except the actual contracting unit, i. e., the myofibril. The level of sarcoplasm may vary within the same muscle fiber.

Ruminants may have considerably more sarcoplasm per muscle fiber than single-stomach animals. The amount of sarcoplasm is directly related to the amount of work or exercise the animal is exposed to.

Sarcoplasm normally contains large amounts of soluble protein, lipids, and a very minute amount of glycogen.

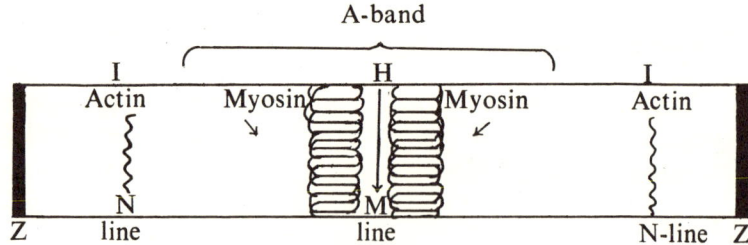

Z = narrow, very dense type of protein that does not let light go through it.
I = small type of protein, light; a fair amount of light goes through it.
A = a heavy area with a light H-zone and a thin M line in the middle. There is (1) heavy myosin with ATPase activity, and (2) light myosin with no ATPase activity.

In contraction, the distance or length of A-band does not change, while the I-bands are shortened. Thus, the change in length is due to the change in I-band.

Muscle Contraction

The nerve impulse causing depolarization of the sarcolema brings about changes on the sarcoplasmic reticulum that allow Ca^{2+} to flow in the sarcoplasmic fluid. This Ca^{2+} activates the ATPase enzyme on the H-meuromyosin and causes the breakage of the Mg^{2+}-ATP complex. When ATPase is activated, it causes the hydrolysis of ATP to ADP+energy. The energy produced causes the sliding of actin over myosin and the opening of the sites of attachment between actin and myosin. The ADP produced is rephosphorylated by creatine phosphate back to ATP. Breakage of the bond between actin and myosin brings about the energy that causes sliding of actin over myosin. When the nerve impulse is lost, Mg^{2+} comes back in again, forming a Ca–Mg complex and, thus, actin and myosin sliding becomes free.

Lipid Component of Muscle

Neutral lipids, phospholipids, and cholesterol—the striated muscle is only 3% lipids but in the connective tissue there is a large amount of neutral lipid.

Water Extractives

1. Organic Matter: glycogen (1%), lactic acid, pyruvic acid, amino acids, urea, creatine, creatinine, anserine, carnosine, and carnitine.
2. Salts: 1%.
3. Water-soluble proteins: natural and derived proteins; extract may contain some gelatin.

Types of Connective Tissue

This tissue is composed of fibrous and amorphous ground substances ranging from fragile to very tough tissue (tendons, ligaments, etc).

White Collagen

This is fibers that are arranged in bundles of filaments that are threadlike, coarse, and considerably strong but are not elastic, such as tendons which are very strong, and collagen which is inelastic.

Yellow elastic fibers

These are abundant in ligaments. They are fine and branched fibers that are randomly distributed, forming a network.

Ground substances

These form the soft jelly-like mass to the tough matrix. They usually are present in soft connective tissues, skin, cartillage, and bone.

Specialized Connective Tissue

Adipose tissue is a specialized type of connective tissue occurring in the subcutaneous regions and around organs such as the kidneys, in the omentum, and between muscles.

Cartillage Tissue is a specialized type of connective tissue that is composed of fibers and gel-like ground substances of chondriotin sulfate.

Bone also is a highly specialized type of connective tissue that is composed of cells which are surrounded by ground substance and fibers. There are tiny salt crystals deposited in the ground substances.

COMPOSITION OF COOKED MEATS

	Beef	Pork	Lamb	Chicken (Raw)	Fish
Water	47–60	40	40–56	66	60–70
Protein	22–27	23	24	20	19–27
Fat	12–27	33	20–35	13	3–13
Carbohydrates	0	1.5	0	0	0
Ash	0.7–1.3	4.1	1.1–1.2	1	1.7–2.0

MEAT SALTS

SALTS:

Cations	Anions
Potassium (Principal one)	Phosphate
Magnesium	Bicarbonate
Sodium	Sulfate

COMPARISON OF AMINO ACID CONTENT OF MUSCLE AND CONNECTIVE TISSUE PROTEIN

Component	Beef Muscle	Collagen	Gelatin	Elastin	Human Epidermis	Human Hair	Sheep Wool	Cattle Horn
Total N	16.0	18.6	...	...	14.2–15.5	15.5–16.9	16.2–16.9	14.8–16.9
Amino N	...	0.46	...	...	1.16	1.17	....	....
Glycine	5.0	26.6	29.3	26.7	6.0–(13.8)	4.1–4.2	5.2–6.5	9.6
Alanine	7.4	10.3	10.0	21.3	....	2.8	3.4–4.4	2.5
Leucine	8.0	3.7	3.2	9.0	(8.3)	6.4–(8.3)	7.6–8.1	7.6–8.3
Isoleucine	6.0	1.9	2.1	3.8	(6.8)	(4.7)–4.8	3.1–4.5	4.3–4.8
Valine	5.5	2.5	2.7	17.7	4.2–(5.6)	5.5–(5.9)	5.0–5.9	5.3–5.5
Phenylalanine	5.0	2.4	2.4	6.2	2.8	2.4–3.6	3.4–4.0	3.2–4.0
Tyrosine	4.0	1.0	1.0	1.5	3.4–5.7	2.2–3.0	4.0–6.4	3.7–5.6
Tryptophan	1.4	...	...	...	0.5–1.8	0.4–1.3	1.8–2.1	0.7–1.4
Serine	6.0	4.3	3.4	0.85	16.5	7.4–10.6	7.2–9.5	6.8
Threonine	5.0	2.3	2.2	1.1	3.4	7.0–8.5	6.6–6.7	6.1
Cystine	1.2	...	0.1	0.35	2.3–3.8	16.6–18.0	11.0–13.7	10.5–15.7
Methionine	3.2	1.0	0.9	trace	1.0–2.5	0.7–1.0	0.5–0.7	0.5–2.2
Proline	6.0	14.4	16.5	13.5	3.2	4.3–(9.6)	5.3–8.1	8.2
Hydroxyproline	1.0	12.8	14.0	1.6	....	....	....	....
Arginine	7.7	8.2	8.6	1.3	5.9–11.7	8.9–10.8	9.2–10.6	6.8–10.7
Histidine	3.3	0.7	0.7	...	0.6–1.8	0.6–1.2	0.7–1.1	0.6–1.0
Hydroxylysine	...	1.2	5.0	...	....	0	0.2	....
Lysine	9.0	4.0	7.5	0.5	3.1–6.9	1.9–3.1	2.8–3.3	2.4–3.6
Asparatic Acid	10.5	7.0	10.8	1.1	(6.4–8.1)	3.9–7.7	6.4–7.3	7.7–7.9
Glutamic Acid	17.0	11.2		2.4	(9.1–15.4)	13.6–14.2	13.1–16.0	13.8

Note: Values in parentheses are for species other than human.

Rigor Mortis

Rigor mortis is stiffening after death. For many years, it was attributed to coagulation of blood or lactic acid or both; however, it is well established now that rigor mortis occurs upon the disappearance of ATP.

After the death of animals, ATP remains, for very few minutes at a constant level; then at some point, it very quickly drops off. Roughly, this point is in the middle of the downward slope of rigor mortis. For ATP's complete depletion, it takes from one to five minutes. Some people believe that rigor mortis occurs when all the ATP has been depleted.

In rigor mortis, feet of myosin and actin are connected, i.e., there is a contracted state where there is covalent chemical bonding between actin and myosin giving rise to actinomyosin. Since there are three actins for every one myosin, it would take a considerable amount of energy to cause the chemical bonding between actin and myosin. This energy comes from ATP in the muscle.

When rigor mortis occurs, myosin and actin combine together, forming actinomyosin.

There are three myosins around each actin, and there are six actins around each myosin (this relationship is important for muscle contraction and for tenderness).

The higher the heat or the holding temperature of the carcass the faster the drop in ATP level and the onset of rigor. Thus, keeping a carcass at temperatures above 10°C may render it less tender.

There usually are about 300 potential sites of feet attachment per molecule of meat protein. When an animal is killed, not all these feet between actin and myosin are attached. However, placing the carcass in temperatures above 10°C, or even beating on it, will cause faster hydrolysis of ATP and, thus, more feet attachment.

Generally, the larger the animal in size, the longer it would take for rigor to set in.

Meat Color

Myoglobin (Mb) is the color pigment found within muscle cells. Hemoglobin (Hb) is the color pigment of blood. Mb and Hb are similar in many respects, except that Hb is the vehicle for O_2 transport to cells, while Mb functions in the temporary storage of O_2 within the cell. Mb's molecular weight is 16,000 to 17,000, compared to 67,000 to 68,000 for Hb. This large difference in molecular weight is due to the fact that each molecule of Mb contains one iron atom, compared to four irons in Hb. When an animal is slaughtered and bled, 90 to 95% of its color pigment is Mb; the other 5–10% is that of Hb of the blood that did not bleed out. On a gram per gram basis, however, both Hb and Mb contain the same amount of color pigment per tissue.

Mb may possess a slightly stronger affinity for O_2 than Hb, which aids in the transfer of O_2 from Hb to Mb in the cell. The amount of Mb also increases with age. Veal meat is estimated to contain 1–4 mg/gram of color pigment in its red tissues, compared to 16–20 mg/gm of the mature cow's. Chicks are believed to contain 0.8–0.9 mg/gm of pigment compared to 1.5–1.8 mg/gm in the mature hen. The same principle makes carbon monoxide (CO) poisoning in children more critical than in adults as children have less Mb per gram of tissue than adults. Hb, too, is less per tissue of children than adults.

Structure of Myoglobin and Hemoglobin

The two pigments are composed of: a) a globular type protein; and b) a porphyrin ring.

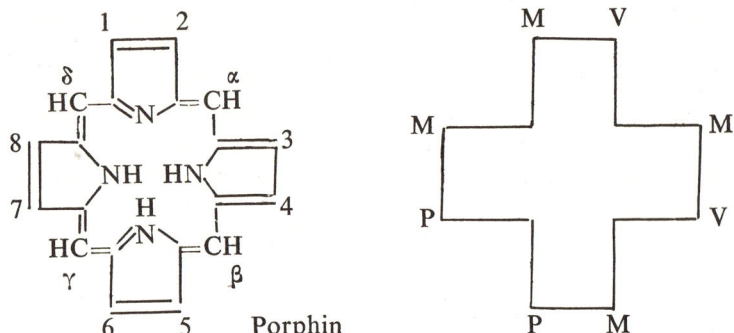

Porphin

Mostly, Mb and Hb undergo similar chemical reactions. Both are water-soluble; their iron is in the ferrous state (Fe^{++}); Mb's color is purple and Hb's color is red.

In the presence of O_2, Mb undergoes a reaction that is called "oxygenation" where it forms oxymyoglobin (OxyMb), that is red in color and the iron of which is still in the ferrous state. For this reaction to occur, the pH must be less than 6, and the partial O_2 pressure should be above 30 mmHg.

$$Mb + O_2 \underset{pH > 6.0}{\overset{(pH < 6.0) \text{ (over 30 mmHg)}}{\rightleftarrows}} OxyMb$$

(purple) Fe^{++} (red, Fe^{++})   (Reaction is reversible)

If the pH reaches 7, mostly the reaction will go towards Mb formation, causing a rather dark color. About 1–3% of the cattle meat remains dark after cutting, "dark cuts," with pH probably around 7 or even a little higher. This high pH and, consequently, the dark color, are due to the presence of lactic acid (arizing from glycolysis). When an animal is under rather prolonged stress, it depletes its glycogen, and lactic acid gets used up with time. However, if the animal undergoes stress or high excitement for 2–3 minutes just prior to slaughter, and is then immediately killed, this results in rather a low

---

* This is the naturally occurring parent compound protoporphyrin where M is methyl, V is Vinyl, and P is propionic acid.

pH as the lactic acid formed from glycogen at the excitement did not have the time to be reused, and low pH and bright red meat are the desirable end results.

If the partial pressure of oxygen is below 30 mmHg, instead of Mb's undergoing "oxygenation" it undergoes "oxidation" and gives rise, in this case, to "met-myoglobin (Met-Mb) which is brown in color and the iron of which is in the ferric state. This reaction is only slightly reversible to the left.

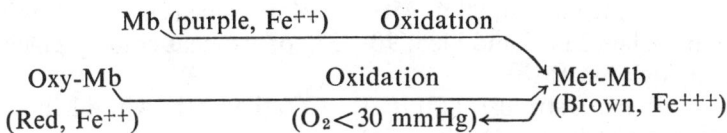

An O_2 partial pressure of 1.5 to 30.0 mmHg favors Met-Mb formation, but above 30 mmHg favors Oxy-Mb formation. As we go from 0°C to 10°C in temperature, the speed of oxidation gets to be 2.8 times faster for each 10° rise. Besides the rise in temperature, bacteria and light also cause "oxidation" of the pigment to Met-Mb. Good sanitation may help in holding meat pigment at the bright red Oxy-Mb for seven to eight days. However, this is not usually the case, as bacteria is to blame for shortening the shelf life of meat.

Mb in the presence of carbon monoxide (CO) forms carboxy-Mb), which is also red in color. CO has stronger affinity for Mb than O_2, and thus forms a much more stable pigment than oxy-Mb. Upon freezing, CO-Mb may stay stable for about six months.

Myoglobin (Mb) increases with the age of animal, with exercise, and at high altitude. The male contains more Mb than the female. An advantage for testing for Hb and Mb is that they are 95% water-soluble, while a disadvantage of this is the loss of red color in meat processing, such as sausage-making, etc. Mb is also salt-soluble and is, to a good extent, heat stable. It may be heated to 50°C for one-half hour and still retain its color.

Usually, the porphyrin ring is the same in all types of Mb, and only the protein part varies.

Mb, as such, is actually purple in color and absorbs at wavelength of 555 mu. Metmyoglobin (Met-Mb) is brown in color and absorbs at 505 and 627 mu. Ultraviolet light, foot candle, bacteria, and temperature all cause the "oxidation" of Mb or oxy-Mb to the Met-Mb brown-colored form.

Cyanide (CN) has strong affinity for practically all forms of the color pigments of meat (one reason CN is a strong poison). In the living animal, 90 to 95% of color pigment is hemoglobin (Hb); when killed and bled, 90-95% of its meat color pigment is myoglobin (Mb).

Hb has a stronger affinity for CO_2 than Mb, while Mb has greater affinity for O_2 than CO_2.

Most of the meat color pigments combine with nitric oxide (NO) forming nitrosyl-met-Mb, which is brown and its iron is in the ferric state. Upon reduction, however, it is converted to nitrosyl-Mb. Ascorbic acid or sodium ascorbate may be used as reducing agents.

Green pigments

These green color pigments occur on meats when there are alterations of the porphyrin ring itself. Such color developments as choleglobin are usually an event where the methene bridge gets partially oxidized; verdoheme is a result of a more severe oxidation of that bridge and the consequence breakage of the ring. When hydrogen sulfide and hydrogen peroxide are present along with ascorbic acid, oxidation may occur as a result of these compounds' adding to the double bonds. An irridescent color which is a rainbow-prismatic effect is another green color development that may occur in meats.

Chlorine (Cl_2) may combine with Mb giving rise to a green-yellowish color. Hypochlorite (HOCl), which is a chlorine-oxidizing agent, is a common sanitation agent used in the meat industry.

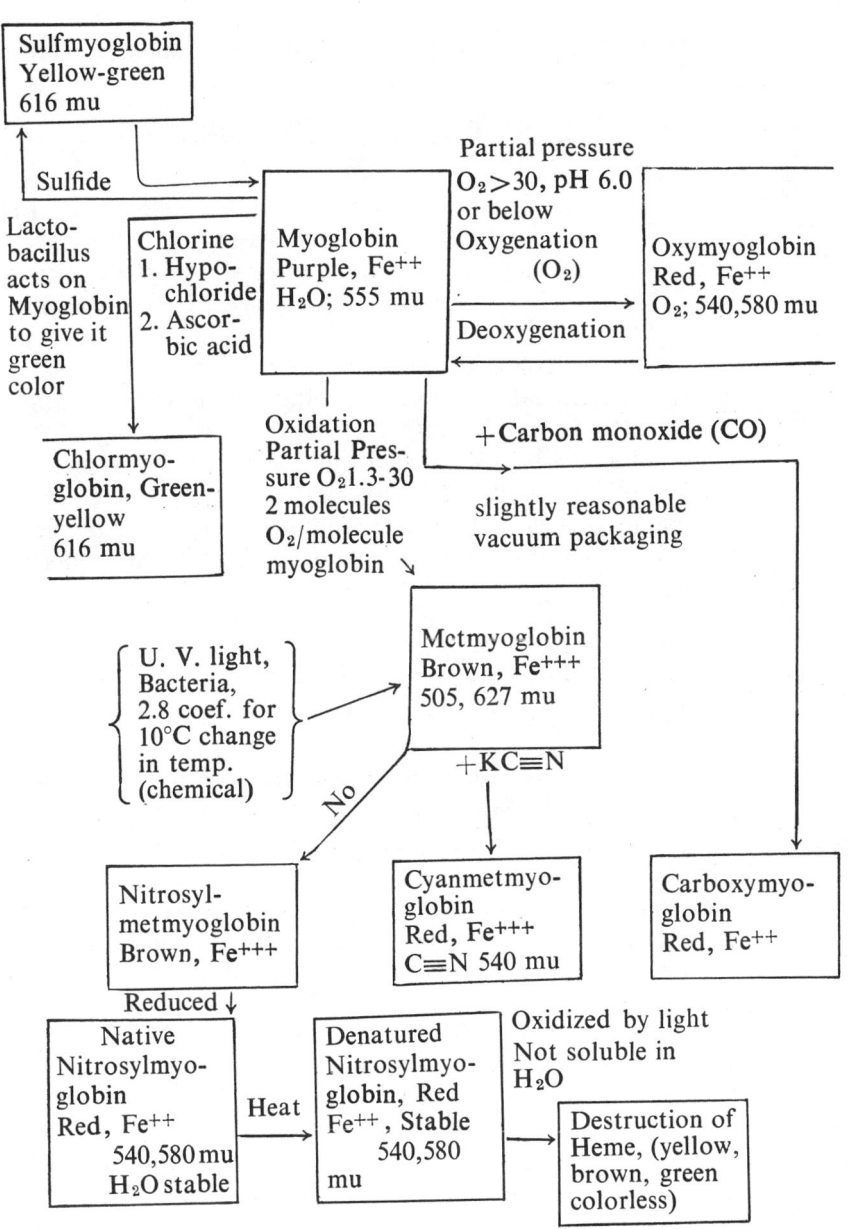

HOCl, besides killing bacteria in meat, imparts a vivid green color to meat.

$$Mb \underset{\longleftarrow}{\overset{HOCl}{\longrightarrow}} \text{Chole-Mb (green)}$$

Ascorbic acid, as well, may cause the development of a greenish color in sausages. Excess nitrites may cause green or brown color development in meats, normally termed "nitrite burn." "Green cores" are caused in sausages by lactobacillus bacteria.

Dark bone color in young poultry is due to their bones lacking adequate calcification and, being porous, upon freezing, Hb concentrates on the surface of bone; upon cooking, it gets denatured and a brown to black color of bones develops.

There is also a hereditary variation of shades of color in fresh ham. This is due to an accumulation of color pigment in specific locations of muscle and the lack of pigments in another part.

The color of meat that we see depends on the myoglobin (Mb) level in that meat. Beef meat generally has higher level of Mb than other kinds of meats, (10–15 mg/g). A dark, mature cut of beef may at times have as high as 20–25 mg/g Mb, which will give it a purplish-red color. Upon exposure to O_2, this, however, is converted to oxy-Mb which then just becomes bright red. Hb and Mb have the same basic reactions. When they are treated with acid, their protein is easily separated from the heme. Once the heme is separated from the globin, it is very easily oxidized, especially if oxygen is available. The oxidation of Mb from the ferrous to the ferric state produces Met-Mb, which is a brown, grey, or dull color.

In curing meat, the nitric oxide (NO) changes the purple color into red nitroso-Mb which, upon heating, turns pink as protein is denatured to hemochromogen.

When meat protein is denatured through cooking, freezing, or adding salt or acid, the heme is disengaged and becomes very susceptible to oxidation. From this point on, color changes and developments take place quite rapidly.

Meat Curing

These methods for curing meat may include dry curing, pickle curing, comminution and mixing, and injection methods.
Curing substances approved by the U.S.D.A. (FDA) are:

Sodium chloride (NaCl)

This salt will help in flavor development, and in the development of a bright red color; it acts as an inhibitor of bacterial growth, and accelerates the onset of oxidative rancidity.

Sodium nitrate (NaNO$_3$)

This salt is added primarily for the maintenance of the red color in meats.

The chemical reactions that (NaNO$_3$) undergoes in meat may be summarized in the following steps:

1. $NO_3 \xrightarrow{\text{bacterial reduction}} NO_2$

2. $NO_2 \xrightarrow{\text{reduction}} HNO_2$ (gives the pH of meat)

3. $HNO_2 \xrightarrow{\text{reduction}} NO$ (brings reduction reactions in meats)

4. $NO + \text{Myoglobin} \longrightarrow NO\text{-Myoglobin}$ (bright red color)

Sugar

Sugar contributes to flavor and color, and provides protection to protein.

Functions of Smoking of Meats

1. Provides a drying effect on meats, as a considerable amount of moisture is expelled in due process.
2. Imparts a desirable and appealing flavor to meats.

3. For color development.

4. Provides a tenderizing effect on meats, and brings about a desirable gloss.

5. When temperature reaches 49°C and above, it exerts a bacteriocidal effect on the meat surface.

6. Smoking meats helps impregnate the surface with antiseptics and germicides.

7. Development of antioxidants from the ensuing smoke, thus protecting the fat content from oxidation.

8. It reduces the nitric oxide content of meats.

Meat Tenderness

Tenderness of meat is highly influenced by the amount of collagen present. The older the animal gets, the more collagen it contains till it reaches physical maturity. As age increases, the animal's meat becomes less tender. This decrease in tenderness is not totally due to the increase in collagen. The amount of collagen present varies from one type of muscle to another. Active muscles contain more collagen than the less active muscles.

Exercise of an animal renders its meat less tender than that of the rested ones. Animals on pasture usually do not provide as tender meat as those held and fed inside.

The amount of collagen per gram of lean tissue, although it remains constant within the single muscle of a breed, varies with the different breeds and the individual animals of the same breed.

The addition of certain salts may increase the tenderness of meat, as they increase the protein-holding capacity of water.

Raising the pH may also increase tenderization of meat, as it causes an increase in the protein's net charge and keeps it away from the isoelectric point. For example, a highly negative charge on protein will attract more moisture to it and make protein more tender.

An infusion of phosphates increases the water-holding capacity of proteins, and thus increases meat tenderness. Injection of 2% salt will result in same thing.

Hereditary factors are supposed to contribute to over 60% of the tenderness of meat.

Of course, age is an important factor in determining tenderness. As mentioned earlier, meat becomes less tender with age, probably due to the increase in collagen.

Partial feet dissociation between actin and myosin results in meat tenderization.

Breaking the peptide linkages between amino acids of meat protein also causes tenderization, and so does the breakage of the electrostatic bonds of protein.

Very rapid freezing of meat helps preserve its tenderness, while slow freezing causes a decrease in tenderness due to partial denaturation of the sarcoplasm.

Enzymes in Meat

Cathespins are in lysosomes of the muscle fiber. There are four types of these enzymes, and they function in the proteolytic digestion of protein. The tropical proteolytic enzyme "papain" that is obtained from the papaya tropical tree is used as meat tenderizer. Bromelin is another proteolytic enzyme that is found in pineapples, and ficin is derived from fig tree juice, both of which tenderize meat through their proteolytic action on protein.

In meat aging, the lysosomal membranes are broken and their hydrolytic enzyme content is released. These lysosomal enzymes, by partially digesting muscle protein, render the meat more tender.

Tenderness of meat is thought to be due to partial dissociation between myosin and actin feet.

Enzymes of Bacteria and Fungi

Enzymes of this source are those such as protease-15 and 11, rhozyme-P-11, rhozyme-A-4, fungal amylases, hydrolase-D, hydrolase-Tp, and pancreatic viokase and trypsin.

Mostly, enzymes produced by bacteria and fungi are very strongly proteolytic, and if allowed to penetrate the surface of meat for over two minutes, they may cause mushy and objectionable characteristics.

Changes Brought in Meats by Cooking

1. Changes of color from red to brown.
2. Drip formation (usually from the fat content of the meat).
3. The collagen content is hydrolyzed to gelatin.
4. Protein is denatured.
5. Fat cells are ruptured and dispersion of fat occurs.
6. There is a loss of some of the vitamins, and possibly of other nutrient contents of meat.

A small amount of hemoglobin (Hb) contributes to meat color, but this mainly is a residual and the principal color pigment in meat is myoglobin (Mb.). Myoglobin is a globular protein with one molecule of globin and one heme; while hemoglobin is composed of one globin and four hemes. Thus, the molecular weight of Mb is 17,000, while that of Hb is 68,000. The amino acid sequence of the Mb and Hb proteins is different, and their solubility is different. Mb gets saturated with O_2 at a lower pressure than does Hb, and their reactions with CO_2 and NO are also different. However, the porphorin groups are the same for the two entities.

Meat Emulsions

Sausages and sausage products compose the majority of meat emulsions formation. They are any meat—chopped, ground, and with seasonings added to it. Ground beef, thus, is not a sausage item, as seasonings are not added to it.

Emulsified sausages are the main area of sausage-making. The two major products of emulsified sausages are hot dogs and bologna. There is the coarse-ground, cooked product with high moisture content, such as "cooked salami," and the coarse-ground semidried product that is nonfermented, such as summer sausages, and, finally, there is the dry sausage product, such as dry salami and pepperoni.

Definition of Emulsions

An emulsion must have three things present: (a) a discontinuous or dispersed phase such as fat, (b) a continuous phase, such as saline or water; and (c) an emulsifying agent, such as a soluble protein. Meat emulsions, for the most part, are oil-in-water emulsions.

An oil-in-water emulsion, if looked at under a microscope, is composed of a fat particle surrounded by a saline or water continuous line which, in turn, is surrounded by soluble protein.

For the soluble protein to be a successful emulsifying agent, it must have water-loving hydrophillic charges, and must possess fat-loving lipophillic groups. That is, an emulsifying agent should have polar charges (for dissolving water-soluble materials) and nonpolar properties (for containment of the fat-soluble material). The size of meat emulsions usually ranges from 0.1–1.0 micron (μ), with large variations.

The soluble protein around the fat is usually in the denatured form (at least the first protein layer around the fat is definitely denatured). This emulsifying agent surrounding the fat lowers the interfacial tension, as well as forming a membrane around the fat particle.

All emulsions are considered stable, and it may take a long or only a little time to break them. A surface active or wetting agent may serve in stabilization of an emulsion as it reduces shear in the system.

An emulsion may remain stable through processing at 145°F, the temperature at which protein gets coagulated, and there is no further concern about emulsion stability. The smaller the fat particle size, the less the surface tension, and the more stable is the emulsion formation. The greater the viscosity of the media, the more stable is the emulsion. This makes it difficult for the fat particle to come to the surface.

Usually, calcium (Ca^{++}) and most of the divalent ions favor the formation of the water-in-oil–type emulsion, while monovalent-type ions such as sodium (Na^+), chlorine Cl^-_2) favor the formation of oil-in-water emulsions. It would take 100 Na^+ to offset the effect of 1 Ca^{++} ion, or it may take 30 Cl^- ions to offset one Ca^{++}. The presence of Ca^{++} in emulsions has a tremendous effect on their stability. The hydrophillic-lipophillic balance (HLB) of the emulsifier also contributes to the stability of the emulsion. A ratio of 8 : 18 of hydrophillic to lipophillic groups would favor the formation of an oil-in-water emulsion, while a ratio between them of below eight will tend to favor formation of water-in-oil emulsions. The addition of fine particles to an emulsion will stabilize it. A salt-soluble protein which is long and narrow, with its length being 200 times its width, is much more efficient in its emulsifying capacity than the water-soluble protein type. It seems that addition of salt causes the unfolding of charges on the protein molecules, i.e., increasing the length-to-width ratio of protein increases the emulsification capacity as it causes the unfolding of charges, especially the water-soluble (polar) ones. Generally, adding salt to protein lowers the isoelectric point, thus rendering it more soluble. Adequate shear action is also necessary for the formation of good emulsion. The temperature of the media should be kept low, as, with temperature rise, the amount of fat emulsified is lowered.

A slow rate of oil flow into the system during emulsification is usually recommended. However, it still remains that control of the temperature in the process is a most critical factor. An increase in temperature may cause fat expansion to a point where the fat particles may surround the protein. Some believe that, as temperature increases, it causes the first layer of protein to become more rigid, and with the continuation of mixing, parts of the protein may start breaking up, allowing fat to escape to the exterior. The breaking point may occur at 70 to 80°F, depending on the amount of shear applied.

Neutral fats or triglycerides are considered to be better emulsified than free fatty acids (FFA), and the shorter the fatty acid (FA) carbon-chain, the better the fat is emulsified. FA with one double bond is considered best for emulsion formation; second to this, a FA with two double bonds, third best is a FA with no double bonds, and poorest for emulsification is any FA that possesses more than two double bonds. It follows that highly unsaturated fats, like linseed oil, make poor emulsions.

The percent of protein that is extractable is important in emulsion formation. Although fat has no effect on protein in this respect, the pH does exert a remarkable effect on the percentage of protein extractable. The higher the pH, the more protein is extracted. The pH of 6, if held constant, is believed to serve well for good protein extractability. Raising the pH keeps it away from the isoelectric point of proteins, thus increasing their solubility. If the pH becomes too high, glucono-delta-lactone is added, where it gets converted to gluconic acid and lowers the pH. Lean meat's being 35% extractable, makes it a very good source for emulsion formation.

Freezing meat reduces its extractable protein by 8%. This makes fresh meat superior in emulsifying more fat. Also, prerigor meat is supposed to have 50% more soluble protein than postrigor meat. This means that prerigor meat would emulsify 50% more fat than postrigor meat. Salt addition increases emulsification up to 10%. Increasing the salt content beyond 10% will decrease the amount of protein extracted. Bull meat is considered the most superior in emulsion formation, as it emulsifies more fat than any other meat. Second to bull meat, is cow's, followed by lean pork, followed by cheek meat, with ox lips being the least capable in emulsifying fat. If a value of 1.0 is assigned to bull meat, cow = 0.9, lean pork = 0.8, cheek meat = 0.6, and ox lips = 0.05. In summary, for good emulsion formation, as far as protein is concerned, the questions that arise are: how much protein is available? how soluble is it? and to what extent is it extractable?

Time, speed of blending, and centrifuge speed must be held constant in emulsification.

PERCENT OF SOLUBLE PROTEIN (EXTRACTABLE) OF VARIOUS MEAT SOURCES

Meat	Extractable Protein(%)
Bull	38
Cow	36
Lean Pork	34
Cheek Meat	25
Ox Lips	10

METABOLISM OF AMINO ACIDS

Making Use of Nitrogen

Nitrogen is the building block for animal, plant, and microorganisms, and although the air we breathe is dominantly nitrogen, we cannot use it, however, in the build-up of our tissues as it needs be reduced to ammonia first. It is only when atmospheric nitrogen is either oxidized or reduced that it becomes useful for animals or plants. Certain microorganisms, specifically Rhizobium, can fix nitrogen in nature. This maybe accomplished through the oxidative decarboxylation of pyruvate with the use of thyamine pyrophosphate (TPP) coenzyme and ferredoxin, the reaction which gives rise to acetyl-Co-A and reduced ferredoxin. Energy from this reaction helps carry on nitrogen reduction to its usable ammonia (NH_3) form.

The usable nitrogen in plants is in the nitrate (NO_3) form. If ammonia (NH_3) is available instead, the plant gets it in the NO_3 oxidized form. Plants then reduce NO_3 to nitrite (NO_2) by the enzyme nitrate reductase. For animal use, the NH_3 form must be the one available, which subsequently gets converted to amino acids (aa), building blocks of tissues. There are three processes whereby ammonia gets trapped or fixed for utilization, a matter of great importance to animals in preventing NH_3 toxicity if present in high levels in the blood. Nonprotein nitrogen sources, such as urea, are provided to ruminants specifically to avoid this NH_3 toxicity. High levels of NH_3 in the blood causes disorders in the nervous system as it freely passes through the highly polar cellular membrane.

Amino acids are the formidable building blocks of all proteins that must be provided for the very survival of animals. Any abnormality in amino acid metabolism may lead to well recognized diseases.

Most amino acids undergo deamination reactions and, thus, are converted to keto acids. These keto acids may enter the glycolytic and TCA cycles or may go into ketone body production. We get rid of the metabolic end products of nitrogen as urea; some other species excrete N_2 as NH_3, while others, like avians, get rid of N_2 as uric acid.

Nitrogen Fixation

Rhizobium bacteria symbiotically fixes nitrogen in plants and legumes. Photosynthetic and aerobic bacteria in soil may also nonsymbiotically fix N_2 in nature. Ammonia (NH_3) may be commercially made by the catalytic reduction of N_2 with high temperature in the presence of ferredoxin. In plants, nitrification takes place where NH_3 is oxidized to NO_2 and, subsequently, to NO_3.

There are organisms that may do the opposite of this, i.e., carry out a denitrification process, where NO_3 is reduced to NO_2 and subsequently to NH_3.

$$NO_3 \xrightarrow[\text{(Fe, Mo, FAD)}]{\text{nitrate reductase}} NO_2 \xrightarrow{\text{nitrite reductase}} NH_3$$

We have little NH_3 in the form of ammonium (NH^+_4) salt in our body, as we fix NH_3 into amino acids or excrete it in the form of urea.

Processes of NH_3 Fixation

1. Through glutamic dehydrogenase reaction, where α-ketoglutaric acid, coming from carbohydrate metabolism through the TCA cycle, gets reduced to the amino acid, glutamic by $NADPH_2$ and NH_3. This reaction thus catalyzes the conversion of a carbohydrate metabolite into an amino acid.

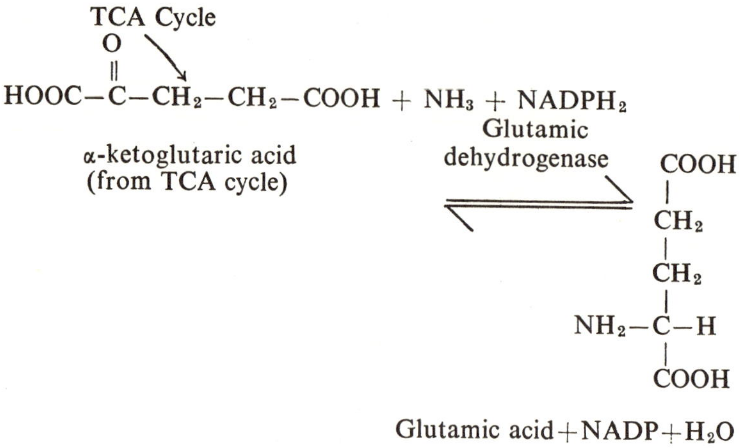

$$\text{HOOC}-\overset{O}{\underset{\|}{C}}-CH_2-CH_2-COOH + NH_3 + NADPH_2 \xrightleftharpoons[]{\text{Glutamic dehydrogenase}} \begin{array}{c} COOH \\ | \\ CH_2 \\ | \\ CH_2 \\ | \\ NH_2-C-H \\ | \\ COOH \end{array}$$

α-ketoglutaric acid (from TCA cycle)

Glutamic acid + NADP + H_2O

Once we have made an amino acid from a carbohydrate in the above reaction, and through transamination reactions, we may get into making other amino or keto acids (except, of course, the essential amino acids, where the carbon skeleton required is not available in this case). The above reaction detoxifies NH_3 and traps it into a useful amino acid.

2. The second process of NH_3 fixation is through glutamic acid's trapping of another mole of NH_3 through the use of

energy from ATP and, thus, making an amide from the amino acid through the aid of glutamine synthetase enzyme.

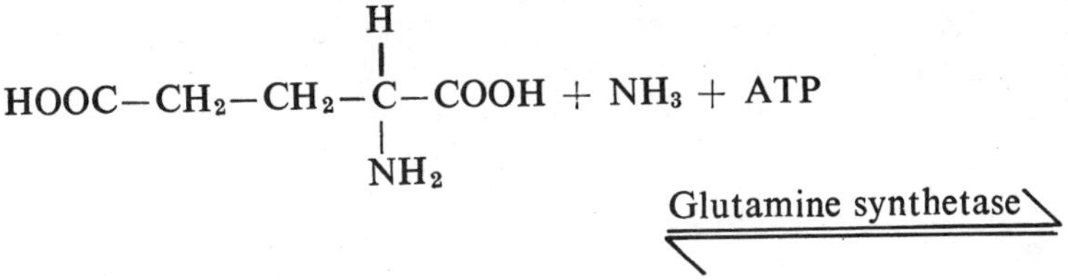

$$HOOC-CH_2-CH_2-\underset{\underset{NH_2}{|}}{\overset{\overset{H}{|}}{C}}-COOH + NH_3 + ATP$$

$$\xrightleftharpoons{\text{Glutamine synthetase}}$$

Glutamic Acid

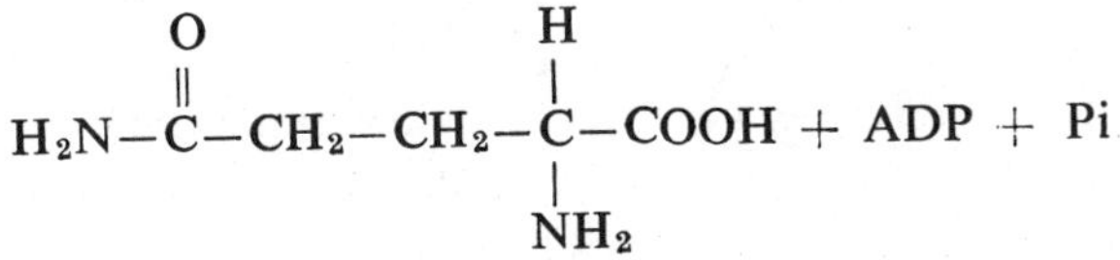

$$H_2N-\overset{\overset{O}{\|}}{C}-CH_2-CH_2-\underset{\underset{NH_2}{|}}{\overset{\overset{H}{|}}{C}}-COOH + ADP + Pi$$

Glutamine

Although the first reaction serves in the immediate removal of NH_3, the second one serves in removal of an excess of it. Glutamine produced here may also serve as a source of NH_3 if it is needed, besides its role in purine, hexoseamine, and mucopolysaccharide synthesis. Like glutamic, it may also undergo transamination reactions and the conversion into keto or amino acids.

3. The third process of NH_3-trapping is through the carbamyl phosphate synthetase reaction. This is an irreversible process that occurs in mammals and where 2ATPs are required for fixation of NH_3, plus carbon dioxide (CO_2)

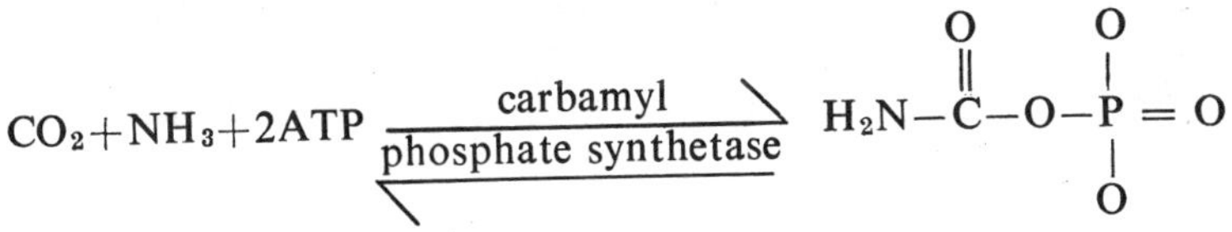

$$CO_2+NH_3+2ATP \xrightleftharpoons{\text{carbamyl phosphate synthetase}} H_2N-\overset{\overset{O}{\|}}{C}-O-\underset{\underset{O}{|}}{\overset{\overset{O}{|}}{P}}=O$$

carbamyl phosphate
$+2NAD+2Pi$

The carbamyl phosphate produced in this reaction goes into the formation of urea (through the urea cycle), and also into pyrimidine synthesis (carbamyl phosphate synthetase is absent in avians and the synthesis of pyrimidines in this species, thus, remains unknown).

The Essential Amino Acids

There are ten essential amino acids which are recognized on the basis of the young, ungrown rat. These amino acids are: valine, leucine, isoleucine, threonine, methionine, phenylalanine, tryptophan, lysine, histidine, and arginine. The full-grown rat does not require arginine. The adult man does not require either arginine or histidine. Avians require glycine in addition to the above ten amino acids. There are certain amino acids that have a sparing effect on others, since these are made from the essential ones. Such are ones like cysteine, which spares the essential methionine since it is made from it. On the same basis, tyrosine spares the essential phenylalanine, and hydroxylysine spares lysine.

The carbon skeleton of certain amino acids that the body cannot make is the essential thing here. The nonessential amino acids, however, can all be manufactured by the body if N_2 and the keto acid or the carbon skeleton are available.

Amino Acids Requirement

The state of growth influences highly the requirement for essential and nonessential amino acids. The growing animal has higher requirements for all amino acids. Disease, lactation, pregnancy, and degree of activity all influence amino acid requirements.

The full-grown man requires eight amino acids as essential in his diet, while the young infant requires histidine and arginine in addition. Nutritional status, optimum growth, and optimum reproduction influence the general requirement for all amino acids as well.

For protein synthesis to take place, all amino acids must be present at the same time.

Digestion and Absorption of Proteins

The first stage of protein digestion starts in the stomach where the proteolytic enzyme, pepsin, secreted by the chief cells, starts breaking the peptide linkages of protein. For pepsin itself to be active, a basic amino acids peptide must be cleaved off from it, thus reducing its molecular weight from 42,000 to that of 35,000, and changing its isoelectric point from 3·8 to 1·5.

Trypsin is another proteolytic enzyme that is secreted in the inactive form, trypsinogen, in the pancreas. It travels to the small intestines in the inactive form, and in the presence of another proteolytic enzyme secreted by the small intestines, enterokinase, a six basic amino acid peptide gets cleaved off from it, thus converting the inactive trypsinogen to the active form, trypsin. This enzyme works in an alkaline pH in the small intestines and cleaves peptide bonds where the carboxyl group comes from a basic amino acid such as lysine or arginine.

Chymotrypsin is another proteolytic enzyme that comes from the pancreas. It is converted from the inactive chymotrypsinogen to the active chymotrypsin form by trypsin. This enzyme works on peptide chains where the carboxyl groups are attributed by aromatic amino acids, such as tyrosine and phenylalanine.

The intestinal peptidases are also proteolytic enzymes that hydrolyze proteins into their amino acid constituents. The cathepsin enzymes are proteolytic ones contained within the lysosomes. When the lysosomal membranes break (as in aging meat or in inflammations), these cathepsins are released and their proteolytic action or their partial digestion of protein may cause either the desirable meat tenderization (in meat aging), or the undesirable inflammation which is usually remedied by the intake of corticoid hormones which presumably correct the lysosomal membrane and keep the cathepsins from leaching out.

There are also commercial proteolytic enzymes that are derived from plant sources. These are such as papain, which comes from the papain tree; bromelin, from pineapple; ficin, from figs; and which all cause digestion of proteins. They have a commercial benefit, such as in meat tenderization, where partial protein digestion is achieved.

Absorption of amino acids in the small intestines is an active process which works against a concentration gradient and which requires energy. It is thought that specific carriers may aid in the transport of amino acids through selective binding with them, thus facilitating their movement across the concentration gradient. The B_6 vitamin may also be involved in the active transport of amino acids. Amino acids are carried in the blood by specific proteins that transport them on to tissues. Tissue amino acids may undergo deamination reactions that convert them to keto acids which get metabolized as carbohydrates or fat. There is a constant turnover of tissue amino acids and the tearing-down and rebuilding of tissue proteins is rapid, especially in active organs like the liver.

General Reactions of Amino Acids

Deamination

The amino acid loses its amino group and becomes more of a carbohydrate than an amino acid.

Transamination

The amino acid exchanges its amino group with a keto acid.

Transamidination

The amino acid donates its guanidine group.

Transdeamination

This is transamination followed by deamination.

Transmethylation

A methyl group is exchanged.

Transcarbamylation

A carbamyl group is exchanged.

Transamination Reactions of Amino Acids

These are reversible reactions through which the amino group of one amino acid is transferred to a keto acid.

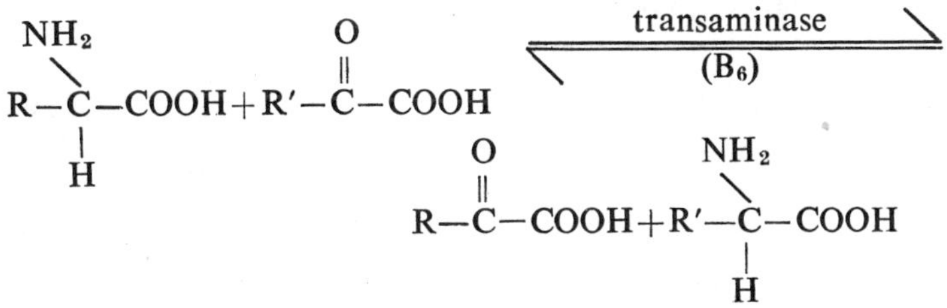

$$\begin{array}{c} NH_2 \\ \diagdown \\ R-C-COOH + R'-C-COOH \\ | \quad\quad\quad\quad\quad \| \\ H \quad\quad\quad\quad\quad O \end{array} \xrightleftharpoons[]{\text{transaminase} \, (B_6)} \begin{array}{c} O \quad\quad\quad\quad\quad NH_2 \\ \| \quad\quad\quad\quad\quad \diagdown \\ R-C-COOH + R'-C-COOH \\ \quad\quad\quad\quad\quad\quad | \\ \quad\quad\quad\quad\quad\quad H \end{array}$$

Transamination of glutamic acid to and from α-ketoglutaric is most important in the living system as it serves as a shuttle between amino acid, carbohydrate, and fat interconversion.

Deamination Reactions of Amino Acids

There are two types of this reaction:

1. Deamination through amino acid oxidase enzyme that directly passes electrons to O_2, giving rise to keto acid, hydrogen peroxide, and ammonia.

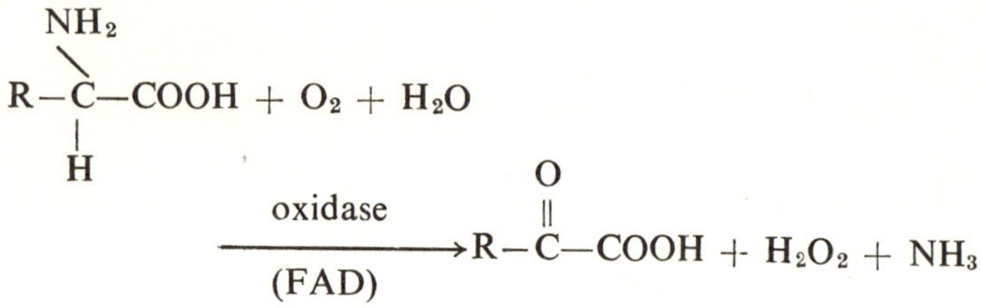

$$R-\underset{H}{\underset{|}{C}}(NH_2)-COOH + O_2 + H_2O \xrightarrow[\text{(FAD)}]{\text{oxidase}} R-\underset{\|}{\overset{O}{C}}-COOH + H_2O_2 + NH_3$$

There are D- and L-amino acid oxidases

2. Oxidative deamination, where the amino group of the amino acid is split off, giving rise to keto acid and ammonia.

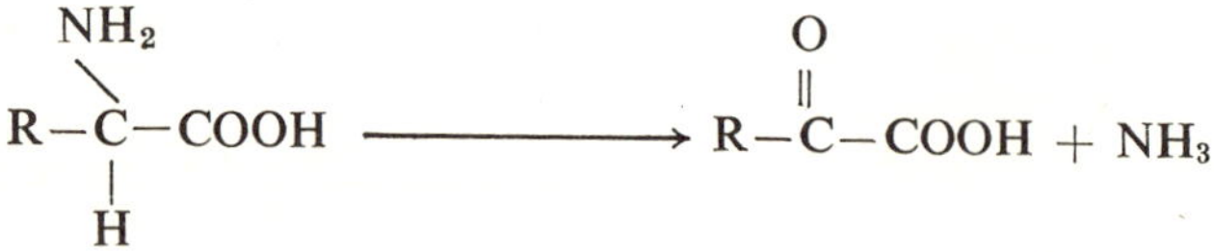

$$R-\underset{H}{\underset{|}{C}}(NH_2)-COOH \longrightarrow R-\underset{\|}{\overset{O}{C}}-COOH + NH_3$$

Dehydrogenation Reactions

Where glutamic dehydrogenase converts glutamic acid to α-ketoglutaric along with the production of NH_3 is a key reaction in the metabolism of amino acids, as it eventually converts proteins to fat and carbohydrates.

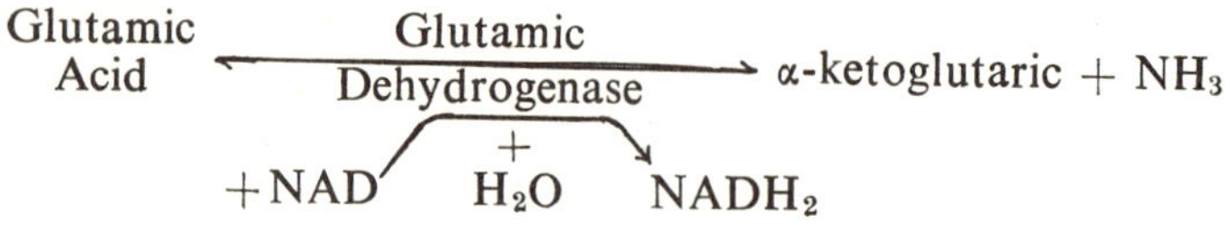

$$\text{Glutamic Acid} \xrightleftharpoons[+NAD \quad H_2O \quad NADH_2]{\text{Glutamic Dehydrogenase}} \text{α-ketoglutaric} + NH_3$$

Amino and Keto Acids Interconversions

Glutamic → Proline:

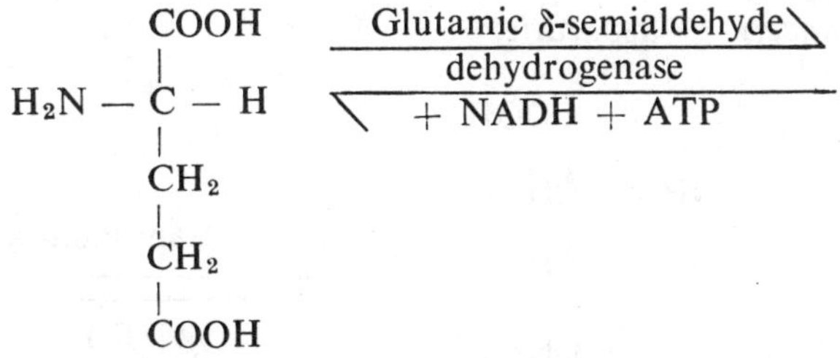

$$\begin{array}{c} \text{COOH} \\ | \\ H_2N - C - H \\ | \\ CH_2 \\ | \\ CH_2 \\ | \\ COOH \end{array} \quad \xrightarrow{\text{Glutamic } \delta\text{-semialdehyde dehydrogenase} \atop + \text{NADH} + \text{ATP}}$$

Glutamic Acid

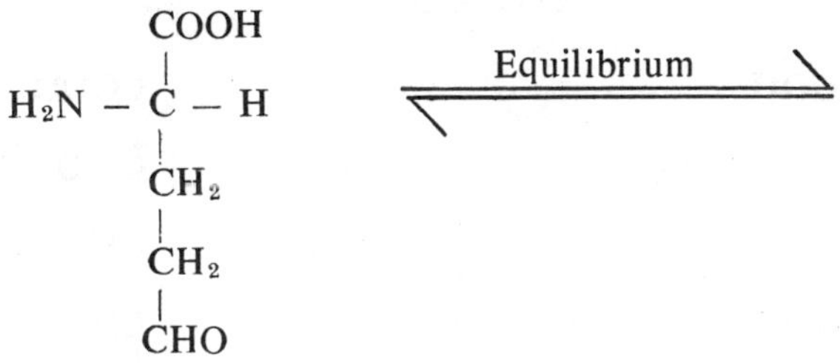

$$\begin{array}{c} \text{COOH} \\ | \\ H_2N - C - H \\ | \\ CH_2 \\ | \\ CH_2 \\ | \\ CHO \end{array} \quad \xrightleftharpoons{\text{Equilibrium}}$$

Glutamic Semialdehyde

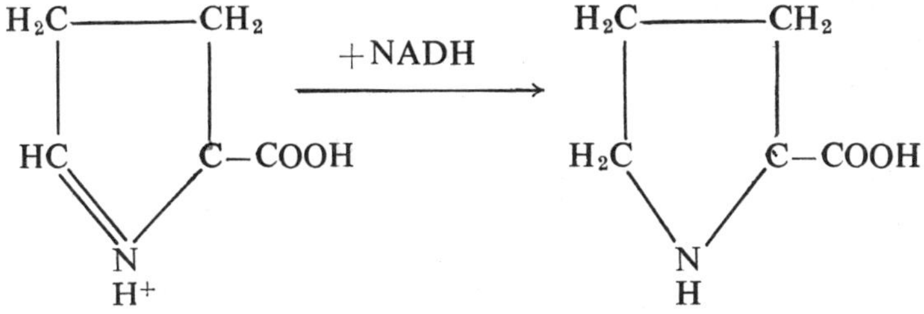

$$\underset{\delta^1\text{-Pyrroline-5-Carboxylic acid}}{\begin{array}{c} H_2C \text{———} CH_2 \\ | \quad\quad\quad | \\ HC \quad\quad C-COOH \\ \diagdown \;\; \diagup \\ N \\ H^+ \end{array}} \xrightarrow{+\text{NADH}} \underset{\text{Proline (Imino Acid)}}{\begin{array}{c} H_2C \text{———} CH_2 \\ | \quad\quad\quad | \\ H_2C \quad\quad C-COOH \\ \diagdown \;\; \diagup \\ N \\ H \end{array}}$$

Proline has a unique characteristic in that it is an imino instead of amino acid, and it gets hydroxylated only after its incorporation into collagen.

Oxaloacetic → asparatic:

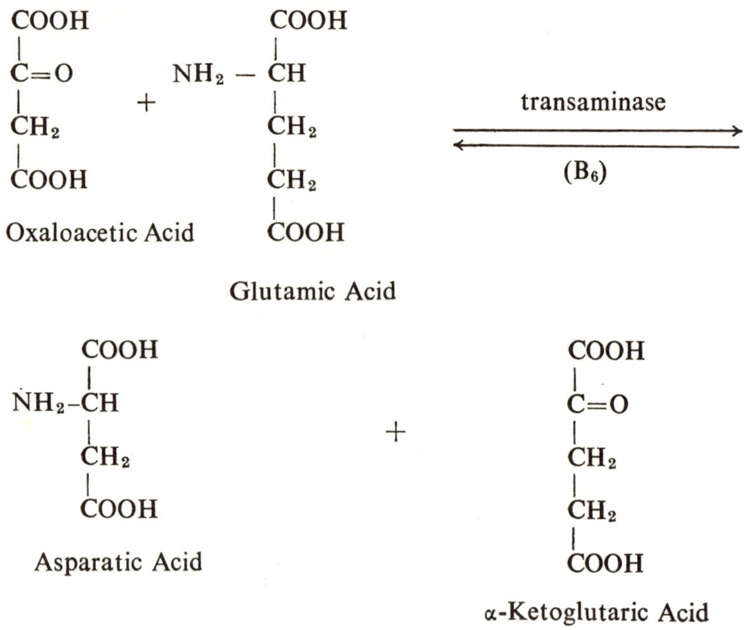

$$\begin{array}{c} COOH \\ | \\ C=O \\ | \\ CH_2 \\ | \\ COOH \end{array} \quad + \quad \begin{array}{c} COOH \\ | \\ NH_2-CH \\ | \\ CH_2 \\ | \\ CH_2 \\ | \\ COOH \end{array} \quad \xrightleftharpoons[(B_6)]{transaminase}$$

Oxaloacetic Acid   Glutamic Acid

$$\begin{array}{c} COOH \\ | \\ NH_2-CH \\ | \\ CH_2 \\ | \\ COOH \end{array} \quad + \quad \begin{array}{c} COOH \\ | \\ C=O \\ | \\ CH_2 \\ | \\ CH_2 \\ | \\ COOH \end{array}$$

Asparatic Acid   α-Ketoglutaric Acid

Pyruvic → Alanine:

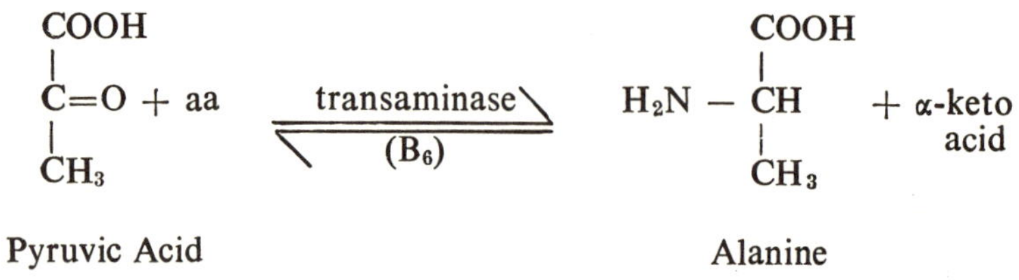

$$\begin{array}{c} COOH \\ | \\ C=O \\ | \\ CH_3 \end{array} + aa \quad \xrightleftharpoons[(B_6)]{transaminase} \quad H_2N-\begin{array}{c} COOH \\ | \\ CH \\ | \\ CH_3 \end{array} + \text{α-keto acid}$$

Pyruvic Acid   Alanine

Decarboxylation of Some Amino Acids

Generally, the decarboxylation of amino acids is catalyzed by B_6 vitamin.

$$\text{Glutamic acid} \xrightarrow[\substack{\text{decarboxylatin}\\(B_6)}]{\nearrow CO_2} \text{gamma amino butyric acid}$$

(which inhibits synapse transmission)

Histidine is decarboxylated to histamine, which is a powerful allergetic agent.

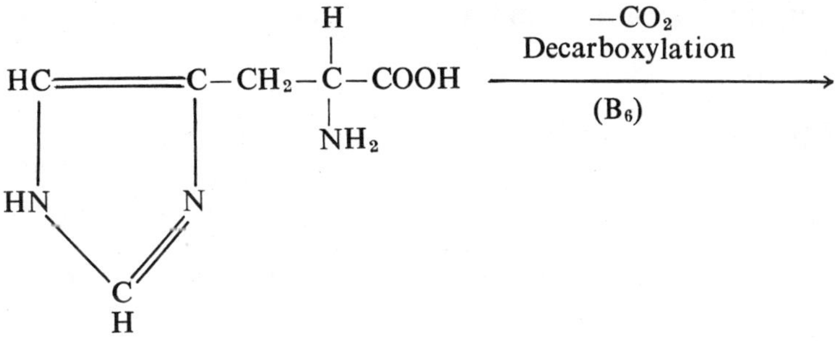

$$\begin{array}{c}\text{HC}=\!=\!=\text{C}-\text{CH}_2-\overset{\overset{\displaystyle H}{|}}{\underset{\underset{\displaystyle NH_2}{|}}{C}}-\text{COOH}\\ |\qquad\qquad| \\ \text{HN}\qquad\text{N}\\ \diagdown\;\diagup\\ \text{C}\\ |\\ \text{H}\end{array} \xrightarrow[\substack{(B_6)}]{\substack{-CO_2\\ \text{Decarboxylation}}}$$

Histidine

$$\begin{array}{c}\text{HC}=\!=\!=\text{C}-\text{CH}_2-\text{CH}_2-\text{NH}_2\\ |\qquad\qquad| \\ \text{HN}\qquad\text{N}\\ \diagdown\;\diagup\\ \text{C}\\ |\\ \text{H}\end{array}$$

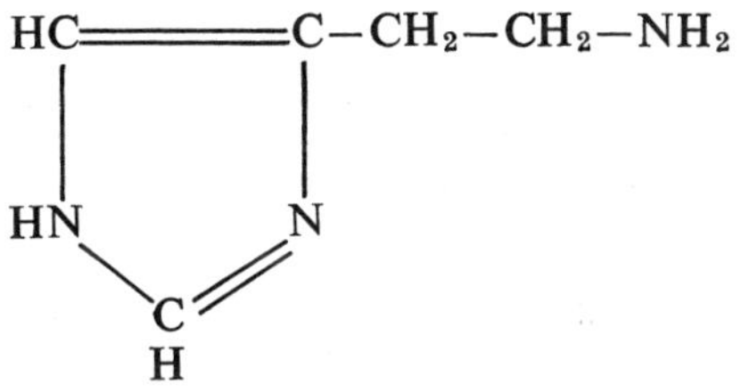

Histamine
(a potent vasodilator)

$$\text{Lysine} \xrightarrow[\text{(B}_6)]{\substack{-CO_2 \\ \text{Decarboxylation}}} \text{cadaverine}$$
(an amine characteristic of decay processes caused by bacteria)

$$\text{Tyrosine} \xrightarrow[\text{(B}_6)]{\substack{-CO_2 \\ \text{Decarboxylation}}} \text{Tyramine}$$
(a vasopressor)

Hydroxylation of tryptophan followed by a decarboxylation reaction gives rise to serotonin which is a vasoconstrictor of smooth muscle. Ornithine is decarboxylated to putrecine which contributes a carbon-amino group to methionine.

Important Reactions of Specific Amino Acids

Some amino acids are precursors of important hormones or vitamins. Tyrosine, for example, after undergoing several reactions in the living systems, gives rise to the important catecholamine hormones.

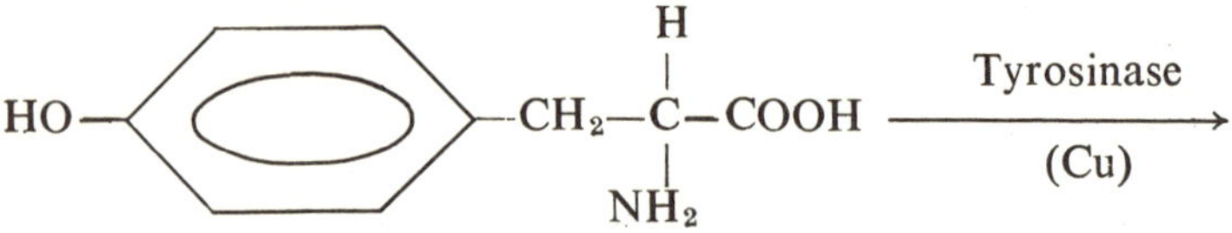

$$\text{HO–C}_6\text{H}_4\text{–CH}_2\text{–CH(NH}_2)\text{–COOH} \xrightarrow[\text{(Cu)}]{\text{Tyrosinase}}$$

Tyrosine

$$\text{(HO)}_2\text{C}_6\text{H}_3\text{–CH}_2\text{–CH(NH}_2)\text{–COOH} \xrightarrow{\substack{-CO_2 \\ \text{decarboxylation}}}$$

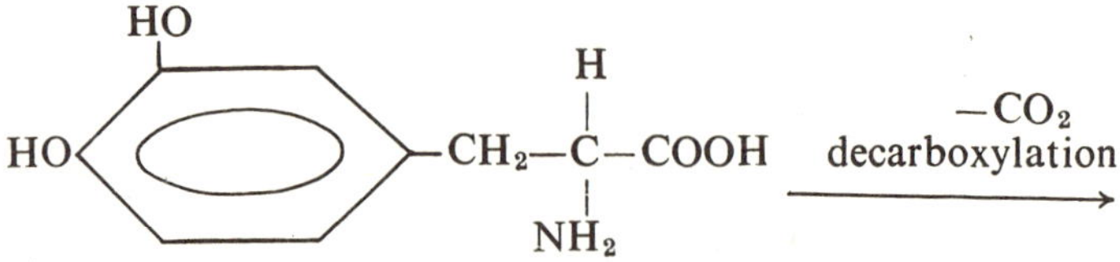

L-Dopa

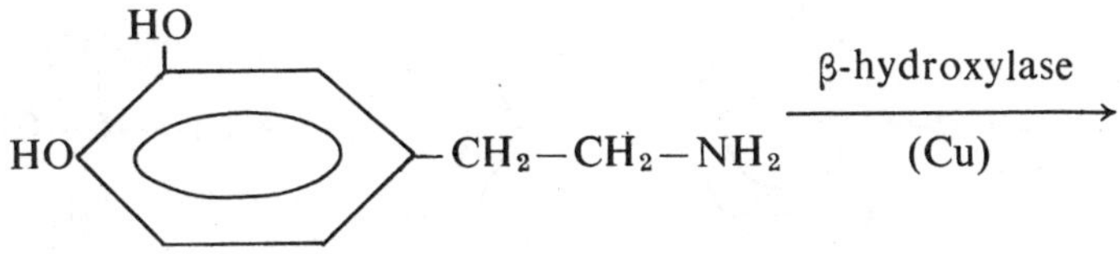

HO—⟨C₆H₃(OH)⟩—CH₂—CH₂—NH₂ →[β-hydroxylase (Cu)]

Dopamine ↘ (melanin)

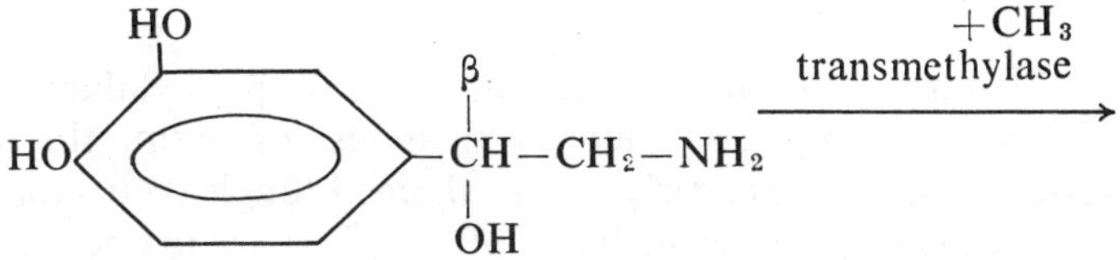

HO—⟨C₆H₃(OH)⟩—CH(OH)(β)—CH₂—NH₂ →[+CH₃ transmethylase]

Norepinephrine

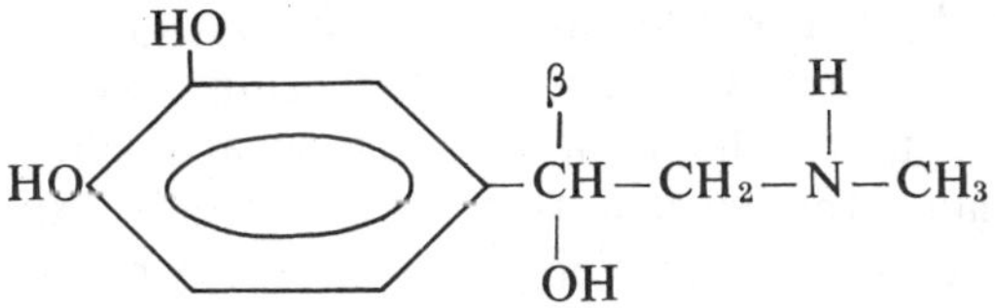

HO—⟨C₆H₃(OH)⟩—CH(OH)(β)—CH₂—N(H)—CH₃

Epinephrine

Tyrosine is also involved in the production of another most important hormone, that is, thyroxine. The iodination of tyrosine in the thyroglobulin molecules in the thyroid gland produces either the tetraiodothyronine or triiodothyronine which are the two thyroid hormones.

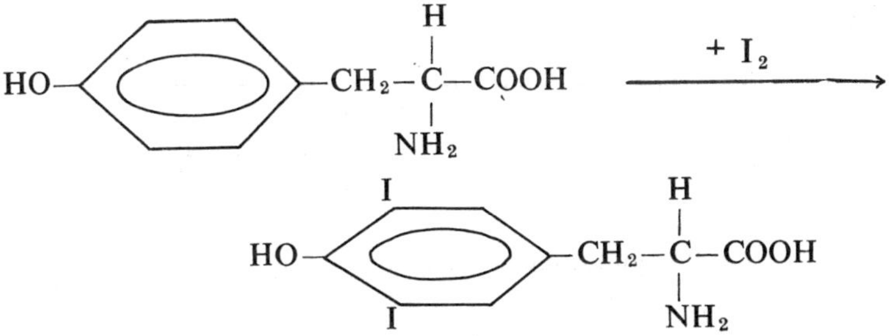

HO—⟨C₆H₄⟩—CH₂—CH(NH₂)—COOH →[+ I₂]

HO—⟨C₆H₂I₂⟩—CH₂—CH(NH₂)—COOH

$$2\times \longrightarrow \text{HO}-\underset{\text{I}}{\overset{\text{I}}{\bigcirc}}-\text{O}-\underset{\text{I}}{\overset{\text{I}}{\bigcirc}}-\text{CH}_2-\underset{\text{NH}_2}{\overset{\text{H}}{\text{C}}}-\text{COOH}$$

Tetraiodothyronine
(thyroxine)

If large amounts of the amino acids phenylalanine and tyrosine are taken and there is an excess of them, the living system may very properly utilize them through their complete metabolism into intermediates of the TCA cycle and into fat metabolism. In this case, phenylalanine must first be irreversibly hydroxylated to tyrosine and the enzyme, mixed-function oxidase, required must be present. Subsequently, tyrosine undergoes transamination, where it loses its amino group and becomes a keto acid, which, in turn, is decarboxylated to homogentistic acid, the benzene ring of which gets oxidized, giving rise to fumaric acid (which goes into energy production through the TCA cycle) and to acetoacetic acid (which goes into fat metabolism and ketone body production).

If the mixed-function oxidase enzymatic reaction is blocked (genetic inheritance), the mental retardation disease, phenyl ketone urea, will ensue due to the subsequent accumulation in the urine of phenyl pyruvic acid.

$$\bigcirc-\text{CH}_2-\underset{\text{H}}{\overset{\text{NH}_2}{\text{C}}}-\text{COOH} \xrightarrow[\substack{\text{mixed-} \\ \text{NADPH}_2 \rightarrow \text{NADP} \\ \text{function oxidase}}]{+O_2}$$

Phenylalanine

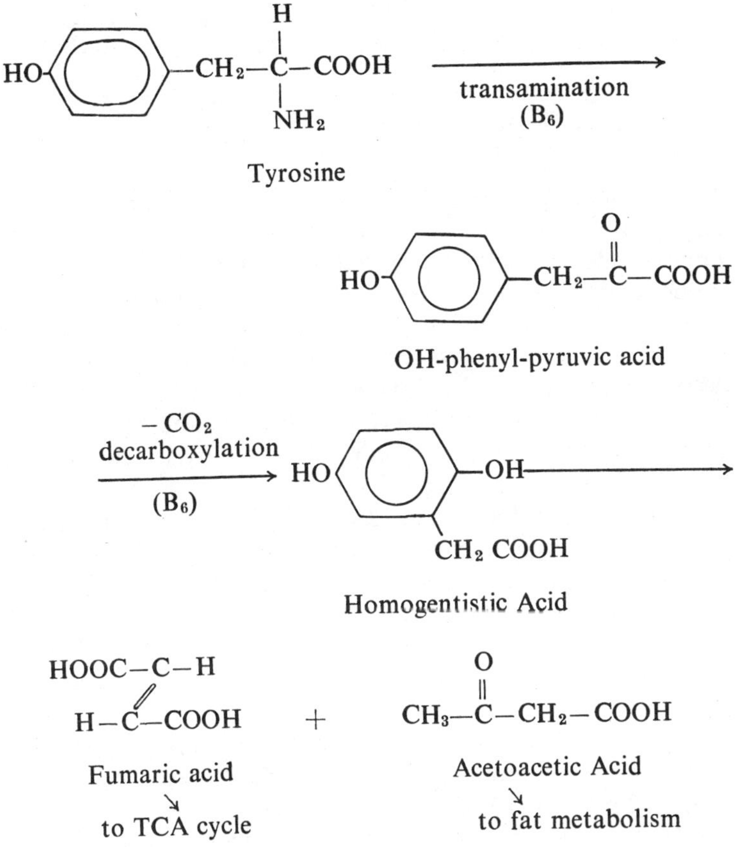

HO—⌬—CH$_2$—CH(NH$_2$)—COOH → transamination (B$_6$)

Tyrosine

HO—⌬—CH$_2$—C(=O)—COOH

OH-phenyl-pyruvic acid

$-CO_2$ decarboxylation (B$_6$) → HO—⌬(—OH)—CH$_2$COOH →

Homogentistic Acid

HOOC—CH=CH—COOH   +   CH$_3$—C(=O)—CH$_2$—COOH

Fumaric acid ↘ to TCA cycle

Acetoacetic Acid ↘ to fat metabolism

Another genetically inherited problem here would be the accumulation of homogentistic acid instead of its subsequent metabolism to fumarate and acetoacetate. The build-up of homogentistic acid causes dark pigmentation in the eyes, tissues, and urine.

Cysteine, through several reactions, gives rise to taurine, which conjugates with bile acids, giving rise to bile salts. When glutamic acid, cysteine, and glycine combine in peptide linkages, they form the tripeptide, glutathione, which is an important component in the maintenance of the cell walls' integrity of red blood cells (RBC). Serine, when phosphorylated, is involved in sphingosine synthesis and, hence, in lipid metabolism.

Glycine

Glycine, the only amino acid with no D- and L-isomers, has some unique functions and properties. It can be made by mammals from carbohydrates through the formation of the amino acid serine first. Avians have the special problem of not being able to make glycine, and, as mentioned earlier, it is an essential amino acid for avians. Besides its role in purine synthesis, glycine is also involved in porphyrin and creatine synthesis. A side chain of tryptophan splits and gets metabolized as alanine, which is converted to pyruvate through deamination.

Alanine, glycine, serine, and cysteine are all convertable to pyruvate through simple step reactions.

Asparatic acid undergoes transamination into oxaloacetic acid and enters the TCA cycle. Glutamic acid, proline, arginine, and histidine are also interconvertible into TCA cycle intermediates, and, consequently, enter into carbohydrate and fat interconversion and metabolism.

Threonine may be converted to α-ketobutyric acid which, in turn, is converted to propionyl-Co-A that is carboxylated to methyl malonyl-Co-A. The latter gets converted to succinyl-Co-A by a mutase and B_{12}, and enters the TCA cycle.

$$CH_3-\underset{\underset{\text{Threonine}}{}}{\overset{\overset{OH}{|}}{CH}}-\overset{\overset{NH_2}{|}}{CH}-COOH \xrightarrow[(B_6)]{\text{deamination}} CH_3-CH_2-\overset{\overset{O}{\|}}{C}-COOH$$

α—keto-butyric acid

+COA, +NH₃, −CO₂
oxidative decarboxylation

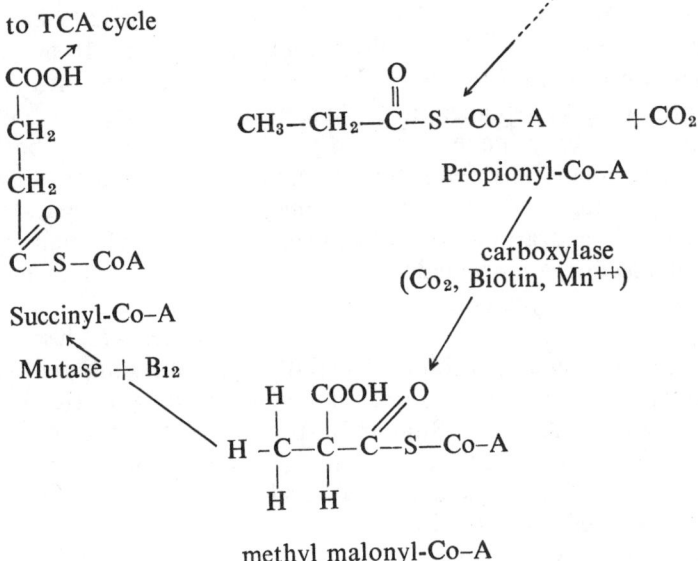

Lysine

Lysine finally gets metabolized to acetoacetate. Like ornithine, its subsequent conversion to glutamic semialdehyde makes it end up as succinate and enter the TCA cycle.

Tryptophan

Tryptophan metabolism ultimately leads to acetoacetate. The B vitamin niacin, is endogeneously synthesized from tryptophan, and we can, in this respect, obtain all the niacin we need if sufficient tryptophan is present in our diet.

Valine, Leucine, Isoleucine

These undergo oxidative deamination and are converted to keto acids which, in turn, undergo oxidative decarboxylation and end up in the TCA cycle. Leucine is a strictly ketogenic amino acid in its end-product metabolism. It finally gives rise to the 2-carbon-unit ketone body that goes into fat metabolism. Isoleucine, in the meantime, gives rise to both acetate and propionate as its end metabolism; thus, it is both a ketogenic and glucogenic amino acid, and so is lysine. Glycine, alanine, serine, valine, threonine, methionine, asparatic acid, proline, arginine, histidine, cysteine, and tryptophan are all glucogenic amino acids in that their end product of metabolism is a carbohydrate.

Arginine, in excess, gets acted on by arginase, which splits it into urea and ornithine. The latter is transaminated to glutamic acid which gets deaminated to α-ketoglutaric that enters the TCA cycle. Similarily, proline gets converted to glutamic, to α-ketoglutaric. Alanine, in excess, gets directly transaminated into pyruvic acid, and thus enters the TCA cycle for energy production.

Excess glycine gets converted to serine, to pyruvic acid, or glycine may be oxidized by glycine oxidase into glyoxylic acid, into oxalate, into formate plus carbon dioxide (the oxalate produced here may cause kidney-stone problems).

Amino acids with hydroxyl groups, such as threonine and serine, do not readily undergo transamination or oxidative deamination reactions. Instead, they undergo dehydration:

$$\underset{\text{Serine}}{HO-CH_2-\underset{\underset{NH_2}{|}}{\overset{\overset{H}{|}}{C}}-COOH} \xrightarrow{\text{Serine dehydratase}} \underset{\underset{\text{+Ammonia}}{\text{Pyruvic Acid}}}{CH_3-\overset{\overset{O}{\|}}{C}-COOH + NH_3}$$

$$\underset{\text{Threonine}}{CH_3-\underset{\underset{OH}{|}}{\overset{\overset{H}{|}}{C}}-\underset{\underset{NH_2}{\backslash}}{\overset{\overset{H}{|}}{C}}-COOH} \xrightarrow{\text{Threonine dehydratase}} \underset{\alpha\text{-Ketobutyric Acid}}{CH_3-CH_2-\overset{\overset{O}{\|}}{C}-COOH}$$

Likewise, the sulfur-containing amino acid cysteine gets acted on by a desulfurase enzyme.

$$HS-CH_2-\underset{NH_2}{\overset{H}{C}}-COOH \xrightarrow{\text{Cysteine desulfurase}}$$

Cysteine

$$CH_3-\overset{O}{\underset{\|}{C}}-COOH + H_2S + NH_3$$

Pyruvic acid

Interconversion of Sulfur–Amino Acids

Methionine, giving up its methyl group, gets converted to homocysteine, which, in turn, gives up its sulfur and ends up as α-ketobutyric acid that gets converted to propionate.

$$CH_3-S-CH_2-CH_2-\underset{NH_2}{\overset{H}{C}}-COOH \xrightarrow{-CH_3}$$

Methionine

$$HS-CH_2-CH_2-\underset{NH_2}{\overset{H}{C}}-COOH$$

Homocysteine

Cystathionine Synthetase (B$_6$)

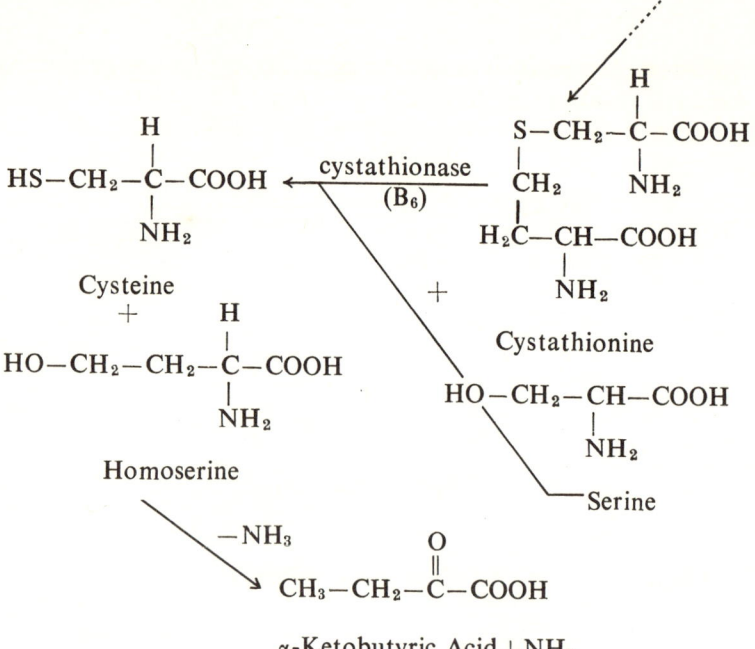

Thus, cysteine and α-ketobutyric acid are produced from methionine and serine.

Urea Cycle

$$NH_3 + CO_2 + 2ATP$$

$$\underset{\text{Carbamyl-ph.}}{H_2N-\overset{O}{\overset{\|}{C}}-O-\overset{O}{\underset{O}{\overset{\|}{P}}}=O} + \underset{\text{Ornithine}}{H_2N-CH_2-CH_2-CH_2-\overset{H}{\underset{NH_2}{\overset{|}{C}}}-COOH}$$

Ornithine Transcarbamylase

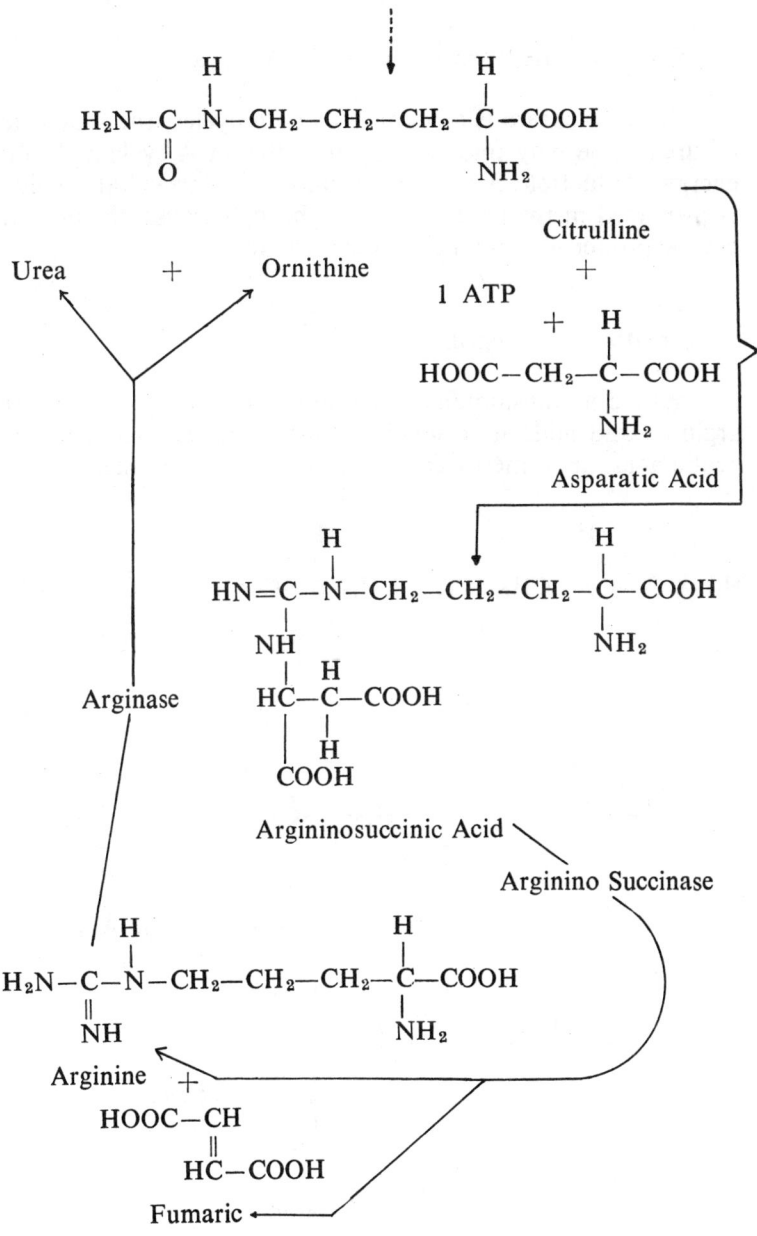

Relationship Between Urea and TCA Cycles

Since fumarate is a product of both cycles, the fumarate of urea cycle may find its way into the TCA cycle and into energy production. Asparatate is another intermediate which is produced in the urea cycle and, through transamination, it can be produced from TCA intermediate.

Formation of Creatine

Where a transamidinase enzyme splits the guanidine of arginine and adds it to glycine, thus giving rise to guanidoacetic acid, upon methylation it gives rise to creatine.

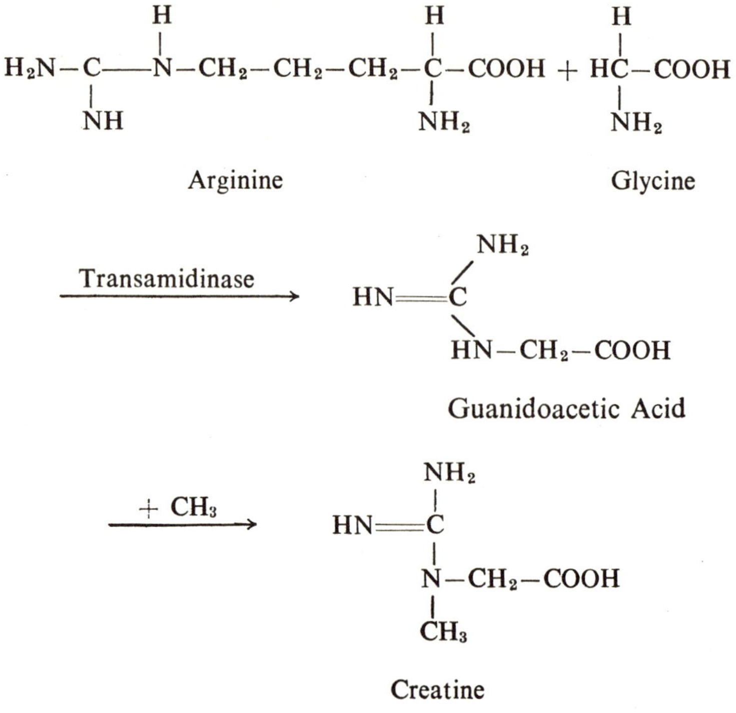

$$H_2N-\underset{NH}{\underset{|}{C}}-\overset{H}{\underset{|}{N}}-CH_2-CH_2-CH_2-\overset{H}{\underset{NH_2}{\underset{|}{C}}}-COOH + H\overset{H}{\underset{NH_2}{\underset{|}{C}}}-COOH$$

Arginine   Glycine

$$\xrightarrow{\text{Transamidinase}} HN=C\underset{HN-CH_2-COOH}{\overset{NH_2}{\diagup}}$$

Guanidoacetic Acid

$$\xrightarrow{+ CH_3} HN=\underset{\underset{CH_3}{\underset{|}{N}}-CH_2-COOH}{\overset{NH_2}{\underset{|}{C}}}$$

Creatine

Creatine is present in muscle as phosphocreatine, a high-energy compound that readily gives off its phosphate to ADP and hence makes ATP, which is required for muscle contraction.

REFERENCES

Altschul, A. M. 1958. *Processed plant protein foodstuffs.* New York: Academic Press.

Bendall, J. R. 1969. *Muscles, molecules, and movement.* London: Heinemann, Ltd.

Blackburn, S. 1968. *Amino acid determination.* New York: Marcel Dekker, Inc.

Davies, R. E. 1965. On the mechanism of muscular contraction. In *Essays in biochemistry.* Vol. I, New York: Academic Press, pp. 29–56.

Fox, S. W., and Foster, J. F. 1957. *Introduction to protein chemistry.* New York: J. Wiley and Sons.

Gibbons, I. R. 1968. The biochemistry of motility. *Ann. Rev. Biochem.* 37:521.

Haurowitz, F. 1950. *Chemistry and biology of proteins.* New York: Academic Press.

Holwill, M. 1967. Contractile mechanisms in cilia and flagella. In D. R. Sanadi, *Current topics in bioenergetics.* Vol. II. New York: Academic Press, pp. 288–334.

Huxley, H. E. 1969. The mechanism of muscular contraction. *Science.* 164:1456.

———, and Brown, W. 1967. Low angle X-ray diagram of vertebrate striated muscle and its behavior during contraction and rigor. *J. Mol. Biol.* 33:383.

Meister, A. 1957. *Biochemistry of the amino acids.* New York: Academic Press.

———. 1965. *Biochemistry of the amino acids.* 2nd. ed. New York: Academic Press.

Neurath, H. 1966. *The proteins.* 3rd. ed. Vols. I–IV. New York: Academic Press.

———, and Bailey, K. 1954. *The Proteins.* New York: Academic Press.

Peachey, L. D. 1965. Sarcoplasmic reticulum and transverse tubules of frog muscles. *J. Cell Biol.* 25:209

Perry, S. V. 1968. The role of myosin in muscular contraction. In *Aspects of cell motility.* London: Cambridge University Press.

Porter, K. R., and Franzini-Armstrong, C. 1965. The sarcoplasmic reticulum. *Sci. American* 212:72.

Sacktor, B. 1965. Energetics and respiratory metabolism of muscular contraction. In *The physiology of Insects.* Vol. II. New York. Academic Press.

Seifter, S., and Gallop, P. M. 1966. The structure of proteins. In *The proteins.* Vol. IV, pp. 372–430.

Smith, D. S. 1965. The flight muscles of insects. *Sci. American* 212:76.

Stracher, A., and Driezen, P. 1966. Structure and function of the contractile protein myosin. In D. R. Sanadi, *Current Topics in bioenergetics.* Vol. I. New York: Academic Press, pp. 154–202.

Chapter 5

FLAVONOIDS

CLASSIFICATION

Flavonoids are the water-soluble pigments in plants. Their general structure is two benzene rings joined together by a 3-carbon bridge. There are four general types of flavonoids:

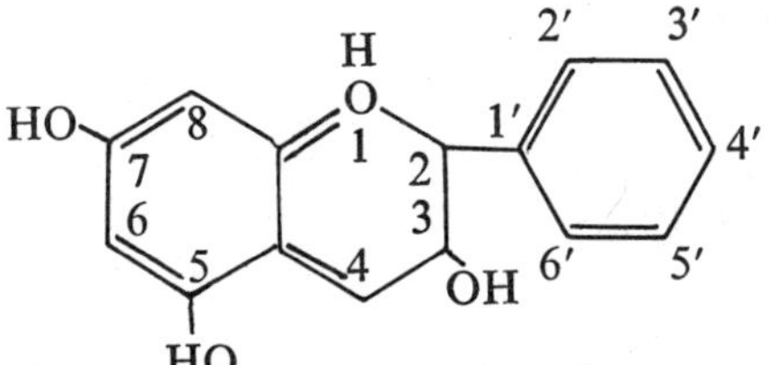

Flavylium

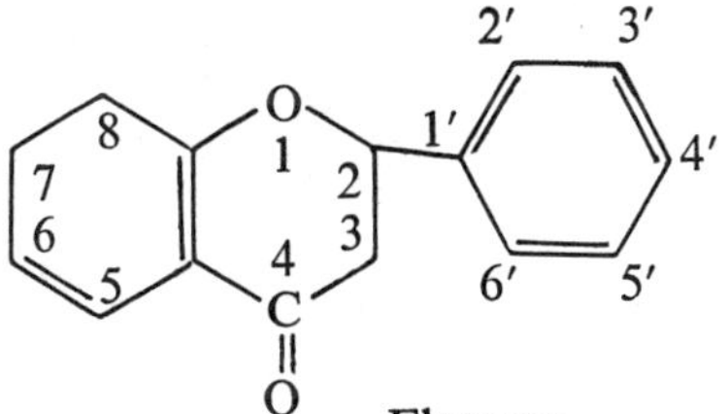

Flavone
(Parent of anthoxanthins)

1. Anthocyanins: are red, blue, and purple pigments of plants and are water-soluble.
2. Anthoxanthins: yellow, water-soluble plant pigments.
3. Catechols and
4. Tannins: brown pigments.

When there is one OH group on the A ring, the pigment is "pelargonidin"; if two OH groups on A ring, the pigment is "cyanidin"; and if three OH groups, the pigment is "delphinidin."

The above flavonoids exist as glycosides, such as anthocyanin (anthocyanidin + sugar).

The anthocyanins, which are blue, red, and purple water-soluble plant pigments, occur in plants as glycosides, which are esters of monosaccharides, although, at times, two sugars may be involved. The carbohydrate part is usually a monosaccharide such as glucose, galactose, or rhamnose, and occasionally it might be a pentose, although this is not often in occurrence. They are all soluble in water, and when the anthocyanins are boiled in mineral acid, they yield anthocyanidin plus a sugar at position 3, 5, or 7. Of these anthocyanins, only three types were identified:

1. Pelargonidin chloride 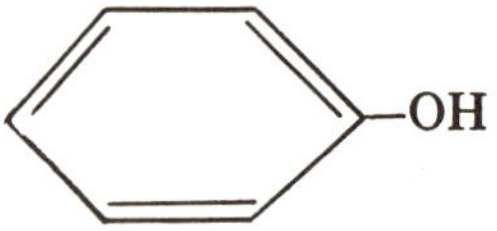
2. Cyanidin chloride 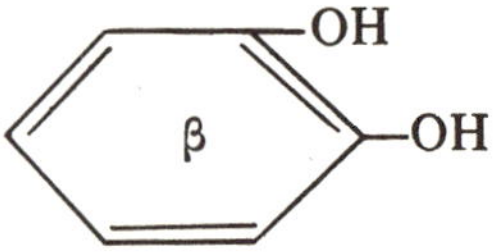
3. Delphinidin chloride

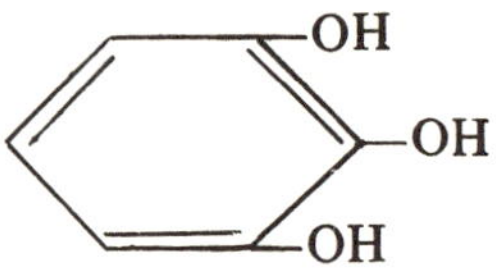

In most cases, the anthocyanins have the same anthocyanidin, but they differ in the types of sugar moiety.

The color intensity is usually determined and controlled by constituents on ring 2. For an example, when ring 2 has only one OH group, color is only a shade of blue; when it has two OH groups, it is darker blue; and when it has three OH groups, the color is darkest blue.

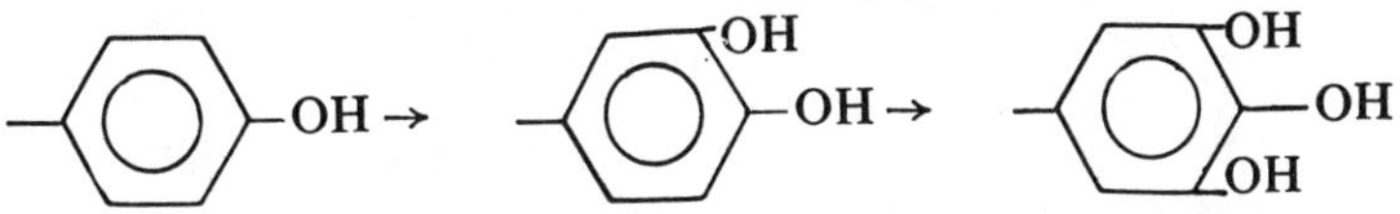

Shade of Blue → Darker Blue → Darkest Blue

Likewise, the number of methoxy groups on ring 2 determines the red color intensity of the anthocyanins.

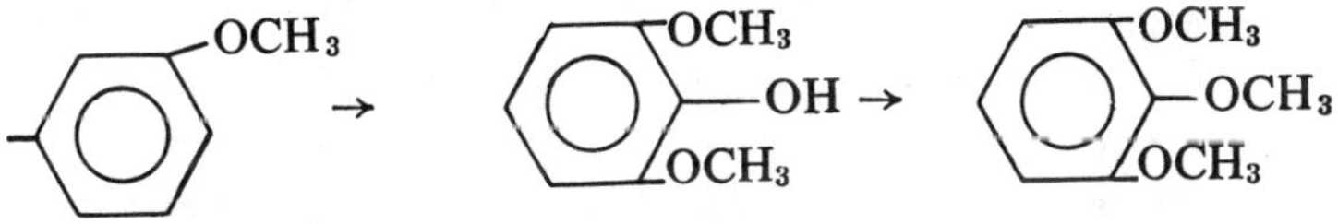

Shade of Red → Darker Red → Darkest Red

The color of anthocyanins also changes with changes in pH. An alkaline medium will result in a blue color, acidic media tend to give a red color, while a neutral pH gives a purple color. This may, in a sense, explain the various colors that flowers take at the various pH's of their environment and at different times of the day.

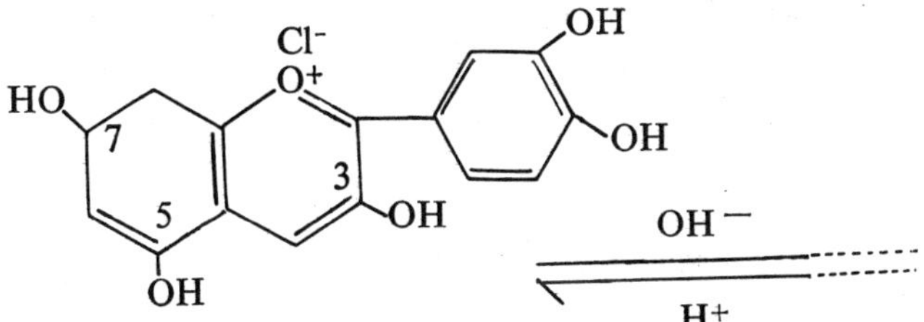

Red (in Acid)

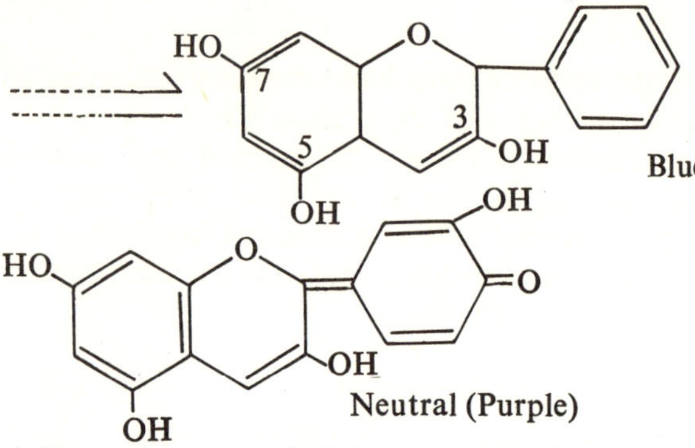

Blue (Alkali)

Neutral (Purple)

Temperature and light, as well, have an effect on the color of anthocyanins. The anthocyanin betain of red cabbage develops at temperatures of between 20 and 30°C, while at 10°C it is hampered.

While the development of these pigments in fruits is at its maximum at the light wavelengths between 360 and 450 mu. This may, also mean that these pigments are related to chlorophyll.

The anthocyanins may provide bad effects such as depolarization of H_2 which forms electrocoupling with metal in tin cans. In canning acid foods, corrosion and the production of H_2 gas occur due to the presence of anthocyanins. Therefore, during the processing of foods that contain these pigments, serious problems may arise due to their presence. If cells of the food ingredient that contain them are left intact, perhaps little change in color and not much trouble be expected. However, if cells are ruptured, leaking of the pigment and the subsequent loss of color occurs. Changes in pH that take place during cooking will definitely affect the color of the processed food. The presence of metal contaminants in most cases changes the food color into a dull gray one.

FLAVONES

The flavones are produced by the reaction of the glycosides and anthoxanthins with dilute acid, along with a sugar moiety.

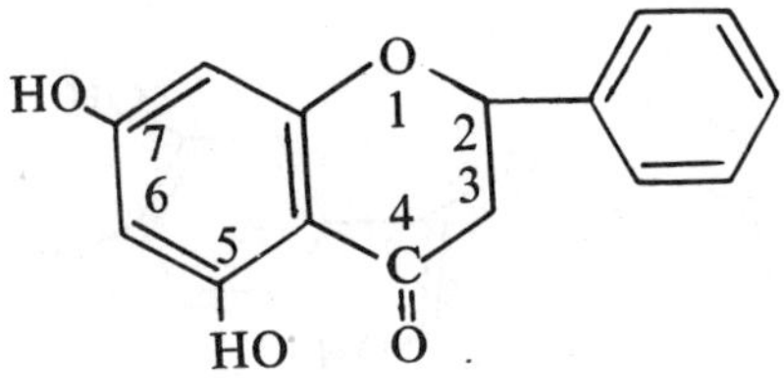

Flavone

Other flavone derivatives are:

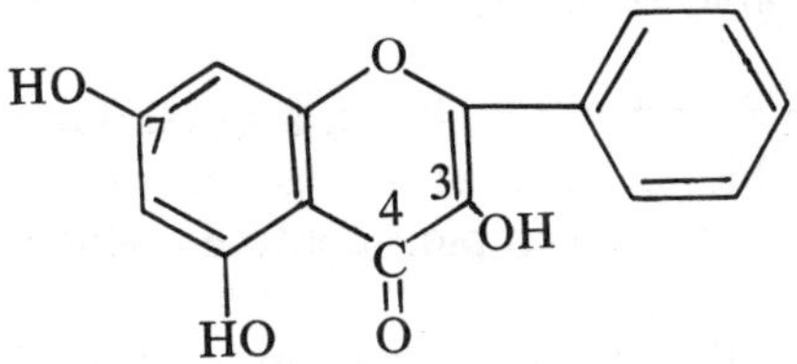

Flavonol (with OH on C 3)

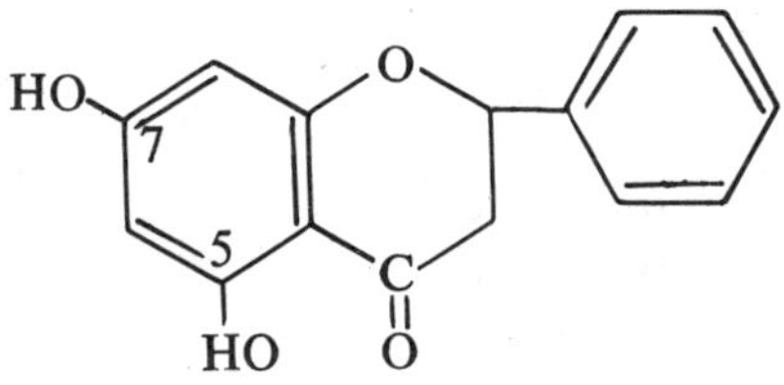

Flavanone

OCCURRENCE OF ANTHOXANTHINS

About 18 flavones, nearly 27 flavonols, 5 flavanonols, and 14 isoflavones were thus far reported as isolated from plants.

Anthoxanthins usually present in foods include quercetin, which is a flavonol that occurs in grains, tomato stalks, and other plants.

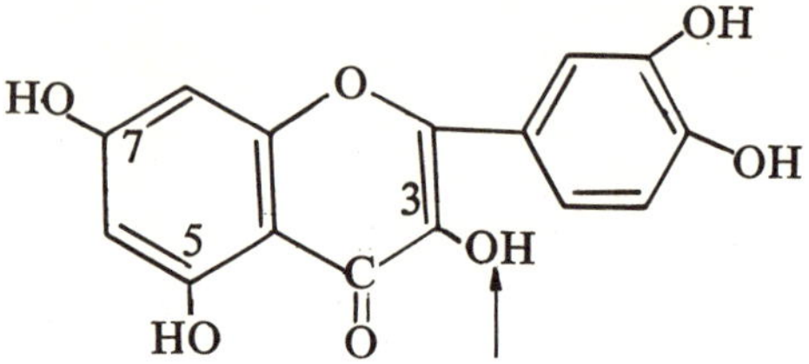

if galactose (golden apples)
if glucose (corn)

Quercetin is also found in tea, hops, apples, onions, and corn. The location and kind of sugar varies with the different anthoxanthins. Rutin, for example, contains glucose and rhamnose besides quercetin.

Apigenin is also a flavone that occurs in foods in the form of a glucoside. It is usually found in flowers, parsley, and in others.

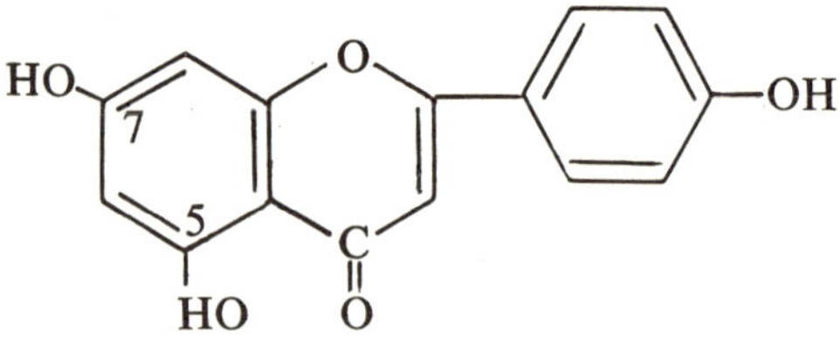

Hesperitin is an aglucose flavanone that is found in foods. It is usually present in citrus fruits like lemons and oranges in the form of 7-rhamnoside or 7-rutinoside (hesperidin). The flavonone form is slightly soluble in water, but is very soluble in alkali. When the pH of the media reaches 11 to 12, the inner ring of hesperitin opens up, thus giving rise to a new, yellow to brown pigment, "chalcone."

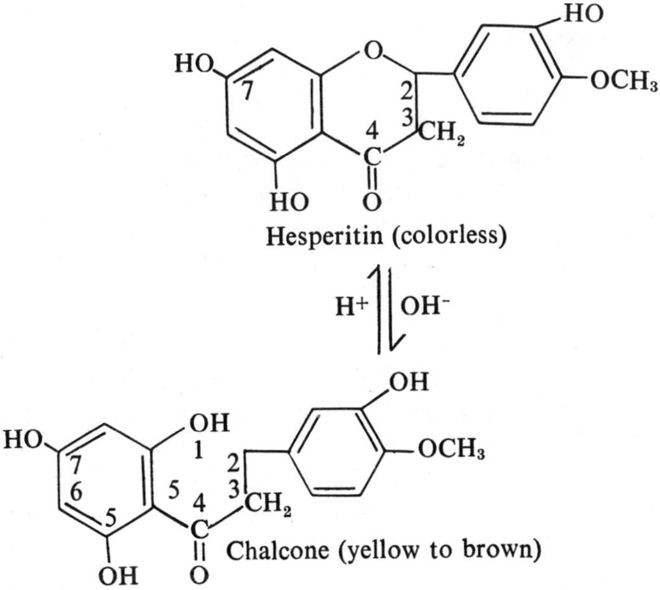

Hesperitin (colorless)

$H^+ \updownarrow OH^-$

Chalcone (yellow to brown)

GENERAL PROPERTIES OF THE FLAVONOIDS

Flavanones are oxidizable with acid and amylnitrite, as in the case of eriodictoyl oxidation to quercetin.

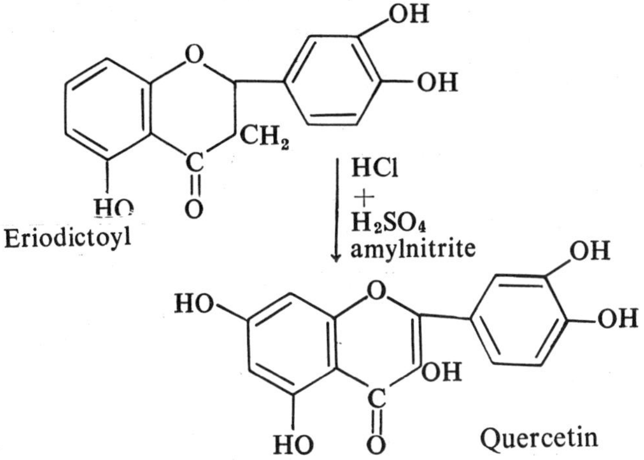

Eriodictoyl

HCl + H_2SO_4 amylnitrite

Quercetin

BIOLOGICAL ACTIVITY OF THE BIOFLAVONOIDS

Hesperidin and its chalcone derivative are believed to manifest a vitamin P activity, as they influence the permeability of blood capillaries to protein. They also are thought to inhibit hyaluronidase enzyme activity. Other flavones are believed to exert a vasoconstriction effect on the heart vessels.

BITTERNESS OF SOME FLAVONOIDS

Naringin is a flavonoid that possesses a bitter flavor. However, when the enzyme naringinase acts on it at pH 3.5 to 5.0 and 20–50°C, it removes the rutinose part from it, thus converting it to the aglucone naringenin, which is not bitter.

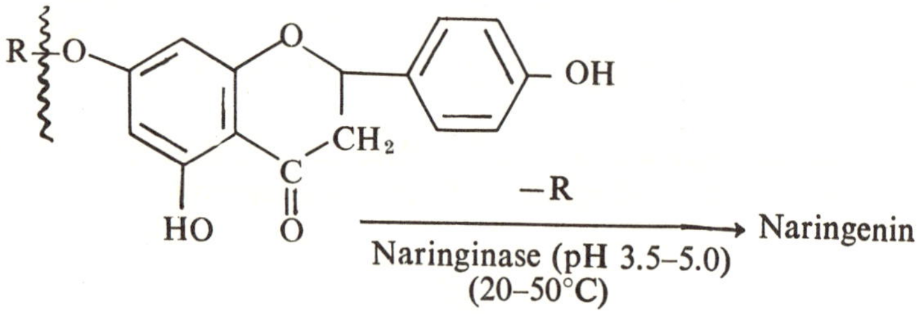

$$\text{Naringin} \xrightarrow[\text{(20–50°C)}]{\underset{\text{Naringinase (pH 3.5–5.0)}}{-R}} \text{Naringenin}$$

R = rutinose

Naringin

TANNINS

Tannins cause darkening in foods. They are astringent in taste, and originally come from plant sources, causing tanning of animal skins.

The best examples of tannins in foods are the catechins and the leucoanthocyanins.

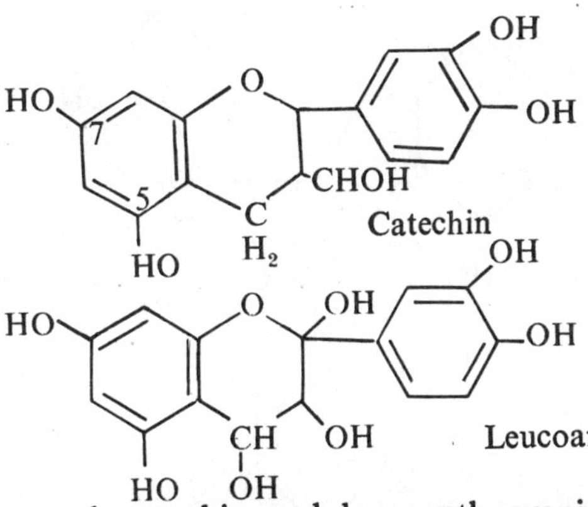

[Structures: Catechin and Leucoanthocyanin]

Both catechin and leucoanthocyanin are derived from tissue of woody plants, like apples, grapes, peaches, pears, and almonds. They also occur in cereals in varying amounts. They are absent in herbs, and small amounts of them will cause astringency.

Examples of tannins frequently occurring in foods are glucosides of digallic acid and tea chatechic.

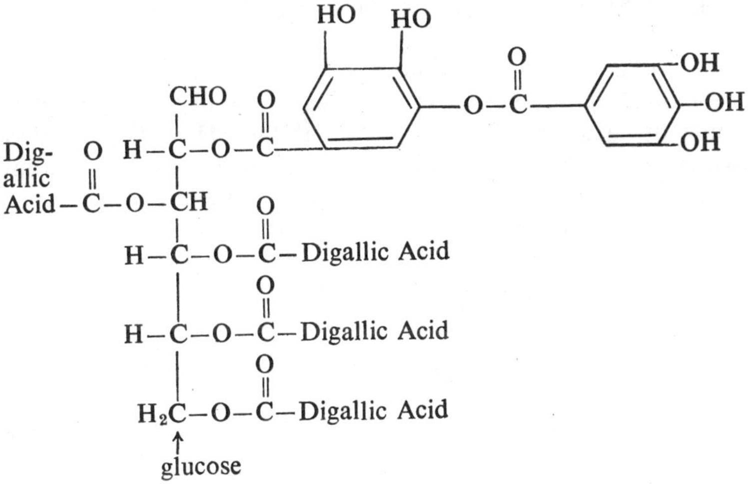

Glucosides of Digallic Acid

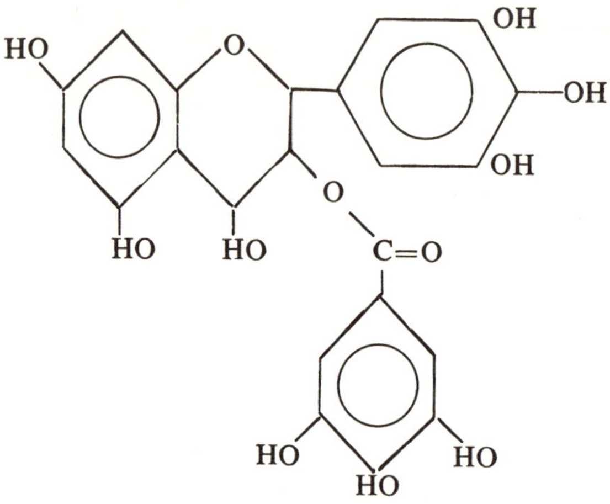

Tea Chatechic

CHANGES THAT OCCUR DURING PROCESSING

Foods that contain anthoxanthins upon cooking may develop a creamy-yellow color. This is due to the slight alkalinity of hard water. In this case, if the pH of the cooking water is lowered (by adding, for example, acid potassium tartarate), the yellow color development due to the presence of anthoxanthins may be avoided.

As with tannins of tea, if brewed in hard water a brown precipitate will develop which is due to the reaction between tannins and the iron content of hard water. Therefore, the green color development in frosted-flavored coffee is due to the reaction between tannins and iron, the iron source in this case being the beaters. The green color of chocolate ice creams, a mixture of canned evaporated milk with coffee, and the darkening of sweet potatoes are all, as well, due to reactions between tannins and iron.

REFERENCES

Bate-Smith, E. C. 1954a. Flavonoid compounds in foods. *Advances in Food Research* 4:262.

———. 1954 b. Flavonoid compounds in foods. *Advances in Food Research* 2:567.

Geissman, T. A. 1955. Anthocyanina, chalcones, aurones, flavones and related water-soluble plant pigments. In K. Peach and M. V. Tracy, *Modern methods of plant analysis.* Vol. 3. Berlin: Springer, p. 450.

Geissman, T. A., and Hinreiner., E. 1952. Theories of the bitterness of flavonoid compounds. *Botan. Rev.* 18:77.

Hall, J. A. 1925. Glucosides of the navel orange. *J. Am. Chem. Soc.* 47:1191.

Highby, R. H. 1938. The bitter constituents of navel and valencia oranges. *J. Amer. Chem. Soc.* 60:3013.

Huang, H. T. 1955. Decolorization of anthocyanins by fungal enzymes. *Agr. Food Chem.* 3:141.

Lawrence, W. J.; Price, J. R.; Robinson, R. M.; and Robinson, R. 1938. A survey of the anthocyanins. *Biochem. J.* 32:1661.

Mayer, F. 1952. The chemistry of natural coloring matters. *Ann. Rev. Biochem.* 21:472.

Robinson, G. M., and Robinson, R. 1934. A survey of anthocyanins. *Biochem. J.* 28:1712.

Sechardi, T. R. 1951. Biochemistry of natural pigments. *Ann. Rev. Biochem.* 20:487.

Thomas, D. W.; Smythe, C. V.; and Albbee, M. D. 1958. Enzymatic hydrolysis of naringin, the bitter principle of grapefruit. *Food Research* 23:591.

Ting, S. V. 1958. Enzymatic hydrolysis of naringin in grapefruit. *J. Agr. Food Chem.* 6:546.

Yang, H. Y., and Steele, W. F. 1958. Removal of excess anthocyanins by enzymes. *Food Technol.* 12:517.

Chapter 6

FLAVORS IN FOODS

INTRODUCTION

Flavor is the complex composite sensation of taste, odor, roughness, smoothness, hotness or coldness, and pungency or blandness, and added to these are the characteristics of sight and sound. However, odor and taste alone are the most important components of flavor. Taste is related to those components in the dissolved state that cause flavor sensations like salty, sour, sweet, and bitter, as they act on the taste buds on the tongue, creating a stimulus that goes to the brain. Sourness is mainly due to the effect of hydrogen ions. If we have equimolar solutions of acetic and hydrochloric acids for tasting, the hydrochloric acid solution will be more sour to the taste as it has a higher concentration of hydrogen ions. If, however, we take both solutions at the same pH, the acetic acid, in such case, will be more sour to the taste than HCl, as the former is more soluble to the taste bud than the latter, and because of the whole anion contribution to sourness.

In the case of salt's contribution to flavor, the low-molecular-weight salts usually cause a salty taste; as the molecular weight of salts increases, however, the salty flavor changes to that of bitterness. Potassium chloride (KCl), sodium carbonate (Na_2CO_3), and lithium chloride (LiCl) are all low-molecular-weight salts and provide a salty flavor. Potassium bromide (KBr) and ammonium bromide (NH_4Br) are of medium molecular weight, and are both salty and bitter. Heavier salts like cesium chloride ($CsCl_2$) and magnesium sulfate ($MgSO_4$) are bitter.

The chemical structure is of paramount importance in determining the flavor effect of a compound. Even a very slight change in the molecular structure of a compound may change it from a sweet to bitter one, and vice versa For example, 4-amino-2-nitropropoxy-benzene, with slight changes in its structure, gives the following contrasting flavors:

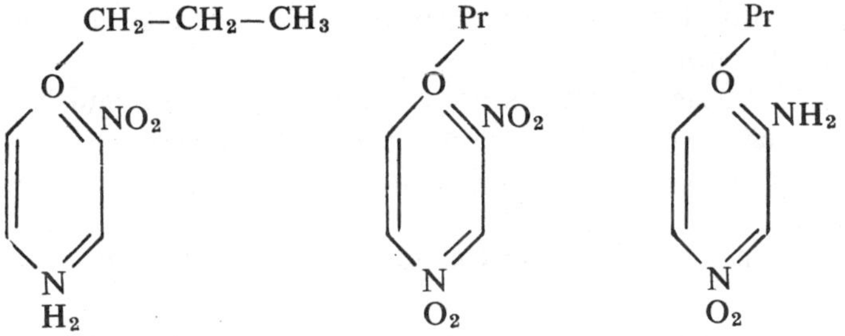

| Tasteless | Bitter | Sweet |

Putting one methyl group in place of hydrogen in the extremely sweet saccharin renders it tasteless.

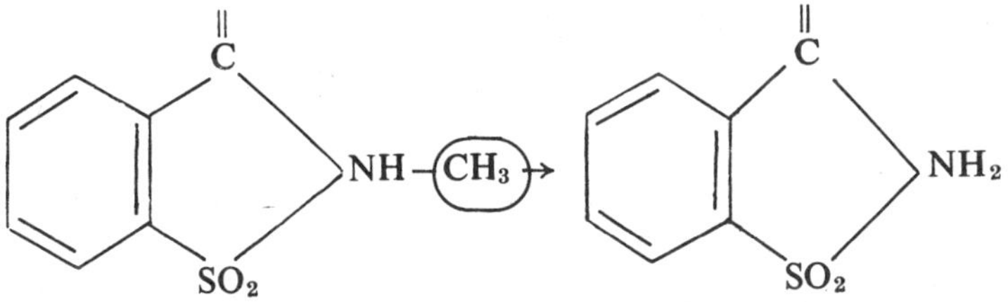

| Tasteless | Sweet |

Characteristics of odor are much more complex than taste. The following are the seven primary odors:

Primary Odor	Chemical Example	Familiar Substance
Camphoraceous	Camphor	Moth balls
Musky	Pentadecano lactone	Angelica root
Floral	Phenylethyl methyl ethyl carbinol	Roses
Pepperminty	Menthone	Mint Candy
Etheral	Ethylene dichloride	Dry-cleaning fluid
Pungent	Formic acid	Vinegar
Putrid	Butyl mercaptan	Bad eggs

A blend of the above seven primary odors make the known primary odors.

The human body is better equipped to distinguish odor than taste. There are millions of olfactory receptor sites that can distinguish odors. These make man highly sensitive to odors.

Man, thus, can detect as low a concentration as one part per billion of N-decanal, 0.2 parts per billion of methyl sulfide, and 0.07 parts per billion of β-ionone. A well-trained performer on flavor identification may be able to identify several thousands of various aromas. Man also possesses a great, long, and uncanny memory for recognizing flavors over very prolonged periods of time. The exact mechanism of how the human body detects odors is not known. It is thought by some that the chemical odor molecule may possess the molecular geometry that fits in specific slots on the olfactory receptor sites, and as they do they elicit the response.

Others tend to favor the membrane-puncture theory, which states that the chemical molecule punctures the receptor site of the membrane causing depolarization of the cell membrane, and Na^+ moves inside and K^+ outside the membrane. The cell membrane depolarization will result in stimulation that goes to brain. The extent of intensity of the flavor stimuli, of course, depends on the extent of puncturing the receptor membrane.

SENSORY EVALUATION OF FLAVOR

There are no clearcut physical methods for flavor evaluation. Thus, human evaluation of flavors is mostly used where a well-trained taster's judgment is trusted in placing scores of evaluations within a specific range.

There are also panels of laboratory or consumer scales where select individuals carefully evaluate products. Usually, in such cases, a small laboratory scale panel is first developed before giving their evaluations to the consumer's use.

There is also the testing-through-differences method, where, for example, three samples are provided, two of which are the same, and the third is different. Here, the individuals match or select the two samples that are the same.

The dilution method of testing flavors is also applicable, where a sample is so diluted that its flavor characteristic is no longer apparent, and the tester manipulates the dilutions until the establishment of the proper dilution that brings up the flavor.

Flavors also may be ranked on a scale that ranges from least to most desirable.

FLAVOR PROFILE

This represents a scientific study of flavor through the vocabulary of five to six people who determine flavor characteristics using the same technology. Here each flavor is represented by a flavor note; then follows the establishment of the flavor notes, and the order of appearance of these notes, and the establishment of the aftertaste effect of the product. It takes an experienced individual to perceive the fullness of flavor in this procedure.

FLAVORS IN VARIOUS FOODS

Milk should be bland and possess no flavor. Any flavor that might be sensed in milk is an off-flavor, with the exception of a sweet, mild sensation that arises from the aromatic characteristic of milk. A rough or astringent flavor may develop with the imbalance of the calcium-protein micelle system, where the protein gets altered.

130 flavor components have been isolated from cheese. These include fatty acids, esters, carbonyls, alcohols, lactones, and thiols. Recently, gas chromatography and mass spectroscopy have been very useful tools in the identification of volatile flavor components of foods.

Buttermilk's flavor is mainly due to diacetyl, fatty acids, and lactic acid. Blue cheese flavor is mainly due to methyl amyl ketone. Flavor of vanilla is due to vanillin, of grapes to methyl anthranilate, of garlic to allicin, and of peppermint to menthol. In the case of bread, the many volatile materials that are produced in the fermentation process of dough-making are lost during baking. Thus, the flavor of baked bread is mainly due to the browning reactions (between sugars and amines), and the flavor components, in this case, are mostly located in the crust. Prolamine is thought of as a precursor of the fresh-bread flavor. The flavor of red meat and chicken is probably due, in part, to the reaction between a glycoprotein fraction that is present in meat and inosinic acid in the presence of fat and heat, a matter that gives rise to the meaty aroma and flavor of cooked meat. Lactic acid, amines, and long-chain, unsaturated carbonyls may also contribute to the flavor of cooked meat.

If beef or pork fats are heated in presence of O_2, meaty flavor develops. If heated in presence of N_2, this flavor does not develop.

Since all fresh fruits contain sugars and amines, practically all undergo browning reactions. A volatile fraction of fruits would contain acetaldehydes, formaldehydes, acetoin, fatty acids, diacetyls, amines, and hydrogen sulfides. Distillate of coffee gives an average of 150 volatile flavor components.

The aroma of peaches is mainly due to lactones, especially γ-undecalactone.

Grapes' flavor is due to methyl anthranilate and ortho-amino benzoic acid; pears, to ethyl 2-trans-4-cis-decadienoate; raspberry, to para-hydroxy-phenyl butan-3-one; grapefruit, to noot kanone; apples, to ethyl butyrates; watermelons, to octanal; onions, to aliphatic disulfides; celery, to alkylidine phthalides; and cucumbers, to aliphatic aldehydes.

Many food flavors are influenced by the circumstances of their growth or their production. For example, milk-feed flavor presents a serious problem for the dairy industry. Many green feeds produce feed flavor in milk such as silage, rye, clover, and onions. Besides feed-flavor development in milk, there is also the lipolysis problem.

Beef produced on citrus bulbs will develop off-flavor milk.

OFF-FLAVOR DEVELOPMENT

Free Fatty Acid Development in Meat When Cooked in O_2, N_2, and Vacuum (Flavor Development)

Fat of	O_2	FFA	N_2	Vacuum
Beef	+	1.5–3.7%	−	−
Pork	+	2.9–5.5%	−	−
Lamb	+	0.3%–0.4%	+	+

Quality of water affects the flavor development of fish, and muddy water fish usually develop an off-flavor.

Poultry fish meal or fish oil develop a fishy flavor upon cooking. Fat of turkeys fed linseed oil is more susceptible to oxidation. Chicken fat, in this regard, is less susceptible than turkey fat because it retains vitamin E (the natural antioxidant) more efficiently. Eggs of poultry fed fish meal will have a fishy flavor as well. The same principle applies; if poultry is fed rancid fat, garlic, or onions an off-flavor not only develops in the meat but in the eggs as well.

Off-flavor development is of even more importance in processed foods. Cooking or processing of foods bring about both desirable and undesirable flavors. Rancid flavor is an undesirable flavor produced in dairy products as a result of hydrolytic decomposition of the fat present. In fats and oils, it is an oxidative type of rancidity. Both, however, are usually undesirable (C-4 to C-10 fatty acids usually produce the off-flavor), although in some cases rancidity does develop a desirable flavor, such as in meat curing, cheese-making, and in other instances. Browning can be both desirable and undesirable.

Some desirable flavors do develop in foods during cooking. This is mainly due to the heat treatment. In poultry, oxidative deterioration occurs and the production of 2-4-decadienal is mainly responsible for the flavor developed.

MILK OFF-FLAVORS

Milk Flavors	Characteristics
Lipolytic flavor	soapy or bitter (fatty acids)
Oxidized flavor	tallowy, oily (carbonyls)
Heated flavor	cooked milk (sulfides)
Sunlight flavor	cabbagy, burnt hair, wet dog (carbonyls and mercaptans)
Sterile concentrated milk and milk powder	stale (aldehydes, fatty acids, ketones, and lactones)

VEGETABLE OFF-FLAVORS

Flavor	Characteristics
Potato Bitter flavor	Solanine $\xrightarrow{+H^+}$ solanidine + glucose + galactose + rhamnose
Potato chips flavor	Oxidized fat and browning
Dehydrates potatoes flavor	Staling—limited fat oxidation and browning

Heating pork, lamb, and beef meat in the presence of O_2 results in the highest production of free fatty acids (FFA) in the case of beef (3.7%), while lamb developed the lowest FFA (0.3 to 0.4%). Also FFA developed in beef and pork are believed to be more highly unsaturated than those of lamb. The increased unsaturated FA in beef and pork will, as well, mean that more carbonyl development will be expected in beef and pork. Consequently, the carbonyl 2-4-decadienal is found in beef and pork but is practically absent in lamb. This, in turn, may explain differences in the flavor of beef and pork from that of lamb.

Fat content usually influences the flavor of meat since, when the unsaturated FA are oxidized, they give rise to carbonyls. These carbonyls, only to a certain level, give desirability to flavor of meat, but they are undesirable at high concentrations. Fat also may serve as a solvent or reservoir for fat-soluble volatile compounds that contribute to flavor.

MILK OFF-FLAVORS

A lipolytic flavor development in milk arises from the action of the enzyme lipase on the triglyceride molecule, producing glycerol and free fatty acids (FFA). The FFA production gives rise to the off-flavor here. Although the lipase enzyme is present in raw milk, the fat globule membrane usually protects fat from it. However, upon agitation, or whatever other means may break the fat globule membrane, lipase enters the fat cell and subsequently hydrolyzes the fat, and produces a soapy, bitter flavor due to the FFA production.

An oxidized flavor in milk, meanwhile, is produced through the action of Cu^{2+} and/or light on milk, and in this case, a tallowy or/and an oily type of flavor develops. This action takes place through a free radical mechanism, through which potent flavor developer compounds like alkanals, alkdienals, and unsaturated carbonyls develop.

A sunlight flavor development in milk occurs as light from the sun decomposes the S-containing amino acid, methionine, using B_2 vitamin, into methional, which in turn is converted to mercaptans. These mercaptans are responsible for a wet dog or cabbagy type of off-flavor, or an oxidized type of off-flavor.

The heated milk flavor develops upon cooking milk at temperatures of 171°F or above. If more intense heat treatment is given milk, the developed heat flavor may approach that of a carmalized flavor. The heat treatment apparently gives rise to H_2S from S-containing amino acids of milk, which is objectionable. The intensity of the H_2S effect may dissipate upon standing the milk at room temperature, as some H_2S gets lost.

The sterile, concentrated milk flavor is a stale flavor like the development of Maillard browning and oxidation. Stale flavor may be due to the development of carbonyls, FFA, or lactones.

REFERENCES

Bennett-Clark, T. A. 1948. Organic acids in plants. *Ann. Rev. Biochem.* 18:639.

Bonner, L. 1950. *Plant biochemistry*. New York: Academic Press.

Harvey, R. B. 1920. The relation between the total acidity, the concentration of the hydrogen ion and the taste of acid solutions. *J. Am. Chem. Soc.* 42:712.

———, and Fulton, R. R. 1935. Relation of pH and total acidity to the taste of tomatoes. *Fruit Products J.* 14:238.

Kastle, J. H. 1898. On the taste and affinity of acids. *Am. Chem. J.* 20:466.

Krebs, H. A. 1943. *Advances in Enzymol.* 3:191.

Ochoa, S. 1952. In J. B. Sumner and K. Myrback. *The enzymes.* New York: Academic Press.

Richards, T. W. 1898. The relation of the taste of acids to the degree of their dissociation. *Am. Chem. J.* 20:121.

Stevens. 1951. *Handbook of experimental psychology.* New York: John Wiley, p. 1148.

Chapter 7

VEGETABLES AND FRUITS

INTRODUCTION

There is no well-defined distinction between vegetables and fruits; however, usually those products consumed as the principal part of the meal are termed as the vegetables, and those eaten for dessert are considered the fruit. Fruits may also be considered as those ripened seeds and adjacent tissues that contain them.

CHEMICAL COMPOSITION OF FRUITS AND VEGETABLES

Most fruits and vegetables have a very high water content (70-98%) with a few exceptions, such as nuts and dates.

Most fruits and vegetables, with the exception of legumes and nuts, are very low in their protein content, ranging from 0.3 to 4.5%. Apples, for example, contain 0.3% protein, while Brussel sprouts contains 4.4%. Legumes, on the other hand, may range from 4% to as high as 24% protein content. Nuts are even higher, as their protein content ranges from 10-25%.

All fruits and vegetables contain some carbohydrates in one form or another. Some may contain cellulose, others may contain pectic substances, while starch is present in practically all fruits and vegetables, regardless of the fact that it gradually disappears during ripening. As the starch content decreases, sugar content increases. Most of the available carbohydrates in fruits and vegetables are composed of glucose, fructose, sucrose, and some starch. These sugars constitute the bulk of the caloric content of fruits and vegetables.

The fat content of most fruits and vegetables is very low, except avocados and nuts.

CLASSIFICATION

Fruits and vegetables are still living things. Their metabolism goes on even after being harvested. The vegetables alone may fall under the following general classifications:

1. The leafy vegetables.
2. Flowers, buds and stems.
3. Bulbs, roots, and tubers.
4. Seeds, as legumes, etc.
5. Vegetable fruits (peppers, tomatoes, etc.).

STRUCTURE OF FRUITS AND VEGETABLES

The parenchymal cells are the chief cells of structure of stems and leafs of fruits and vegetables. They are uniform cells of polygonal or cubical shape. They have thin walls and contain air space between cells. The cell walls of fruits and vegetables are fibers of cellulose, and the cells are held together by a cementing substance which, in the young cell, is the pectic substance. However, as the cell gets older, the cellulose gets imbibed with lignin, thus gaining in toughness and undesirability as a food, and even a woody texture may develop upon cooking.

Material inside the cell is protoplasm which is composed of many other substances. This is usually a viscous or gelatinous in nature. Material inside the cell, with the exception of the nucleus, is composed of the cytoplasm. Inside the cytoplasm, there are a number of plastids, which become more pronounced in cells of green leafy vegetables. These are chloroplasts, where the photosynthesis process is initiated and transiet starch is synthesized.

Other chloroplasts inside the cell are the oil-soluble colored pigments, which are the chromoplasts. The leucoplasts are granules of starch storage in the cytoplasm, and, as with maturing, some chloroplasts are transformed to leucoplasts. There are also the vacuoles where water-soluble salts or some other soluble materials are present. As the plant ages, these vacuoles group together in one large vacuole.

The parenchymal cells are the principal part of the food plant. In addition, there are the conducting cells of xylem and phloem, which are large, tubular cells. These cells may thicken with lignin at thickening points, but they are mainly made of cellulose. These cells may give the fruit or vegetable a stringy character.

The supporting cells, on the other hand, are long, pointing cells that are not numerous in young edible plants. Their cell walls are made of cellulose which thickens as plant ages and gets encrusted with lignin. Supporting cells of an old plant can be pretty tough.

The protective cells are specialized parenchymal cells that secrete cutin or contain suberin. These cells are pressed together and are very tough. They generally occur on the epidermis or the outermost cover layer of the plant. Cutin and suberin render plant tissue impermeable to water, thus protecting it from losing its moisture content. Also, they protect plants from attacks of microorganisms, insects, and mechanical injuries. Moreover, cutin, the waxy material secreted by cells, provides a protective waxy membrane around the cell surface of fruits or vegetables. The type of material cutin is made of is quite complex in nature. Cutin of apples, for example, contains long-chain 16-and 18-carbon fatty acid with one, two, and three double bonds, fatty alcohols of 18–26 carbons, ketones, esters, ethers, hydrocarbons, aromatic compounds, and pseudoesters. Suberin, meanwhile, is present in the inner lining of the cell and is plentiful in leaves and stems of plants. This material is insoluble in 80% H_2SO_4.

TURGOR OF FRUITS AND VEGETABLES

This turgor depends on the rigidity at the cell wall of fruits and vegetables. Changes in the moisture content of cells will greatly influence turgor; thus, forces of osmosis and changes in cell volume are determining factors of turgor. On this basis, the turgor of fruits and vegetables is highly dictated by the concentration of osmotically active substances present in the vacuoles, whether in true solutions or in a dispersion, the permeability of protoplasm, and by the elasticity of the cell walls.

Cooking fruits and vegetables brings about many changes in cellular structure that relate to turgor. Denaturation of protein, which consequently changes the cell wall permeability, death of cells, vacuoles' loss of their active membranes, and swelling and gelatinization of starch granules to an extent that might fill up the cell and give new firmness to food are all changes in turgor which are brought about by cooking.

GENERAL CHANGES THAT OCCUR DURING COOKING OF FRUITS AND VEGETABLES

The soluble pectic material of fruits and vegetables is generally increased during cooking as the protopectin is partially hydrolyzed. As the soluble pectic material increases, the middle lamillae area shrinks. Cooking potatoes results in mealiness, which is a desirable effect. Sloughening is an undesirable effect of cooking that might be remedied by adding Ca^{2+} salts, which help bridge the pectic acids.

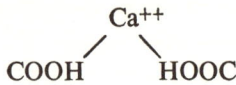

Addition of calcium chloride, calcium citrate, ca-sulfate or calcium phosphate to approximately 700 ppm helps firm up canned potatoes, peaches, etc. Needless to say, the above mentioned mealiness is not related to pectic substances.

Although calcium salts may help firm up, for example, young peas, still these salts will have no firming effect on mature old peas, as the hemicellulose and polypentans have covered, in this instance, protopectin, and have prevented its solubilization. The cellulose part in the cell wall may somewhat thin out during cooking, but to the greatest effect, it is not changed in due process.

Cooking fruits and vegetables also changes the intercellular air spaces. With heating, this air space expands and cells are rendered more permeable. The cell saps are converted into interspaces and there results changes in appearance and juiciness of food as it becomes more translucent. If syrup is added in the process (apple processing), the air space is replaced in such instance by the syrup, which is a desirable effect.

Cooking fruits and vegetables also helps give off the volatile acids that they contain. If a lid is left on the pot, these volatile acids are compressed back into the cooking media. This attributes a favorable flavor to the broth. When fruits and vegetables that contain chlorophylls are cooked, the chlorophyll is converted to pheophytin and the cooked food changes its color to brownish instead of green.

Cooking fruits and vegetables may, in instances, produce a sour flavor and speed up hydrolysis of sulfur-containing glycosides.

Organic acids present in foods cause the hydrolysis of sulfur glycosides which, in turn, gives rise to the obnoxious odor of volatile sulfur compounds. The organic acids are generally present in the cell sap of the fruit or vegetable, and they mostly are water-soluble volatile acids. The most common acids in fruits and vegetables are malic and citric. Volatile acids do exert an effect on both color and flavor of fruits and vegetables. A sour flavor is mostly attributed to volatile acids. The level of organic acids present in fruits and vegetables varies with their ripeness and variety, and with the season. Succinic, isocitric, lactic, and benzoic acids are common in fruits and vegetables. Acetic and caproic acids contribute to flavor in certain fruits; butyric acid to the flavor of carrots; acetic and propionic to the flavor of molasses; octanoic, valeric, and N-nonoic to the flavor of cocoa.

VOLATILE SULFUR COMPOUNDS AND FLAVOR

Volatile sulfur compounds make important contributions to food flavor. Some sulfur flavors are brought about during processing foods, such as, after crushing cells, enzymes are liberated, or upon the hydrolysis of foods, etc. Cooking, in many instances, produces a volatile sulfur-compounds flavor.

The flavor of many prominent spices is due, to a good extent, to the sulfur compounds of their precursors.

Onion and garlic contain allyl cysteine which is a precursor for the compound "allicin" that contributes to the characteristic flavor of onion and garlic. Allicin is obtained upon enzymatic action after crushing the tissue cells.

$$CH_2=CH-CH_2-\overset{\overset{O}{|}}{S}-CH_2-\overset{\overset{NH_2}{|}}{CH}-COOH$$

Allyl Cysteine

$$CH_2=CH-CH_2-\overset{\overset{O}{|}}{S}-S-CH_2-CH=CH_2$$

Allicin

Garlic flavor is due to both allyl propyl di- and tri-sulfides, while the onion flavor is due to allyl isothiocyanates and allyl propyl disulfides. The allyl isothiocyanates also contribute to the flavor of mustard oil and radishes.

Cabbage and cauliflower, upon cooking, evolve hydrogen sulfide and organic sulfur compounds that contribute to their flavor. However, if cooking is carried on for a prolonged time, excess hydrogen sulfide and organic sulfur compound production may bring about obnoxious odors.

Hydrogen sulfide is also generated during canning corn; since it cannot escape, it becomes part of the canned-corn flavor. The characteristic "cooked flavor" of many foods may be attributed to the volatile sulfur compounds present or developed during cooking.

REFERENCES

Boswell, V. R. 1937. Improvement and genetics of tomatoes, peppers, and eggplants. *U. S. Yearbook of Agriculture for 1937*, 176.

Cruess, W. V. 1938. *Commercial fruit and vegetable products.* 2nd ed. New York: McGraw-Hill.

Hohl., L.A.; Swanburg, J.; David, J.; and Ramsay, R. 1947. Cooling of blanched vegetables and fruits for freezing. *Food Research* 12: 484.

Hollinger, M. E. 1944. Ascorbic acid value of sweet potato as affected by variety, storage and cooking. *Food Research* 9: 76.

———, and Colvin, D. 1945. Ascorbic acid content of okra as affected by maturity, storage, and cooking. *Food Research* 10: 255.

Holmes, A. D. 1948. Variation in composition of winter squashes. *Food Research* 13:123.

Jacobs, M. B. 1951. *The Chemistry and Technology of Food and Food Products.* 3 volumes. New York. Interscience Publishing Co.

Joslyn, M. A., and Stepka, W. 1949. The free amino acids of fruits. *Food Research* 14:459.

Krehl, W. A., and Cowgill, G. R. 1950. Vitamin content of citrus products. *Food Research* 15:179.

Landford, C. S. 1942. Studies of liberal citrus intake. *J. Nutr.* 23:409.

MacDowell, L. G. 1950. Research in the citrus industry. *J. Southern Research* 2 (No. 2): 7–8.

McLaughlin, L.; Tarwater, M.; Lowenberg, M.; and Koch, G. 1931. Vegetables in the diet of preschool children. *J. Nutr.* 4:115; *Nutr. Abs. Rev.* 1:232.

Maynard, L. A., and Nelson, W. L. 1951. Foods of plant origin. *J. Am. Med. Assoc.* 136:1043; Chapter XXV of *Handbook of nutrition,* 2nd ed. American Medical Association.

Smith, S. L. 1935. Vegetables in the diet. *J. Home Econ.* 27:73 and 146.

Todhunter, E. N. 1939. Vitamin values of garden type peas preserved by frozen-pack method. II. Vitamin A. A. *Food Research* 4:587.

Watt, B., and Merrill, A. L. 1950. *Composition of foods: raw, processed, and prepared.* U. S. Dept. Agriculture, Agriculture Handbook No. 8. Washington, D.C.: Superintendent of Documents, 147 pages.

Chapter 8

MILK

INTRODUCTION

The legal definition of milk is the lacteal secretion, practically free from colostrum, obtained by the complete milking of one or more healthy cows, which contains not less than 8.25 percent milk solids (not fat) and not less than 3.25 percent milkfat.

The gross composition of milk is:

Water	87%
Carbohydrates	5.1%
Lipids	3.9%
Proteins	3.3%
Ash	0.7%

Milk is fluid and its water content is mostly in the free form, although a small amount may be present as hydrate, and another, still smaller amount of water may be present tightly bound on the surface of protein.

Components of milk that exist in true solution form include lactose and salts, protein in a colloidal state, and fat, which exists in oil-in-water emulsion in tiny globules of 0.1–22 microns in diameter. When milk is homogenized, these globules change diameter to 0.1-2.0 micron.

The lipid components of milk are:

> Neutral triglycerides 98–99%
> Phospholipids 0.2–1.0%
> Sterols 0.25–0.4%

VITAMINS A, D, E, AND K CONTENT

Milk in its natural form contains a large quantity of vitamin A, whether in the free vitamin form or in the carotene percursor form.

The principal carbohydrate in milk is lactose, which is the only sugar present in milk and is found only in milk (found in all mammalian milk, except sea lions).

There are traces of glucose and galactose in milk, probably as products of the metabolism of lactose or its degradation.

PROTEIN CONTENT

Proteins of milk are present in a colloidal state. Casein is 80% of the milk proteins, and is composed of 45–65% α-and kappa-caseins. 20–28% β-, and of 3–7% γ-casein. Milk proteins also include the lactalbumins: lactoglobulin, 7–12%, lactalbumin 2–5%, and serum albumin 0.7–1.3%. The lactoglobulin of milk protein is 0.8–1.7% euglobulin and 0.6–1.4% pseudoglobulin.

The ash content of milk is the residual component after milk is inserinated. The white ash material remaining contains most of the minerals of milk. Many of these minerals do not exist in the free salt form. For example, two-thirds of the calcium and phosphorus of milk exists in the colloidal form. Approximately one-fourth of the magnesium content may also exist in the colloidal form, and 10% of the citrate content.

Since salts dissociate into positive and negative ions, thus the form of the salt occurring in the natural system depends, to a great extent, on the pH of the milk. At the usual pH of milk, which is 6.6–6.8, salt ionic species exist as positive and negative ions. Protein, also, of course, exists as positive and negative ions, and even may be classified as salts. Citrate in milk may exist to 0.2% as free citrate and the balance as ionic with phosphates, calcium, and magnesium, a complex that may be very important in providing protein stability of milk.

Milk varies in its composition with different species. Human milk, for example, contains 7% of the sugar lactose, compared to 5% in cow's milk. This makes human milk superior to animal milk for the nursing child, as he requires this high-energy source for his initial period of growth. Cow's milk, however, contains 3.5% protein compared to 1.5% in humans.

Since the lactose level is high in human milk, it follows that the mineral content has to be low (0.2%), since milk should be isotonic with blood, i.e., the osmotic pressure of blood and milk has to be similar.

After the human infant passes the critical period of 180 days, in which he doubles his weight, the protein content of human milk starts to increase. Along with this increase in protein, the ash content increases too, but not in a linear fashion as in the case of protein.

Although the protein and fat content of milk may vary within the same species, there is very little change in the lactose and ash content in order to maintain isotonicity between milk and blood.

First-drawn milk after calving drastically differs from that drawn after seven days. The first is high in protein, due to increased albumin and globulin fractions that are needed to give the cow immunity against disease at that stage. Lactose of the first-drawn milk is 1% lower and the ash content is up by 1% to cause an isotonic condition. After the passage of seven days from the time of first-drawn milk, its composition returns to normal.

NONCOMBUSTIBLE ASH CONTENT OF MILK

Calcium	0.123
Phosphorus	0.095
Magnesium	0.012
Sodium	0.060
Chlorine	0.120
Potassium	0.141
Sulfur	0.030

VITAMIN CONTENT OF MILK

Milk may be considered high in its vitamin A, riboflavin, and thiamin content, as it contains 500–3000 International Units (I.U.) of vitamin A per quart, 1.5 mg/quart riboflavin, and 0.35–0.45 mg/quart of thiamin. Milk, however, is naturally low in its vitamin D content (5–15 I.U.), and although it may contain most of the B vitamins, and fat-soluble ones, yet, these all may be present in low quantities in relation to daily requirement.

ENZYMES IN MILK

With the exception of lipase, most enzymes in milk do not have substrates. Prior to pasteurization, lipase may cause limited lipolysis; however, it is easily inactivated with heat. Vigorous agitation releases lipase, which hydrolyzes fat into glycerol and free fatty acids (FFA). Phosphatase is present in milk, but is destroyed during pasteurization. As a matter of fact, its presence is used as an indicator of pasteurization, where the enzyme is allowed to act on the substrate, disodium phenyl phosphate, converting it into a phenol compound that takes a blue color.

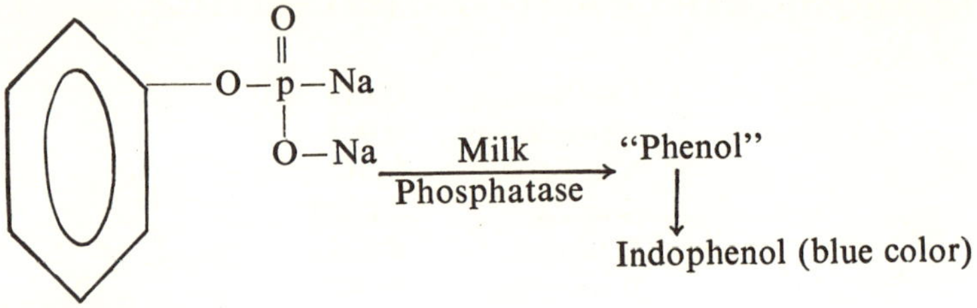

$$\text{Disodium phenyl phosphate} \quad \xrightarrow[\text{Phosphatase}]{\text{Milk}} \quad \text{"Phenol"} \longrightarrow \text{Indophenol (blue color)}$$

Disodium phenyl phosphate

Enzymes present in milk are:

1. Lipase
2. Phosphatase
3. Aldolase
4. Protease
5. Peroxidase
6. Catalase
7. Xanthine Oxidase
8. Carbonic Anhydrase
9. Amylase

REFERENCES

Associates of Rogers. 1935. *Fundamentals of dairy science*, 2nd ed. Reinhold Publishing Corporation.

Barry, J. M. 1961. In Kon, S. K., and Cowie, A. T. (eds) *Milk: the mammary gland and its secretion*. New York: Academic Press, Vol. I. p. 389.

Folley, S. J., and McNaught, M. L. 1961. In Kon, S. K., and Cowie, A. T. (eds). *Milk: the mammary gland and its secretion*. New York: Academic Press, Vol. I, p. 441.

Hansen, R. G., and Carlson, D. M. 1961. In Kon, S. K., and Cowie, A. T. (eds.) *Milk: the mammary gland and its secretion*. New York: Academic Press, Vol. I. p. 371.

Jarvis, B. W. 1930. Milk sugar in infant feeding. *Am. J. Diseases Children* 40:993.

Koehler, A. E.; Rapp, I.; and Hill, E. 1935. The nutritive value of lactose in man. *J. Nutr.* 9:715.

Leloir, L. F., and Cardini, C. E. 1961, In Kon, S. K. and Cowie, A. T. (eds.) *Milk: the mammary gland and its secretion*. New York: Academic Press, Vol. I, p. 421.

Chapter 9

CEREAL CHEMISTRY

INTRODUCTION

Grains can be stored for a prolonged time, and they still maintain their fertility as they contain approximately 15% moisture that may sustain a rapid metabolic activity in the stored state. An increase in temperature will also increase the activity of mold that is present in grain. This increase in mold growth, consequently, will cause damage to grains as it causes "bin burning" and increased respiration and biological activity.

The physical separation of parts of the grain seeds, such as separation of bran from flour, may cause the rupture of starch granules and a subsequent hydrolysis of starch by the amylases.

The lipid content of grains varies with the blend, variety, climate, and the soil. Neutral triglycerides are the dominant fat in grains; however, there are also present polar lipids (as phospholipids), and sterols.

The hydrophobic and hydrophillic character of the wheat protein structure, along with the small amount of polar lipids that is present, have a great influence on bread-making performance of wheat flour.

VITAMINS AND MINERALS

The separation process of bran from grains causes the white flour to be deficient in the vitamins and minerals that were already in the grain. Thus, white flour in the U. S. is usually enriched with thiamine, riboflavin, niacin, and with the mineral iron, all of which were in the original wheat grain.

Also, certain amounts of calcium and vitamin D may be added to white wheat flour, although some nutritionists complain that usually the amount of calcium added exceeds the level normally present in grains. Wheat flour occasionally is fortified with the basic amino acid lysine (the natural wheat content of lysine is low). Lysine may be directly added to the dough or, instead, added in skim-milk powder. The latter case results in a subsequent loaf-depressing characteristic, unless the milk was given a prior high heat treatment. The loaf-depressing characteristic may also be prevented by adding hydrogen peroxide to milk. It is assumed, in this case, that the hydrogen peroxides cause the oxidation of the sulfhydryl groups in the whey protein.

GAS PRODUCTION IN DOUGH-MAKING

The yeast *sacromyces servaecae* is the agent responsible for gas production in dough. The amount of gas produced depends on the amount and strain of yeast present. These yeasts also contribute to flavor development in bread. There are cultures of microorganisms in the baking industry that produce sour dough. These cultures, besides CO_2 also produce organic acids. As CO_2 causes levelling and dough to rise, the organic acids develop the sour flavor. Microorganisms that produce acids and sourness are those like acetobacter, lactobacillus casei, lactobacillus acidophilus, and lactobacillus bulgaricus. The first produces acetic acid, and the latter three, lactic acid. For yeasts to grow, they require a soluble carbohydrate source, nitrogen, calcium, and phosphate. For a sugar, yeasts favor glucose or fructose or the disaccharides sucrose or maltose; however, many yeasts do not use lactose.

Since the sugar present in dough is sucrose, it is subsequently metabolized rapidly by yeast. If the level of sucrose, however, exceeds 10%, it will retard yeast growth and, thus, CO_2 production in dough.

In dough, the α-amylases act on amylose and amylopectin, producing limit dextrin; while β-amylases produce 2-carbon units. The ungerminated cereals do not have α-amylases, but have β-amylases, and the germinated cereals contain both types of enzymes. Thus, the level of amylases is regulated by adding malted wheat or barley so that the yeasts may have sufficient maltose, which is required for their growth.

If amylose is low in the dough, dextrin or soluble starch may be added to improve CO_2 gas production.

During baking, starch imbibes water and gets gelatinized, and during the early stages, the β-amylases attack the gelatinized starch until reaching the temperature of 150-175°F, where β-amylases are destroyed.

Besides the amylases, malted wheat and barley contain protease enzymes that will also exert influence on the dough characteristics.

Boiled mashed potatoes along with water may be incorporated into dough for the purpose of flavor development and for rapid growth of yeast.

GAS RETENTION IN DOUGH

Retention of gas by dough is usually influenced by the mechanical action applied to the dough, by proteolytic action, level of oxidation occurring, addition of milk, salts, and by the level of fat present.

As dough is exposed to mechanical action, there is first a loss of the elasticity of dough, then an increase in elasticity, and finally a decrease.

The proteolytic enzyme activity somewhat weakens the strength of the gluten content in wheat flour dough, since the presence of too much gluten gives stickiness to dough. Nevertheless, only a limited proteolytic activity is desired. If too much proteolytic activity is present, the structure of dough may fail to hold up as desired.

The level of oxidation in the dough influences its buckiness, structure, and gas-bubble development. Addition of an oxidizing agent increases the strength of the gluten structure. Usually, addition of 1% $KBrO_4$ is sufficient for maintenance of the desired strength of gluten. Should strengthening of gluten become excessive, this may be overcome by additional mechanical work or by adding a reducing agent.

Heating to 180°F for thirty minutes may serve as a remedy for the milk-adding-caused, loaf-depressing characteristic. Calcium and sodium salts in hard water may exert a similar effect.

Acid may increase gluten's strength to a point, and fat in a small quantity increases the dough's ability to incorporate or hold CO_2.

REFERENCES

Cowgill, G. R., and Anderson, W. E. 1932. Laxative effect of wheat bran and "washed bran" in healthy men: a comparative study. *J. Am. Med. Assoc.* 98:1866.

Cowgill, G. R., and Sullivan, A. J. 1933. Further studies on the use of wheat bran as a laxative: observations on patients. *J. Am. Med. Assoc.* 100:795.

Fantus, B.; Kopstein, G.; and Schmidt, H. R. 1940. Roentgen study of intestinal motility as influenced by bran. *J. Am. Med. Assoc.* 114:404.

Hassid, W. Z., and McCready, R. M. 1941. The molecular constitution of enzymically synthesized starch. *J. Am. Chem. Soc.* 63:2171.

Langworthy, C. F., and Deuel, H. J., Jr. 1920. Digestibility of raw corn, potato, and wheat starches. *J. Biol. Chem.* 42:27.

Meyer, K. H. 1942. Recent developments in starch chemistry. *Advances in Colloid Science.* Vol. I. New York: Interscience Publishers Inc., p. 143.

Norris, F. W., and Preece. J. A. 1930. The hemicellulose of wheat bran. *Biochem. J.* 24:59.

Posternak, T. 1951. On the phosphorus of potato starch. *J. Biol. Chem.* 188:317.

Reichert, E. T. 1913. The differentiation and specificity of the starches in relation to genera and species. Carnegie Institution of Washington, Publication No. 173.

Rose, M. S.; MacLeod, G.; Vahlteich, E. M.; Funnel, E. H.; and Newton, C. L. 1932. The influence of bran on the alimentary tract. *J. Am. Dietet. Assoc.* 8:133.

Taylor, T. C. 1926, 1929, 1931, 1933. Constitution of starches. *J. Am. Chem. Soc.* 48:1739; 51:294, 3431; 53;3436; 55:258.

Wolfrom, M. L., and Sugihara, J. M. 1950. Carbohydrate chemistry. *Ann. Rev. Biochem.* 19:67.

Part II

THE CHEMISTRY OF VITAMINS

Chapter 10

VITAMIN A

INTRODUCTION

Vitamin A was long known by the Egyptians by its deficiency symptoms thousands of years ago.

It is also recognized for its growth, survival, and dark adaptation effects.

Vitamin A does not naturally occur, as such, in food stuffs. It is derived by the body from the carotenoid pigments. There are α, β, γ carotenes and cryptoxanthins that may give rise to vitamin A, but the principle one is β-carotene from which Vitamin A is derived by the body.

When β-carotene is cleaved in the middle, there are two molecules of vitamin A produced. Vitamin A may be in the forms of aldehyde, alcohol, and acid. The conversion of the carotene pigments may not give 100% vitamin recovery. In the ruminants, for example, a good proportion of the carotenoid pigments are destroyed by the microorganisms of the rumen. In the case of humans, the conversion of carotenes to vitamin A occurs in the liver and it must be in the form of transisomers to have any biological activity.

The original source of vitamin A, that is, the plant carotenoids, depends, of course, on the plant's maturity stage, extent of processing after harvest, and the level of oxidation that occurred in the pigments. Thus, a young, lush, and fairly mature plant is expected to be a rich source of the carotenoids.

To protect vitamin A itself from oxidation or destruction, it may be esterified with palmitic or acetic acids or by forming a gel around it which excludes O_2.

VITAMIN A-PRECURSORS AND FORMS

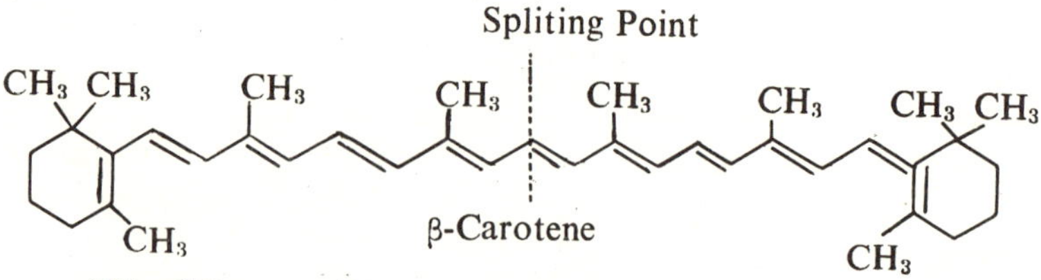

β-Carotene (Splitting Point)

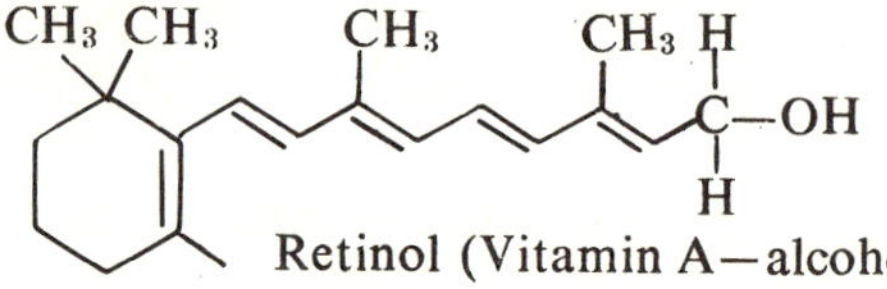

Retinol (Vitamin A—alcohol)

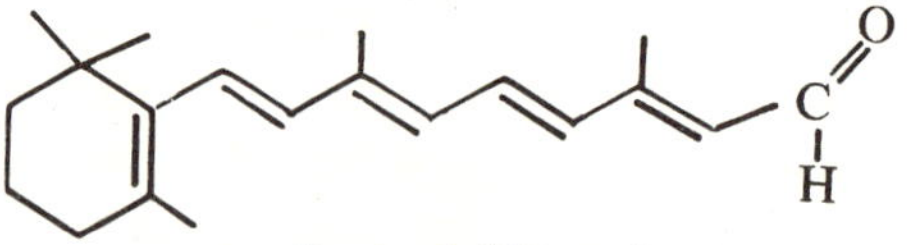

Retinal (Vitamin A—aldehyde)

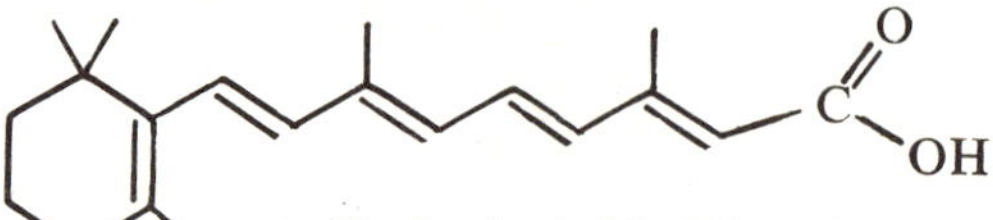

Retinoic Acid (Vitamin A—acid)

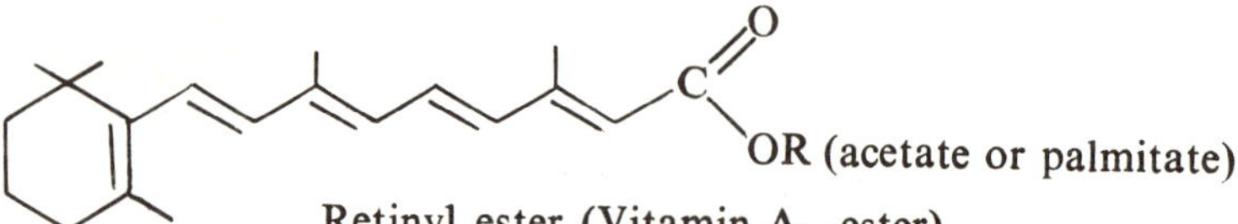

Retinyl ester (Vitamin A—ester), OR (acetate or palmitate)

FACTORS IN CONVERSION OF CAROTENES TO VITAMIN A

Since both the carotenoids and vitamin A are fat-soluble, a small amount of fat in the diet is necessary for the subsequent bile salt production that is needed for emulsification and, thus, getting the fat-soluble vitamin in an absorbable form. A large amount of fat in the diet may have an adverse effect on the conversion of carotenes to vitamin A.

It is thought that thyroxine is involved in the conversion of carotenes to vitamin A. Thus, it should be expected that individuals on the high side of thyroid hormone secretion would have an increased rate of carotene conversion to vitamin A and more of the vitamin stored in the liver than a hypothyroid individual. Needless to say, a hyperthyroid individual has a higher requirement for vitamin A, as his general metabolism on all cellular levels is higher.

Nitrates are known to inhibit thyroxine; thus, it follows from the preceding argument that nitrates will impair β-carotene conversion to vitamin A. Nitrates may also enhance the rate of mobilization of the vitamin in the liver, thus decreasing the storage. The nitrate salt is thought by some to exert an inhibition directly on vitamin A absorption.

Since heat decreases the thyroid hormone's secretion, it may, consequently, have an impairment effect on conversion of the pigments and absorption of the vitamin itself.

A low protein level in the diet may impair conversion to and uptake of vitamin A as, in such a case, the enzymes for conversion may be lacking.

There are also species differences as to level of conversion of the carotenes to the vitamin A form. Holstein cattle, for example, are known to be good converters of carotenes to vitamin A as compared to the Jersey type, the milk of which has a yellowish color that is due to the presence of unconverted carotenes. There is no carotenoid detection, to any great extent, in the plasma of sheep, an indication of the pigment's conversion in the intestinal walls in this case.

Vitamin E may improve absorption and levels of vitamin A, for, with its antioxidant property, it probably exerts a protection effect against oxidation of vitamin A.

The glucocorticoids are thought to deplete vitamin A from the liver, as they may adversely influence the carotene conversion to the vitamin. Therefore, people under stress, where lots of ACTH is secreted from the pituitary, activating the glucocorticoid secretion in the adrenals, may be expected to have problems with vitamin A.

A deficiency of vitamin A in animals most probably means that the specific subject has lost the ability to synthesize the specific enzyme that converts carotenes to the vitamin form. Intake of carotene as a remedy in such cases is useless. As a matter of fact, it may prove harmful, as carotene pigments may turn the skin a yellow color or even, in extreme cases, toxicity may occur.

In case of vitamin A deficiency, the vitamin itself should be given.

TESTS FOR VITAMIN A

Car-Price test (chemical method): This is a spectrophotometric test that develops a color if vitamin A is present. All forms of vitamin A would develop a color in this test. However, for our purpose, the aldehyde form of the vitamin is biologically the most active and important. For night blindness, it is only the aldehyde form wherein the remedy lies. Although the above method may detect and quantify the amount of vitamin A present, it lacks specificity, of course.

ABSORPTION AND UPTAKE OF VITAMIN A

The enzyme, β-carotenase, acts in the intestinal mucosa on β-carotene, converting it to the aldehyde form of vitamin

A (retinal). Retinal is convertible to retinol, the alcohol form, also in the intestinal mucosa. Retinol may exist in the ester form, such as retinyl palmitate, also in the intestines (the ester form is the primary one used in the liver). The ester form is hydrolyzed to the alcohol form, recombines in another ester form, perhaps with acetate, enters the lymphatic system as chylomicrons, becomes lymph-soluble, goes through the thoracic duct to the blood, and then to the liver where it is stored in the retinyl ester form. When the need arises, the ester form in the liver is hydrolyzed by a specific enzyme to the retinol-alcohol form. This alcohol form may go to the blood and be carried through the system in such a form, or it may be oxidized in the liver to the retinoic acid form which is usually not for storage in the liver, but is conjugated into the retinoyl glucuronide which leaves the liver and goes into bile. This form may reenter the gut from the bile and get reabsorbed again, thus completely cycling.

The retinoic acid form may maintain growth, but has no other biological functions known.

The uptake of vitamin A is energy-dependent and requires ATP for its absorption.

As mentioned earlier, the vitamin itself may be given exogeneously if required. It may be injected in the alcohol form, or as an acetate or palmitate ester. Although any one of the three forms will restore the plasma or liver level of vitamin A, the alcohol form is the fastest acting, followed by the acetate, followed by the palmitate ester forms. These forms may be given or injected either in fat-soluble or water-soluble form. The water-soluble form probably increases the rate of absorption of the vitamin at the injection site. The vitamin may be injected into muscle in an emulsified state.

BIOLOGICAL FUNCTION OF VITAMIN A

The only specific biochemical function of vitamin A known in the living system is its role in vision. The retina of

VITAMIN A ABSORPTION

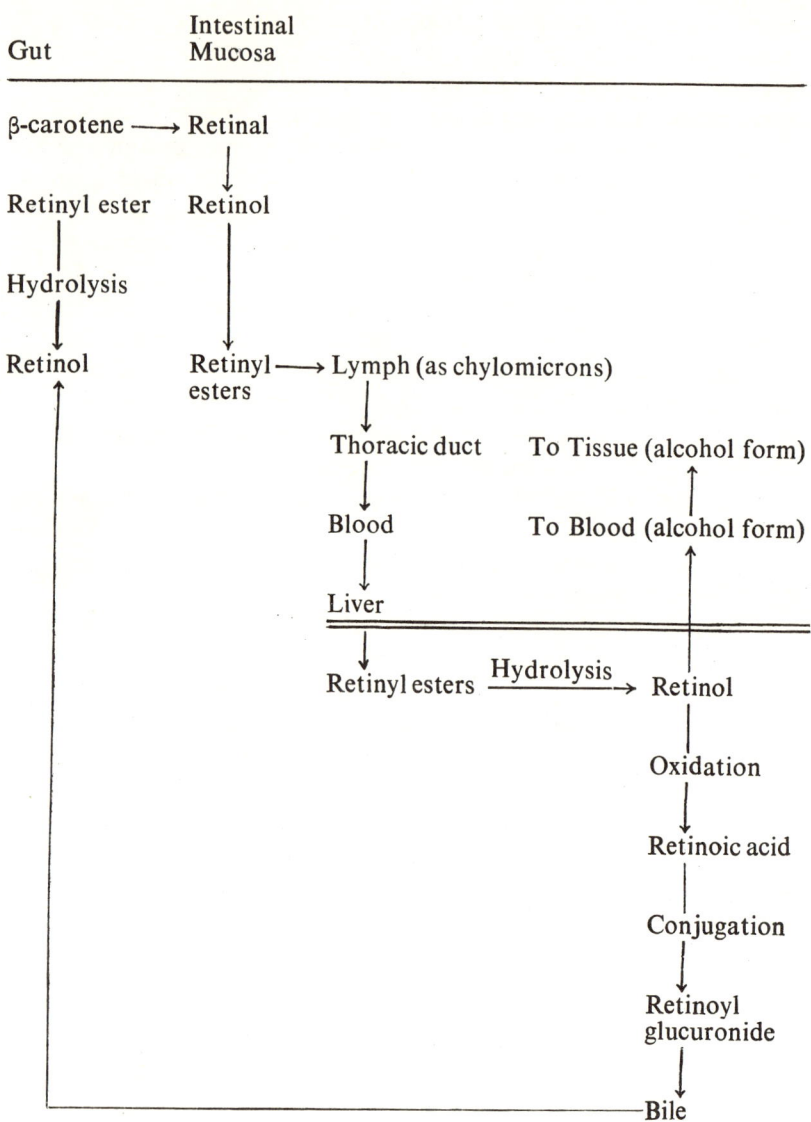

the eye has two different light receptors: the rods, which are sensitive to low-intensity light; and cones, which are sensitive to bright light and colors. Both rods and cones are altered in vitamin A deficiency. The effect on rods explains the inability to see objects in a dim light, i.e., the rods cannot, in this case, perceive in the light. This is night blindness.

MECHANISM OF VITAMIN A INVOLVEMENT IN VISION

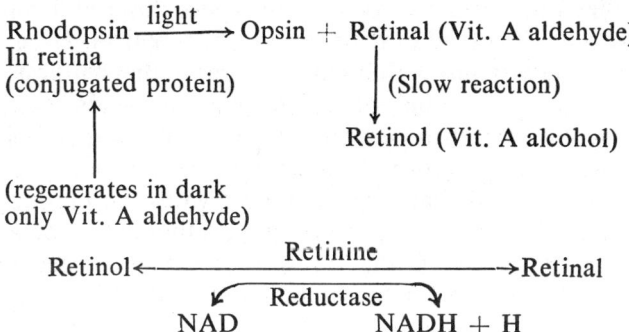

When light hits the eye, the protein rhodopsin is converted to opsin and the aldehyde form of vitamin A is produced (only form active in vision). Retinal undergoes a slow reaction to retinol, the alcohol form. This alcohol form undergoes the reversible oxidation-reduction reaction, through the aid of reductase, into the aldehyde form (retinal). This is a dark-regeneration of retinal.

Vitamin A is believed to play a role in maintaining the cell membrane integrity. This makes it an integral part of the cell membrane. Some tend to think it might act as a coenzyme; however, this has not been proven.

Vitamin A is believed to be involved in growth of bone (excessive vitamin A inhibits bone growth). An excess of the vitamin also causes keratinization of the skin.

In vitro work, has shown that an excessive dose of vitamin A caused mitochondria swelling, red blood cell (RBC) and lysosome rupturing, and finally cellular death.

In vivo work showed an excessive dose of vitamin A as causing a fragility of the lysosomal cell membranes. This suggests, of course, an antagonistic relationship between the glucocorticoids and vitamin A, as the glucocorticoids are known to heal or correct the lysosomal cell membranes.

Membranes of cells are believed to require an optimum amount of vitamin A for their normal function. The vitamin is believed by some to cause an increase in the incorporation of choline and certain fatty acids into phospholipid formation and is, thus, involved in their general metabolism. Here, specifically, the vitamin is thought to increase the synthesis rate and turnover of choline, oleic, and palmitic acids which, in turn, increase the turnover rate of phospholipid metabolism.

Mucus cell secretion is thought to be increased by vitamin A intake, and that the vitamin activates the conjugation of sugars with protein. Specifically, it is thought to be a necessary intermediate in the transfer of the monosaccharide, mannose, to protein for the synthesis of glycoproteins.

VITAMIN A'S ROLE IN BONE

An excess of vitamin A is believed to cause rapid growth of bone through rapid resorption in the cartilage which will result in bone fracture. This subsequent depletion of the cartilage matrix due to the loss of mucoprotein is the cause of bone fracture. The vitamin A deficiency, on the other hand, also alters bone growth, although in a different way. Bone, in this case, becomes very short and very thick.

The faulty bone overgrowth causes contraction on the spinal column and, consequently, an impinging on vision nerves and the loss of sight.

The high bone resorption caused by excess vitamin A may be prevented by the hormone calcitonin, which causes the decrease of blood calcium levels by inhibiting bone resorption.

Like vitamin A, large doses of vitamin D increase bone resorption. However, a high-level intake of both A and D together has shown that vitamin A at that level inhibits the effect of vitamin D.

VITAMIN A VERSUS THYROXINE

Vitamin A and thyroid hormones are thought to be antagonists. Thyroxine stimulates metabolism, while vitamin A is thought to decrease it.

VITAMIN A AND GLYCOGEN SYNTHESIS

Vitamin A indirectly affects synthesis of glycogen through its effect on glucocorticoids' synthesis, as the glucocorticoids cause gluconeogenesis, i.e., deamination of amino acids and production of glucose from protein. Since glucocorticoid synthesis is impaired in vitamin A deficiency, thus, gluconeogenesis, in such a case, is impaired. The point of inhibition is thought to be where 11-deoxycorticosterone is converted to corticosterone, as shown below:

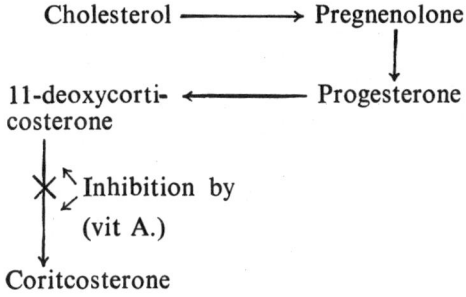

SYMPTOMS OF VITAMIN A DEFICIENCY

Vitamin A deficiency was reported as causing an increase in iron retention.

It causes a great increase in urine output and a change in urine composition; where phosphorous excretion is high and calcium excretion is low, the pH of urine becomes considerably lower. Probably the altered adrenal function discussed earlier, in turn causes an alteration in parathyroid hormone activity. In vitamin A deficiency, there is an increase in glucocorticoids secretion. These hormones, in turn, cause the breakdown of proteins. Retention of sodium and loss of potassium are manifestations of increased aldosterone activity in this case. The protein catabolism already mentioned will also produce more urine output.

Vitamin A deficiency is also thought to alter cerebrospinal fluid pressure in the spinal column. The pressure is believed to drop sufficiently in all animals except man. In man, it is believed that pressure goes up instead. On the other hand, excessive vitamin A intake causes the cerebrospinal fluid pressure to go up, except in man, where it drops in this case. The problem of changing the cerebrospinal fluid pressure is not that of excretion or overexcretion of fluids, but of malabsorption of fluids.

Vitamin A deficiency may have a profound effect on reproduction, and a severe case may cause sterility. The aldehyde and alcohol forms of the vitamin are the only ones involved in reproduction and in vision. If the acid form of the vitamin is given, the animal may grow well, but he may become both blind and sterile. In a severe Vitamin A deficiency, the germinal epithelial cells are degenerated.

Vitamin A's role in reproduction lies in the fact that a keen deficiency in the vitamin inhibits corticoids synthesis. The inhibition is thought to occur at δ^5-3β-OH steroid dehydrogenase, the enzyme that catalyzes androstenedione's conversion to testesterone. The vitamin's deficiency here causes degeneration of the germinal epithelial cells of the testes.

A vitamin A deficiency results in an increased adrenal growth probably because of the decrease in the ascorbic acid level, as vitamin A deficiency causes a reduction in gulonolactone oxidase, the enzyme responsible for ascorbic acid synthesis.

Vitamin A deficiency causes a decrease in ATP-sulfurylase enzyme activity. Since this enzyme controls the first rate-limiting reaction in the synthesis of sulfated mucopolysaccharides, it's lack may cause severe problems in the formation of tendon and collagen.

A deficiency in vitamin A may cause problems in protein synthesis. The vitamin is believed to play a role in the incorporation of uridine into nuclear RNA, which usually leads to protein synthesis.

Vitamin A requirements are higher in the young because of their increased cellular division and tissue cell differentiation. A high carbohydrate diet for cattle necessitates a higher requirement for vitamin A.

In most species studied, there has been no vitamin A storage detected in the fetus. Just enough to supply fetal demand for the vitamin was transferred to it. Upon birth, the colostrum is quite rich in the fat-soluble vitamins, and the need for vitamin A is sufficiently satisfied at this early stage.

SUMMARY OF DEFICIENCY SYMPTOMS OF VITAMIN A

1. Eye night blindness, tear formation in eye, dry mucosa, keratinzation of cornea, blindness, lack of mucus secretion.

2. Increased cerebrospinal fluid pressure, except in man.

3. Ataxia: muscular noncoordination, staggering, and convulsions.

4. Bone overgrowth, compression on nerve, long bone thickening and shortening.

5. Epithelial cell changes in the alimentary, urinary, respiratory, reproductive, and occular tracts, where there is loss of mucoprotein and there is keratinization.

6. Atrophy of reproductive organs, decrease of spermatogenesis, decrease of δ^5-3β OH-steroid dehydrogenase activity, reduced glucocorticoid and glycogen levels.

7. Retardation of ascorbic acid synthesis.

8. Increased red blood cell level.

9. Body fluid imbalance, polyurea, edema, decreased renal plasma flow, decreased urea clearance, high PO_4, low Ca^{2+} in urine, Na^+ retention high and K^+ loss high indicating hormonal upset.

10. Decrease in protein synthesis, mucoprotein goblet cell formation.

11. Decrease in sulfation of mucopolysaccharides, and sulfation of adrenal steroids.

12. Nuclear RNA synthesis is depressed.

13. General senses (taste, hearing, etc.) are depressed.

14. Reduced feed intake.

15. Finally, death.

SUMMARY OF EXCESSIVE VITAMIN A SYMPTOMS

1. Lysosomal cell-membrane breakdown.
2. Low cerebrospinal fluid pressure, in man only.
3. Thickening of long bones, and general bone fracture.
4. Inflamed skin due to dekeratinization.
5. Rupturing of red blood cell membranes.
6. Decreased metabolic rate of lipids.
7. Swelling of mitochondria.
8. Eye-lid sealing.
9. Loss of hair and weight.
10. Finally, death.

REFERENCES

Batchelder, E. L., and Ebbs, J. C. 1944. Some observations of dark adaptation in man and their bearing on the problem of human requirement for vitamin A. *J. Nutr.* 27:295.

Bessey, O. A., and Wolbach, S. B. 1939. Vitamin A physiology and pathology. Chapter II of *The Vitamins*. American Medical Association.

Caldwell, A. B.; MacLeod, G.; and Sherman, H. C. 1945. Bodily storage of vitamin A in relation to diet and age. *J. Nutr.* 30:349.

Clausen, S. W. 1939. The pharmacology and therapeutics of vitamin A. Chapter III of *The Vitamins*. American Medical Association.

Goodwin, T. W. 1959. The biosynthesis and function of the carotenoid pigments. *Advances in Enzymology* 21:295.

Heilbron, I. M.; Jones, W. E.; and Bacharach, A. L. 1944. The chemistry and physiology of vitamin A. In *Vitamins and Hormones*, Vol. II, p. 155.

Johnson, R. M., and Baumann, C. A. 1947. The effect of thyroid on the conversion of carotene into vitamin A. *J. Biol. Chem.* 171:513.

Josephs, H. W. 1939. Vitamin A: relation of vitamin A and carotene to serum lipids. *Bull. Johns Hopkins Hosp.* 65:112.

Joyce, A. 1953. The effect of heat treatment on some plant carotenoids. *Proc. Symp. Color in Foods*, National Academy of Science, November, 1953.

Karrer. P., and Jucker, E. 1950. *Carotenoids*. New York: Elsevier.

McCollum, E. V. 1939. *The newer knowledge of Nutrition*. 5th ed. New York: Macmillan.

Moore, T. 1950. Fat-soluble vitamins. *Ann. Rev. Biochem.* 19:319.

Shantz, E. M., and Brinkman, J. H. 1950. Biological activity of pure vitamin A_2. *J. Biol. Chem.* 183:467.

Sherman, H. C., and Lanford, C. S. 1951. *Essentials of nutrition*. 3rd ed. New York: Macmillan.

Wald, G. 1951. The chemistry of rod vision. *Science* 113:287.

Wolbach, S. B., and Bessey, O. A. 1951. Vitamin A deficiency and the nervous system. *Arch. Path.* 32:689.

Young, G., and Wald, G. 1940. The mobilization of vitamin A by the sympathico-adrenal system. *Am. J. Physiol.* 131:210.

Yudkin, J.; Robertson, G. W.; and Yudkin, S. 1942. Vitamin A and dark adaptation. *Lancet* 245:10; *Nutr. Abs. Rev.* 13:450.

Chapter 11

VITAMIN D

INTRODUCTION

Vitamin D is another fat-soluble vitamin that we are only recently beginning to learn about. It is known as the antirickets vitamin for its role in bone formation. Vitamin D_2 is derived from plant sources through the action of ultraviolet light (U.V.) on ergosterol. Vitamin D_3 is in animals and is derived from 7-dehydrocholesterol in the skin upon irradiation with U.V. light. An active vitamin D_3 must possess an OH group at C-3, a double bond between C-5 and -6, another double bond between C-7 and -8 and the middle ring must be open. There is also a double bond at the point of opening at C-9.

Absorption of vitamin D is similar to that of vitamin A, i. e., it occurs all along the small intestines. Like the rest of the fat-soluble vitamins, it requires bile salts for its emulsification into a micillar form prior to its absorption. Since lactic acid aids in the absorption of Ca^{2+}, and since Ca^{2+} and vitamin D are interrelated, lactic acid, thus, may be considered as of importance in vitamin D's absorption.

In the blood, the vitamin gets protein bound with a_1-, or a_2-globulins. Like vitamin A, it is stored in the liver, but not to the same extent, however.

Similarly to vitamin A, it gets conjugated in the liver with glucuronic acid, and, as well, it goes to bile and may reenter the intestines and gets reabsorbed, as shown below:

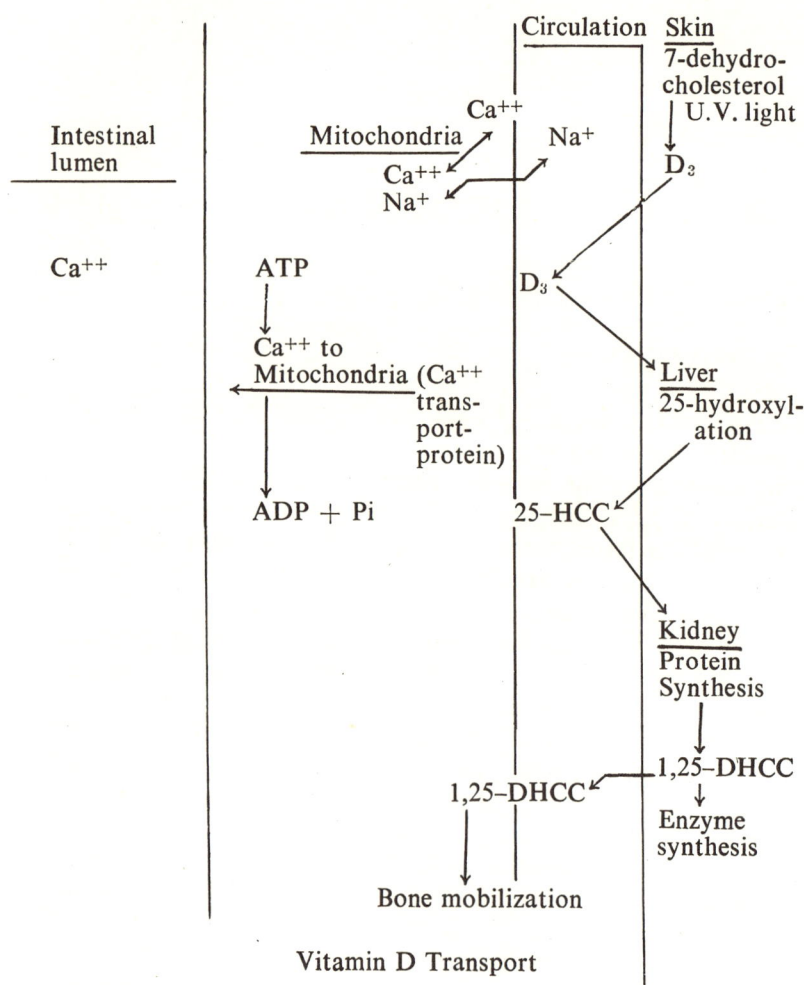

Vitamin D Transport

STRUCTURES OF THE VARIOUS FORMS OF VITAMIN D

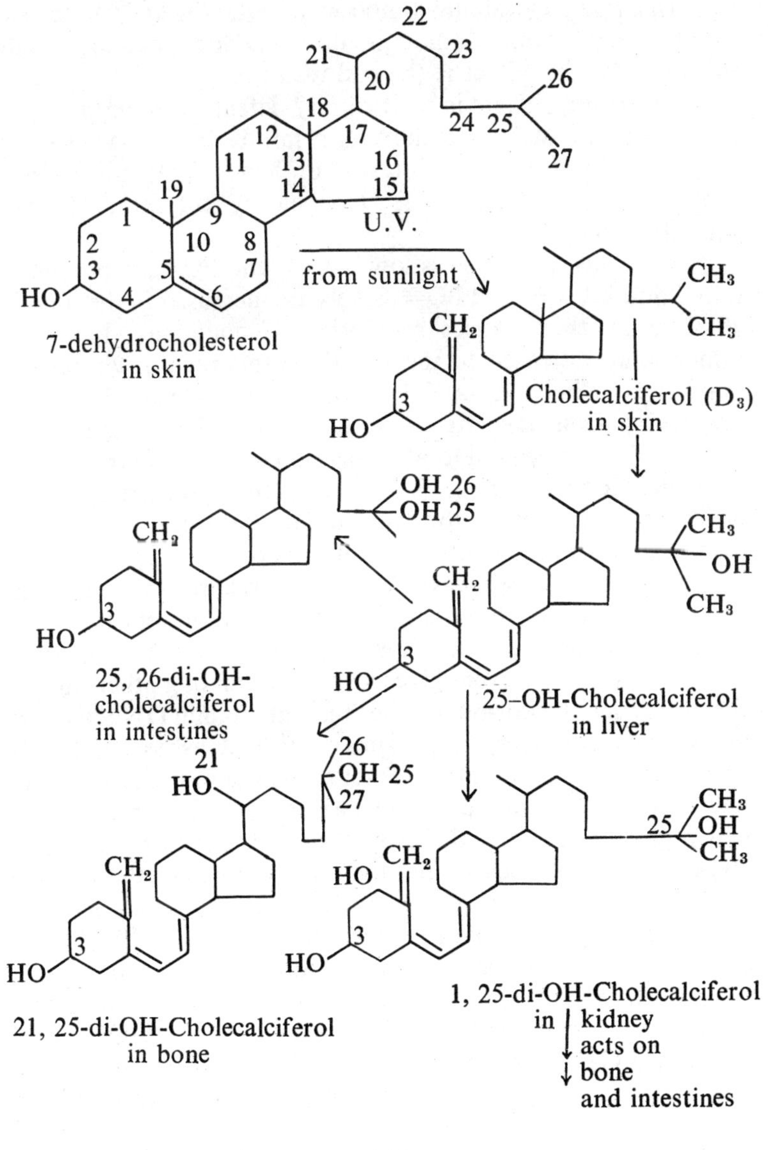

7-dehydrocholesterol in skin

U.V. from sunlight ↓

Cholecalciferol (D_3) in skin

25-OH-Cholecalciferol in liver

25, 26-di-OH-cholecalciferol in intestines

21, 25-di-OH-Cholecalciferol in bone

1, 25-di-OH-Cholecalciferol in kidney
↓ acts on
↓ bone and intestines

FUNCTIONS OF VITAMIN D

The main metabolic function of vitamin D is to cause calcification of bone. Although vitamin D does not accomplish this directly, this effect is the end result of its action.

Therefore, rickets is the primary defect of vitamin D deficiency, as in such case there is a failure in supplying Ca^{2+} and $PO\equiv_4$ to the calcification sites on bones. Thus, the disease itself is not due to vitamin D deficiency but rather to the Ca^{2+} deficiency.

In rickets, the blood is low in Ca^{2+} and $PO\equiv_4$. Upon vitamin D intake, Ca^{2+} and $PO\equiv_4$ levels are increased in the blood. Vitamin D, thus, causes resorption of bone, i.e., the bone minerals are released to the blood. In this regard, vitamin C may exert a helping hand to vitamin D by increasing Ca^{2+} absorption from the intestines. As Ca^{2+} is absorbed, it pulls the $PO\equiv_4$ along behind it. Mobilization of minerals from bone and into blood is a long-known function of the parathyroid hormone. It is believed, however, that the parathyroid hormone cannot do its work unless vitamin D is present.

After irradiation of skin by U.V. waves from sunlight, the product vitamin D_3 in the skin is transported, protein-bound in the blood, to the liver where it gets hydroxylated at the C–25 position. This specific event gives vitamin D, among other things, the properties of a hormone. The fact that vitamin D is activated and stored by the liver, secreted in a controlled fashion, is carried protein-bound in blood to a target tissue where it exerts its effect, whether on bone or the intestines, are all characteristics which grant vitamin D the properties of a hormone. The 25-OH-cholecalciferol leaves the liver and travels in the plasma to the kidney where it gets a further hydroxylation, on C–1 this time. The new dihydroxylated form of 1, 25-dihydroxycholecalciferol gets secreted into blood, and on to bone and the intestines. This dihydroxylated form of vitamin D is the most potent and active form of the vitamin. The 25–OH form may cause Ca^{2+} transport, but the 1,25-dihydroxy form does it at a much faster rate. A specific protein is required for vitamin D's action, and the 1, 25-dihydroxy form induces the synthesis of this protein. It does this through induction of a specific enzyme synthesis that is involved in mobilization of minerals from bone, or by itself acting as an active metabolite.

There are other metabolites of vitamin D that are found in the liver. The 25, 26-dihydroxycholecalciferol, although it is not as active as the other form, still acts on intestinal transport of Ca^{2+}. In the meantime, there is even another form, the 21, 25-dihydroxycholecalciferol, which works on mobilization of minerals from bone and does not work on mineral transport in the intestines.

Mechanism of Ca^{2+} Transport

Since calcium is transported against an electrochemical gradient in the intestines, energy is, thus, required for its transport across the intestinal mucosa. As Ca^{2+} is absorbed, phosphorous follows up.

After intake of vitamin D, ATPase activity was reported as being detected in the intestinal brush borders. This Ca^{2+}-dependent adenosine triphosphatase enzyme breaks down ATP, and, thus, releases energy which is necessary for calcium uptake.

When the blood calcium level goes down, the 1, 25-di-OH form of vitamin D is secreted in the blood (from the kidney), travels to bone, mobilizes Ca^{2+} from bone into the blood or helps bring about the synthesis of a specific Ca^{2+}-transport protein. ATP is brought into the cell, and ADP and Pi are formed; Ca^{2+} goes to the mitochondria, which has great affinity for it and transports it to other sites. Ca^{2+}, finally, is taken out of cell membrane pores, while Na^+ moves in to its place.

Vitamin D is believed to activate the alkaline phosphatase enzyme, which may play a role in Ca^{2+} transport.

The parathyroid hormone, which increases Ca^{2+} levels in plasma and decreases $PO\equiv_4$, cannot influence intestinal absorption of Ca^{2+} in the absence of vitamin D. The parathyroid hormone has to have vitamin D's aid in the absorption of Ca^{2+}, and not vice versa.

In certain conditions where vitamin D is not efficiently absorbed, problems in the general absorption of fat arise. This is called seliac disease, where the pancreas is bypassed. Injury of the liver may prevent the formation of 25-OH-cholecalciferol, and vitamin D's role in plasma Ca^{2+} utilization is impaired.

The genetically inherited disease "familial hypophosphatemia" is characteristic of a decrease in vitamin D_3 turnover, and results in extremely low plasma phosphate.

Also, in chronic renal failure, the kidney may fail in synthesizing the 1, 25-di-OH-cholecalciferol form of vitamin D, and hypercalcemia ensues.

Strontium mimics calcium utilization and inhibits 1,25-di-OHCC formation and intestinal Ca^{2+} absorption. Thus, strontium from atomic fallout may cause its own kind of rickets.

SOURCES OF VITAMIN D

Although vitamin D is not present in the growing plant, hay, for instance, upon drying and curing, the longer it is left outside, the more vitamin D makes its presence shown. In due process, as a plant loses carotenes, it gains vitamin D. In this case, i.e., in plants, the form D_2 is present, which is not effective in preventing rickets in man. However, fish oil is considerably high in vitamin D_3, the form utilized by man and which prevents rickets.

The form D_3 is, as well, effective in all animals. If the D_3 form is given to a high-producing cow that has milk fever and, thus, a very low blood Ca^{2+}, D_3 will prevent the milk fever in such case.

RELATIONSHIP BETWEEN VITAMIN D AND OTHER CHEMICALS

An overtreatment of U.V. light will destroy vitamin D; so will considerable amounts of peroxides from the auto-oxidation of unsaturated fats. A large intake of trace minerals may destroy both vitamins A and D, presumably by oxidizing them.

RELATIONSHIP OF VITAMIN D WITH OTHER HORMONES

There is believed to be an antagonist relationship between vitamin D and the glucocorticoid hormones. A large dose of a corticoid hormone would quite rapidly mobilize vitamin D and rid the body of it.

DEFICIENCY SYMPTOMS OF VITAMIN D

Since absence of vitamin D means a deficiency in calcium, and since the end result of vitamin D is to calcify bone, therefore, in the case of the vitamin's deficiency, the alkaline phosphatase level significantly increases. On the same basis, a tremendous overpolyferation of cartilage occurs in vitamin D deficiency. The deficiency of calcium may also be followed with a deficiency of $PO\equiv_4$, and this may mean a decrease in red blood cells.

Deficiency of this vitamin may result in keen muscular weakness.

SUMMARY OF VITAMIN D DEFICIENCY SYMPTOMS

1. Rickets onset.
2. Red blood cells decline.
3. Increase in serum alkaline phosphatase.

4. Serum calcium elevation.
5. Serum phosphorous decline.
6. Overpolyferation of cartilage.
7. Weakness of muscular activity.
8. Defects in teeth calcification and bone.
9. Growth impairment.

TOXICITY SYMPTOMS OF VITAMIN D

Since the function of vitamin D is to mobilize calcium and phosphates from bone into blood and soft tissues, in the case of the vitamin's toxicity, large amounts of Ca^{2+} in soft tissues will calcify organs such as the heart, kidney, blood vessels, etc. In such a case, symptoms develop such as loss of appetite, thirst, nausia, vomiting, weight loss, etc.

REFERENCES

Blunt, K., and Cowan, R. 1930. *Ultraviolet light and vitamin D in nutrition*. Chicago: University of Chicago Press.

Follis, R. H., Jr.; Jackson, D.; Elliot, M. M.; and Park, E. A. 1943. Prevalence of rickets in children between two and fourteen years of age. *Am. J. Disease Children* 66:1.

Luce-Clausen, E. M. 1939. Clinical aspects of ultraviolet therapy. Chapter XXIX of *The Vitamins*. American Medical Association.

MacLeod, G., and Taylor, C. M. 1944. *Rose's foundations of nutrition*. 4th ed. New York: Macmillan.

McCollum, E. V. 1939. *The newer knowledge of nutrition*. 5th ed. New York: Macmillan.

Nelson, E. M. 1939. The determination and sources of vitamin D. Chapter XXV of *The Vitamins*. American Medical Association.

Park, E. A. 1939. The use of vitamin D preparations in the prevention and treatment of disease. Chapter XXVII of *The Vitamins*. American Medical Association.

Review 1942. Vitamin D and tooth decay. *Nutr. Rev.* 1:5.

Review 1943. Massive doses of vitamin D in prevention of rickets. *Nutr. Rev.* 2:337.

Review 1945. Toxicity following massive doses of vitamin D. *Nutr. Rev.* 3:313.

Review 1952. The enzymatic action of vitamin D. *Nutr. Rev.* 10:25.

Sherman, H. C., and Stiebeling, H. K. 1930. The relation of vitamin D to deposition of calcium in bone. *Proc. Soc. Exptl. Biol. Med.* 27:663.

Smith, M. C., and Spector, H. 1940. Further evidence of the mode of action of vitamin D. *J. Nutr.* 20:197.

Wolf, I. J. 1943. Prevention of rickets with single massive doses of vitamin D. *J. Pediat.* 22:396.

———1944. Further observations on the use of single massive doses of vitamin D in the prevention of rickets. *J. Pediat.* 24:167.

Yudkin, J. 1943. The relative efficacy of calcium carbonate and calcium phosphate in preventing rickets in rats. *Biochem. J.* 37:543.

Zucker, T. F.; Hall, L.; Mason, L.; and Young, M. 1933. Growth-promoting rachitogenic diets for rats. *Proc. Soc. Exptl. Biol. Med.* 30:523.

Chapter 12

VITAMIN E

INTRODUCTION

Vitamin E is a fat-soluble vitamin. It is required for normal reproduction in rats and mice, and thus is a fertility factor in these animals, but not in man.

Most responsible for the vitamin's reactivity is the hydroxyl group on carbon-6, as it can easily donate one electron in a free-radical reaction mechanism, the very behavior that gives vitamin E its antioxidant property. The length of the side chain of C-1 to C-14 is also of great importance as it dictates the vitamin's mechanism of action. It is believed that for the vitamin's absorption to occur, the chain is cleaved between C-6 and C-7.

There are four forms of vitamin E: α, β, γ, and δ, tocopherols. The most reactive form is the α form, followed by β, followed by γ, the δ form being the least active. When considering the vitamin as an antioxidant, the order of its reactivity is reversed from the above. The α form, in this case, is the least reactive and the δ is the most reactive. For its protection against oxidation by minerals and auto-oxidized fatty acids, vitamin E, and likewise the other fat-soluble vitamins, is esterified into the acetate or palmitate fatty acid esters. Like vitamin A, vitamin E's side chain may have double bonds and, as A and D, it may be changed into an "active metabolite" for biological activity.

SOURCES OF VITAMIN E

Lettuce is an excellent source of vitamin E, especially the green part of the plant.

Wheat germ is a very good source of the vitamin.

Vegetable oils, especially if they have not undergone extensive processing are good sources of the vitamin.

Eggs, if hens are fed the proper feed, are usually good source of vitamin E, as well.

Colostrum of milk has a high content of all the fat-soluble vitamins.

ABSORPTION AND UTILIZATION OF VITAMIN E

Like other fat-soluble vitamins, vitamin E needs bile salts for the micelle formation and its subsequent absorption in the small intestine. If the vitamin is present in the alcohol form, it may get absorbed as such. If it is present in the ester form, it has to be hydrolyzed at the absorption site or soon after. It leaves the intestinal walls, enters the lymphatic system in a chilomicron form, and is transported along with proteins that render them in a soluble state. Storage is primarily in the liver, although, since the vitamin is lipid-soluble, large amounts of it may be stored in adipose tissue. Most tissues of the body, thus, contain the vitamin to some extent.

FUNCTIONS OF VITAMIN E

Vitamin E may act as a storage site for phosphates (the rapid source of energy). A vitamin E deficiency may cause a

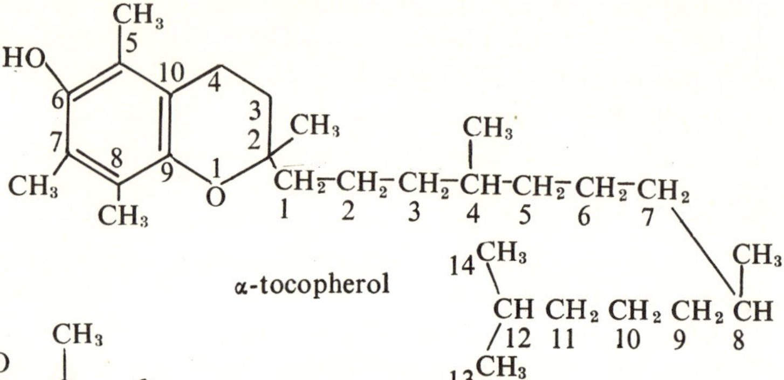

α-tocopherol

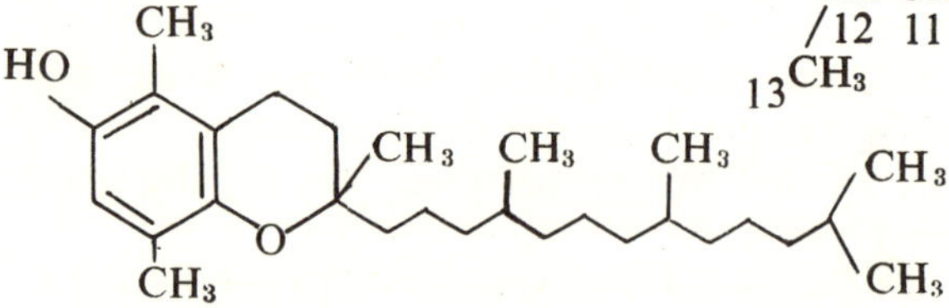

β-tocopherol

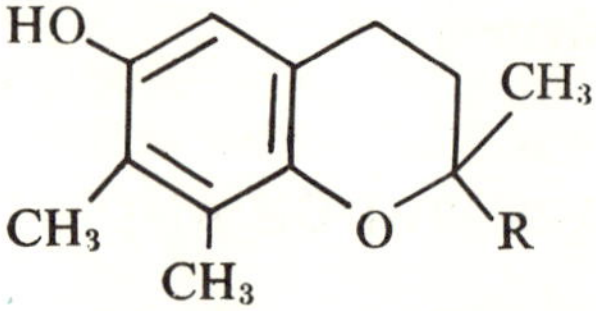

γ-tocopherol

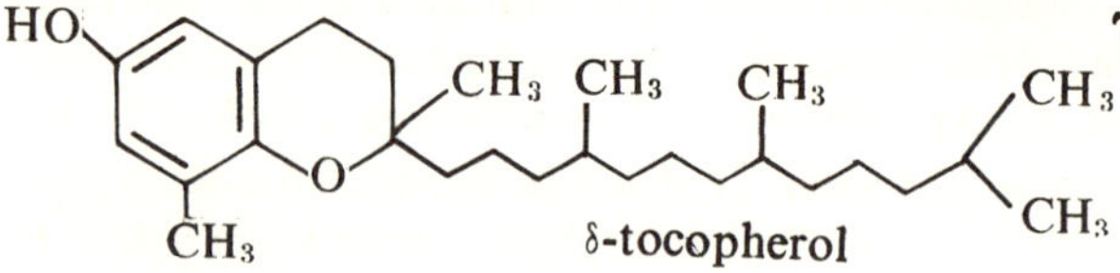

δ-tocopherol

metabolism of α-tocopherol:

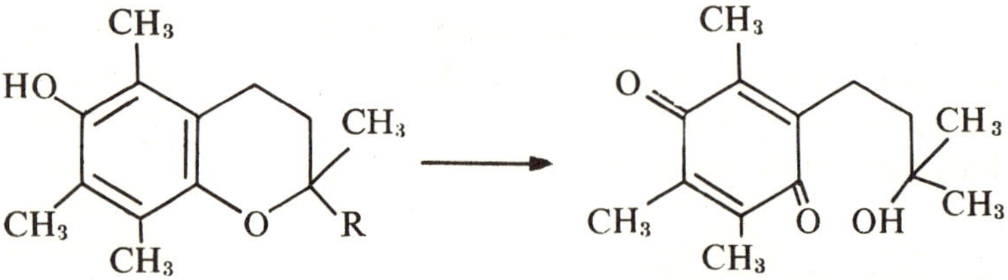

α-tocopherol → α-tocopherylquinone (end product in metabolism)

Forms and Structures of Vitamin E

deficiency in ascorbic acid synthesis. The vitamin may play a role in ubiquinone synthesis, probably through operation on the electron transport system. This suggestion was made as diets low in vitamin E result in low ubiquinone synthesis, and as the structure of the two compounds is similar.

Vitamin E is thought by some to influence thyroid function, and is involved in red blood cell maintenance. However, vitamin E is definitely involved in lipid metabolism as it exerts an antioxidant effect that protects fat from oxidation and destruction. In this respect, it neutralizes the free radicals formed in the peroxidation of lipids. Its possession of an odd number of electrons and its readiness to donate or accept electrons renders vitamin E potent in neutralizing the free radicals produced from the autoxidation of fats. Since the phospholipids of the cellular membrane contain unsaturated fatty acids that are susceptible to oxidation, vitamin E has the specific and important role of protecting the cell membranes.

Vitamin E, along with Se, is necessary for prevention of stiff lamb disease which is a muscular destrophy. Se may play a role in the absorption and retention of vitamin E.

Vitamin E's role in purine synthesis may be at the point where hypoxanthine is converted to xanthine by xanthine dehydrogenase. These purines, in turn, are involved in heme synthesis, the impairment of which results in anemia.

Vitamin E is required for prevention of death and resorption of the fetus in the female rat. It also prevents sterility in rats and mice.

In New Zealand, vitamin E and Se deficiency has been reported as causing reproductive failure in cattle and sheep.

The follicles of the thyroid gland were reported to become very inactive in vitamin E deficiency.

The vitamin has also been reported to cause the synthesis of arachidonic fatty acid from linoleic, and it may influence the chain elongation of yet other fatty acids. Besides its protection role on membrane phospholipids, vitamin E is thought to influence the incorporation of SO_4 into the lipoproteins of membranes, thus helping to make sulfolipids which are an integral part of the cellular membranes.

Vitamin E is, as well, believed to increase the synthesis of the immune cells from the stem cells of bone marrow. In the vitamin's deficiency, the RBC production in bone marrow is impaired. The exact involvement of the vitamin is in the making of the heme portion. It seems the heme-type enzymes are affected by the vitamin, enzymes such as δ-aminolevulinic acid synthetase (in bone), δ-aminolevulinic acid dehydratase (in liver), and the catalases. Vitamin E deficient baby pigs born with anemia-iron deficiency, when given a shot of iron, will die of iron toxicity. This because they could not form the heme portion of porphyrins. If vitamin E, however, were given before or shortly after the iron injection, toxicity would have been prevented.

Hemolytic anemia would favorably respond to a vitamin E injection, and a rapid rise in hemoglobin would occur. The vitamin, as well, may prolong the life of RBC in the circulation.

SUMMARY OF THE METABOLIC FUNCTIONS OF VITAMIN E

1. The main known and best-established function of vitamin E is its role as an antioxidant.
2. It is involved in tissue respiration, and may be in aiding cytochrome c-reductase.
3. It may be involved in the normal phosphorylation process.
4. It may be involved in ascorbic acid synthesis.
5. It may have a role in sulfur amino acid metabolism.
6. It has a role in purine base metabolism.
7. Vitamin E may play a role in ubiquinone synthesis, thyroid activity, red blood cells (RBC) maintenance, lipid metabolism, and in hemoglobin synthesis.

VITAMIN E DEFICIENCY SYMPTOMS

Encephalomalacia which is a prominant vitamin E deficiency symptom, is found primarily in fowl and is prevented by vitamin E intake. Encephalomalacia is manifested by ataxia where lipids undergo auto-oxidation and, as the membrane phospholipids are oxidized, the cellular permeability is upset. In this case, there develops hemorrhages of cerebellum, edema, and destruction of parts of brain. The symptoms of encephalomalacia arise from the fact that, in this disease, the walls of the capillaries become permeable to plasma proteins, specifically the plasma albumins. As the albumins pass out through the capillary walls, they pull water out with them, causing a tremendous edema to occur in tissues. Intake of vitamin E alone does not prevent encephalomalacia if the diet is deficient in selenium (Se).

Nutritional muscular destrophy is due to vitamin E deficiency and is accompanied by sulfur amino acid deficiency. This disease causes degeneration in muscles in chicks. Se along with vitamin E are both needed in the prevention of this disease.

Vitamin E deficiency results in reproduction failure in lower animals, and thyroid function is depressed in man.

The vitamin's deficiency also causes exudative diathesis, depression of lipid synthesis (decrease in SO_4 incorporation into cellular lipoproteins), iron toxicity in pigs, liver necrosis, and kidney degeneration.

FACTORS DETERMINING VITAMIN E'S INTAKE

The amount of unsaturated fatty acids (FA) in the diet is a major factor in determining vitamin E requirements. Autoxidation of these FA gives rise to peroxides that destroy vitamin E. An intake of fish oil, for example, produces an immediate response of extreme vitamin E deficiency in chicks.

Processing foods may change all or most vitamin contents of foods. Pelleting of feeds creates a large surface area and results in oxidation of the lipid-soluble vitamins to a good extent.

The presence of minerals, such as iron and other trace minerals, may oxidize vitamin E and the other fat-soluble vitamins. This is why, in practice, the trace minerals are added to a mix at the last minute.

Additives, such as nitrites or chlorine dioxide, etc., when added to flour for bleaching, for instance, may destroy the vitamin E content.

Carbon tetrachloride is a vitamin E antagonist. Unlike vitamin A, vitamin E's intake in high level does not produce toxicity or lethality.

REFERENCES

Bradway, E. M., and Mattill, H. A. 1934. The association of fat-soluble vitamins and antioxidants in some plant tissues. *J. Am. Chem. Soc.* 56:2405.

Butt, H. R. 1951. Fat-soluble vitamins A, E, and K. *J. Am. Med. Assoc.* 143. Chapter XI of *Handbook of Nutrition.* 2nd ed. American Medical Association.

Cuthbertson, W. F.; Ridgway, R. R.; and Drummond, J. C. 1940. The fate of tocopherols in the animal body. *Biochem. J.* 34:34.

Drummond, J. C.; Noble, R. L.; and Wright, M. D. 1939. Relationship of vitamin E to the endocrine system. *J. Endorinol.* 1:275.

Emerson, G. A., and Evans, H. M. 1939. Restoration of fertility in successively older E-low female rats. *J. Nutr.* 18:501.

Harris, P. L., and Kujawski, W. 1950. *Annotated Bibliography of Vitamin E.* New York: National Vitamin Foundation, Inc.

Harris, P. L.; Quaiff, M. L.; and Swanson, W. J. 1950. Vitamin E content of foods. *J. Nutr.* 40:367.

Joffe, M., and Harris, P. L. 1943. The biolgical potency of the natural tocopherols and certain derivatives. *J. Am. Chem. Soc.* 65:925.

Mackenzie, C. G.; Levine, M. D.; and McCollum, E. V. 1940. The prevention and cure of nutritional muscular dystropy in the rabbit by alpha-tocopheral in the absence of water-soluble factor. *J. Nutr.* 20:399.

Mason, K. E. 1940. Minimal requirements of male and female rats for vitamin E. *Am. J. Physiol.* 131:268.

Mason, K. E. 1944. Physiological action of vitamin E and its homologues. In *Vitamins and Hormones*. Vol. II. p. 107.

Mattill, H. A. 1939. Vitamin E. Chapter XXX of *The Vitamins*. American Medical Association.

New York Academy of Sciences. 1949. *Vitamin E*. A monograph. *Annals N. Y. Acad. Sci.* 52:63.

Olcott, H. S., and Mattill, H. A. 1943. Vitamin E. i, 2. *J. Biol. Chem.* 104:423.

Review 1942. Muscle destrophy and vitamin E deficiency. *Nutr. Rev.* 1:7–8.

Review 1943. Vitamin E and muscle physiology. *Nutr. Rev.* 1:308.

Review 1943b. Further studies on the natural tocopherols. *Nutr. Rev.* 1:371.

Review 1943c. Cod-liver oil and the production of vitamin E deficiency. *Nutr. Rev.* 1:381.

Review 1950. Retrolental fibroplasia and vitamin E. *Nutr. Rev.* 8:116.

Review 1945. Vitamin E as physiologic antioxidant. *Nutr. Rev.* 3:17.

Chapter 13

VITAMIN K

INTRODUCTION

Vitamin K is primarily involved in blood coagulation, and this is where it got its name. It probably activates the synthesis of four specific proteins that are needed for blood clotting. The isoprene units of the side chain of vitamin K are important for its function and activity. These isoprene units range in number from four to nine. The form having four units is usually given 100% on the activity scale.

Vitamin K has a special importance in cardiac disease where it is expected to prevent clotting.

The menaquinones are vitamins of the K_2 type that are produced by the microorganisms of the rumen or the intestines of animals. Man also is capable of providing his own need of vitamin K through the aid of his intestinal microorganisms, especially those in the large intestines. Avians depend on plant sources for their vitamin K source, since the vitamin's absorption occurs very far down along the digestive tract, and since the avian tract is not long enough for the vitamin's synthesis and absorption.

Vitamin K_3 is a synthetic form of the vitamin. It possesses very high biological activity when it is in a sodium-bisulfite form; this renders it highly water-soluble and, thus, highly reactive. Its absorption, however, does not vary from the other forms.

Like the rest of the fat-soluble vitamins, Vitamin K must be in a micellar form with fat in order to be absorbed in the intestines.

Therefore, some fat is needed for its absorption. The metaquinone types get their side chain converted to four isoprene units in the liver for absorption purposes. If the side chain is longer than four isoprene units, the liver shortens it to only four, and, thus, renders the vitamin more active. Storage of the vitamin and its action in influencing protein synthesis are found mainly in the mitochondria of the liver.

A main role of vitamin K is the production of prothrombin in the liver. It takes the vitamin only about four hours to cause prothrombin synthesis. This, however, is achieved *in vivo*, and vitamin K would not cause blood clotting *in vitro*.

It is also believed that vitamin K is involved in the formation of a specific messenger-RNA.

In synthesis of prothrombin, vitamin K is thought to aid in the addition of a carbohydrate moiety to the polypeptide chain of prothrombin. The carbohydrate is believed to be a glucoseamine and mannose.

All factors necessary for blood clotting are in the plasma, and clotting occurs through an intrinsic system.

In case of injury, the system produces thromboplastin, which is an extrinsic component of tissue, and vitamin K's role here is in the activation of prothrombin, or it determines the time of conversion of prothrombin to thrombin.

If calcium in the blood is tied up, with oxalate or citrate, for example, clotting does not occur, and, as we see, the vitamin's role is inhibited.

Aside from its influence in blood clotting, it may serve as a cofactor in oxidative phosphorylation. When rats were fed liver that was sterilized with ultraviolet light, they became vitamin K deficient.

Since vitamin K is located in the mitochondria, where oxidative phosphorylation occurs, and since U.V. irradiation of mitochondria inhibits oxidative phosphorylation, thus adding vitamin K back to the irradiated samples, restores the oxidative phosphorylation process to normal.

VARIOUS FORMS OF VITAMIN K

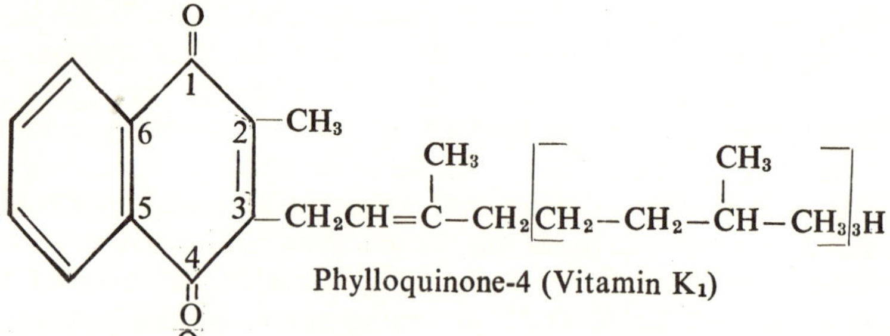

Phylloquinone-4 (Vitamin K_1)

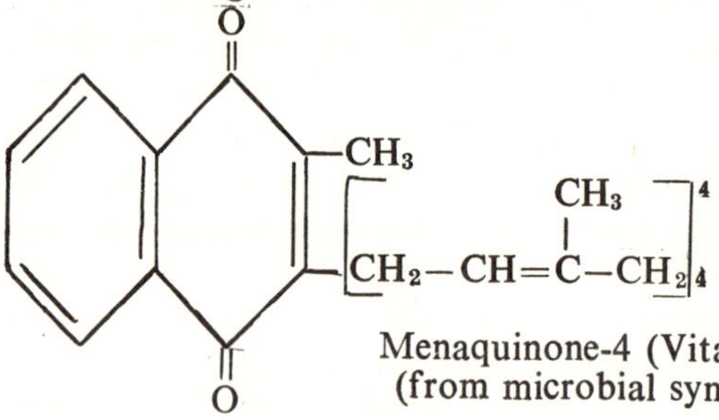

Menaquinone-4 (Vitamin K_2)
(from microbial synthesis)

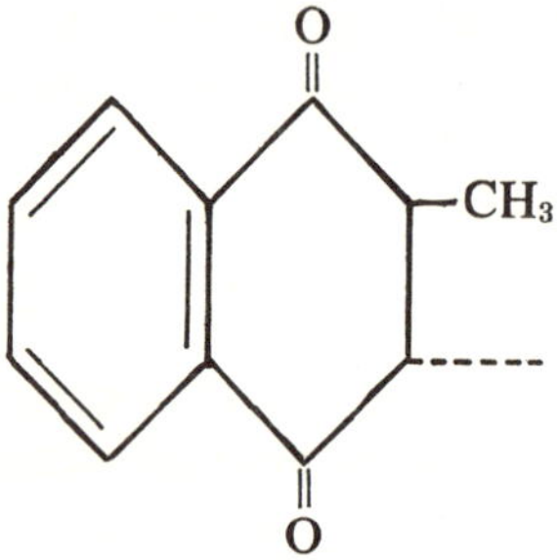

Menadione (Vitamin K_3)

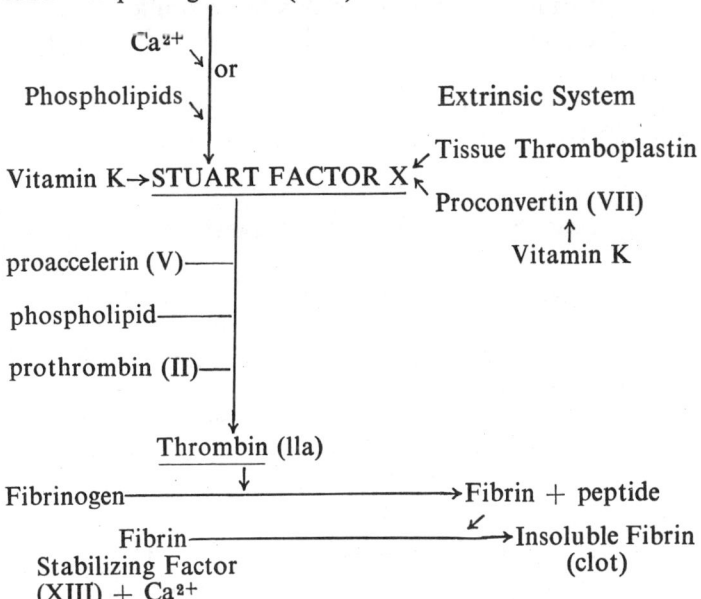

Menadione Sodium bisulfite (a synthetic form)

THE INTRINSIC SYSTEM OF VITAMIN K ACTION

Hageman Factor (XII)
Plasma thromboplastin antecedent (XI)
Vitamin K ⟶ Plasma thromboplastin component (IX)
Antihemophilic globulin (VIII)

FACTORS AFFECTING POOR CLOTTING TIME OR VITAMIN K DEFICIENCY

1. Improper fat absorption or impairment of bile salts formation will result in vitamin K deficiency or poor clotting time.
2. Liver damage will result in inhibition of synthesis of the specific proteins that are required in causing clotting.
3. Level and activity of the intestinal bacteria that endogenously synthesize vitamin K are very important in providing the vitamin for the subject.

VITAMIN K ANTAGONISTS

1. Dicoumoral: This compound competitively inhibits vitamin K. Cattle, sheep, and other animals that graze on spoiled sweet clover develop a vitamin K deficiency as the dicoumoral from such clover inhibits the vitamin.
2. Certain sulfa drugs inhibit vitamin K because of structural similarities.
3. A high intake of vitamin A may prove antagonistic to vitamin K.

DEFICIENCY SYMPTOMS OF VITAMIN K

Hemorrhage, massive bleeding, and death may result in keen vitamin K deficiency. A diet high in sulfur or spoiled sweet clover may produce vitamin K deficiency that might be manifested in animals through restlessness, loss of appetite, abnormal breathing, and, of course, the bleeding factor. Menadione (vitamin K_2) in the sodium bisulfite form would cure these symptoms.

Like vitamin E, vitamin K does not easily become toxic at high levels of intake. However, a very high intake of menadione may produce toxicity symptoms in many animals and in man, such as vomiting and high albumin level in the urine, along with porphorin of hemoglobin.

REFERENCES

Brinkhous, K. M. 1940. Plasma prothrombin; vitamin K. *Medicine* 19:329

Butt, H. R., and Snell, A. M. 1941. *Vitamin K.* Philadelphia: Saunders Co.

Dam, H. 1942. Vitamin K, its chemistry and physiology. *Advances in Enzymology* 2:285.

Dam, H. J. Glavind, and Karrer, P. 1940. Biological activity of the natural K vitamins and of some related compounds. *Helv. Chim. Acta.* 23:224.

Doisy, E. A.; Binkley, S. B.; and Thayer, S. A. 1941. Vitamin K. *Chem. Rev.* 28:477.

Fieser, L. F.; Tishler, M.; Sampson, W. L.; and Woodford, S. 1941. Vitamin K activity and structure. *J. Biol Chem.* 137:659

Bellman, L. M.; and Shettles, L. B. 1940. Factors influencing plasma protherombin in the newborn infant. I-III *Bull. Johns, Hopkins Hosp.* 65:138.

Klose, A. A., and Almquist, H. J. 1940. Synthesis of vitamin K. *J. Biol. Chem.* 132:469.

Kornberg, A.; Daft, F. S.; and Sebrell, W. H. 1944. Mechanism of production of vitamin K deficiency in rats by sulfonamides. *J. Biol. Chem.* 155:193.

McKee, R. W.; Binkley, S. B.; MacCorqodale, D. W.; Thayer, S. A.; and Doisy, E.A. 1939. Isolation of Vitamin K_1 and K_2. *J. Biol. Chem.* 131:327.

Pohle, F. J., and Stewart, J. K. 1940. Observation on plasma prothrombin and effects of vitamin K in patients with liver or biliary tract disease. *J. Clin. Investigation* 19:365.

Review 1942 Specificity of hemorrhagic preventive factors. *Nutr. Rev.* 1:52.

Scarborough, H. 1940. Nutritional deficiency of vitamin K in man. *Lancet,* 1:1080.

Smith, H. P.; Ziffren, S. E. Owen, C. A.; and Hoffman, G. R. 1939. Clinical and experimental studies on vitamin K. *J. Am. Med. Assoc.* 113:308.

Snell, A. M. 1939. Vitamin K: its properties, distribution and clinical importance. *J. Am. Med. Assoc.* 112:1457.

Thayer, S. A.; McKee, R. W.; Binkley, S. B.; and Doisy, E. A. 1940. Potencies of vitamin K_1 and of 2-methyl-1, 4-naphthoquinone. *Proc. Soc. Exptl. Biol. Med.* 44:585.

Waddell, W. W., Jr., and Guerry, D. 1939. Effect of vitamin K on the clotting time of the prothrombin and the blood, with special reference to unnatural bleeding of the newly born. *J. Am. Med. Assoc.* 112:2259.

Wilson, S. J. 1944. Qualitative and quantitative studies on the antithrombic activity of blood serum and plasma. *Am. J. Clin. Path.* 14:307.

Chapter 14

VITAMIN C

INTRODUCTION

Ascorbic acid, although a water-soluble vitamin, it does not fall within the B-complex vitamins. It is essential for man, monkey, guinea pig, and birds, as they lack the ability of synthesizing the enzyme L-gulono-lactone oxidase that converts L-gulonolactone to L-ascorbic acid. Ascorbic acid is involved in oxidation-reduction reactions as it is reversibly converted to monodehydro ascorbic acid by losing one electron, and to the dehydroascorbic acid form by losing a second electron. This oxidation-reduction property renders ascorbic acid extremely unstable in solutions, and at times even under frozen storage or dry conditions. Foods that are rich in vitamin C content may lose a considerable amount of it upon storage. Methylation at the C-2, C-3, or C-6 position of ascorbic acid has been thought of as means of stabilization of the vitamin. Adding palmitate at the C-3 position makes ascorbic acid a fat-soluble vitamin and renders it available as the palmitate breaks away in the body. Sulfation at the C-3 position makes an important derivative of ascorbic acid for steroids metabolism. This derivative is water-soluble.

Vitamin C is absorbed in the small intestines, and the plasma level is a fairly good indication of the level of intake.

There is probably no storage site for the water-soluble form of the vitamin and, ascorbate, especially the C-3 sulfated form, may be carried protein-bound in the blood.

Specific tissues have a high need for ascorbic acid for their metabolism and, consequently, they contain high levels of the vitamin. Such tissues, like the adrenal glands, which contain perhaps the highest concentration of the vitamin, and the brain, which is second highest in ascorbic acid concentration, do require the vitamin for their metabolism. The free form cannot get from the blood to the brain, and it is thought that the C-3 sulfated form gets protein bound and travels to the brain where it gets desulfated. The pituitary and hypothalamus are also high in ascorbic acid content.

Since it is a water-soluble vitamin, ascorbic acid has a very short half-life in the living system. Its half-life in man is about sixteen days; in guinea pigs it is four days. The total pool size of the vitamin in man is around 1.5 grams and, on a vitamin C deficient diet, he loses it at the rate of three percent per day. In ninety days, this pool size drops to 300 mg, where scurvy develops. The scurvy onset is rapid and can be lethal, as the nervous system is affected (brain is high in ascorbic acid content). As symptoms of the vitamin's deficiency hit rapidly and cause emotional changes and finally death, it, thus, becomes difficult to adequately study the disease's symptoms.

Since in man and other primates about five percent of vitamin C is converted to the oxalate form, a large intake of the vitamin may result in the development of kidney stones as calcium-oxalate precipitate develops in the kidney tissues, especially since man lacks the lactonase enzyme that converts ascorbic acid to CO_2.

OXIDATION-REDUCTION SYSTEM OF ASCORBIC ACID

Ascorbic acid is oxidizable by a group of various enzymes, such as ascorbic acid oxidases, cytochrome oxidases, phenolases, peroxidases, and laccases.

Metal ions, hemochromogens, and quinones also oxidize ascorbic acid.

Most systems that reduce ascorbic acid are located in plants, bacteria, and yeast. Dehydrogenases, with their NADH cofactor, exchange electrons back and forth in oxidation-reduction reactions with ascorbic acid as shown below:

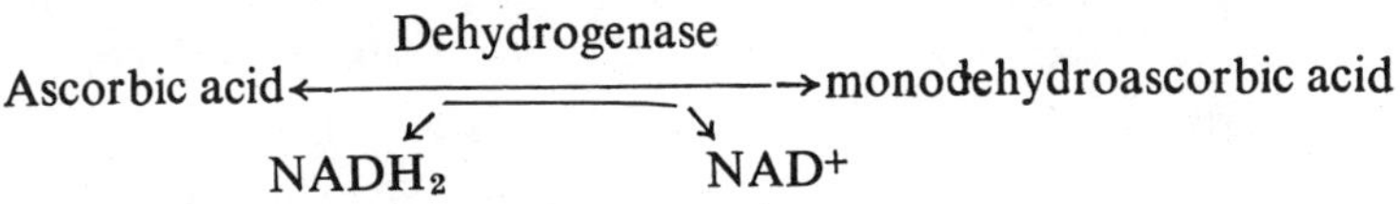

$$\text{Ascorbic acid} \underset{NADH_2}{\overset{\text{Dehydrogenase}}{\longleftrightarrow}} \text{monodehydroascorbic acid} \quad NAD^+$$

Oxidation-Reduction Reactions of Ascorbic Acid

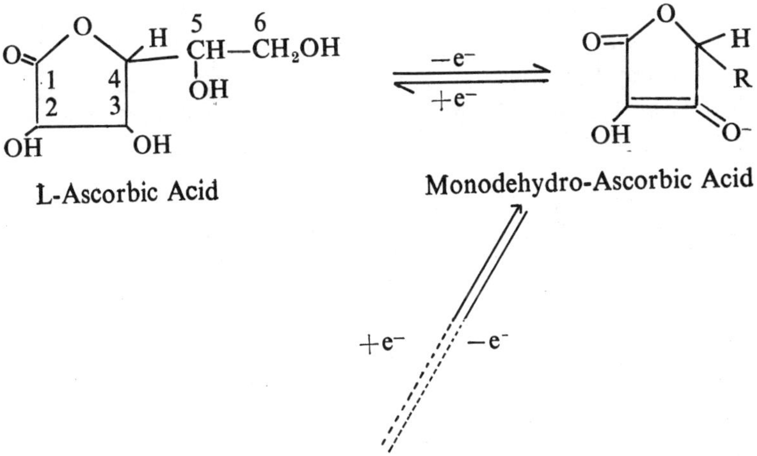

L-Ascorbic Acid $\underset{+e^-}{\overset{-e^-}{\rightleftharpoons}}$ Monodehydro-Ascorbic Acid

$+e^- \quad -e^-$

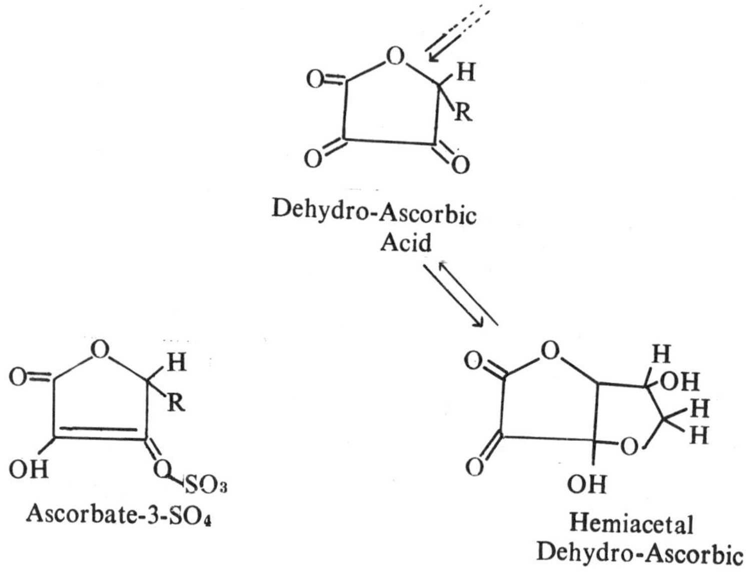

[Dehydro-Ascorbic Acid structure]

Dehydro-Ascorbic Acid

[Ascorbate-3-SO₄ structure]

Ascorbate-3-SO$_4$

[Hemiacetal Dehydro-Ascorbic structure]

Hemiacetal Dehydro-Ascorbic

OTHER BIOCHEMICAL FUNCTIONS OF VITAMIN C

Ascorbic acid is also involved in tyrosine metabolism, where tyrosine is converted to p-OH-phenyl-pyruvic acid that, in turn, is converted to homogentistic acid. The step where p-OH-phenyl-pyruvic is converted to homogentistic acid is catalyzed by the enzyme p-OH-phenyl-pyruvic acid oxidase, which is activated by ascorbic acid:

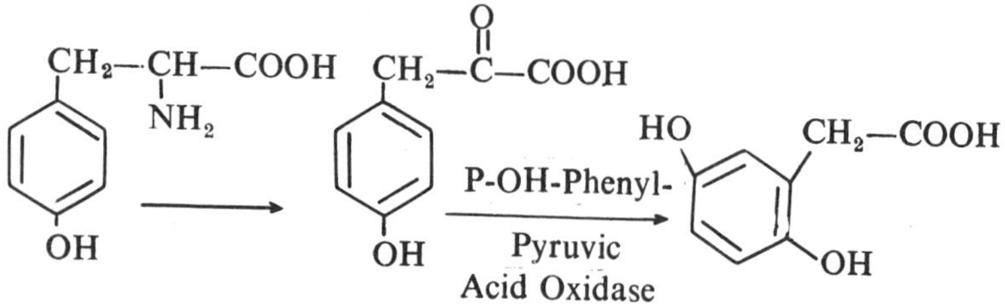

Tyrosine → P-OH-Phenyl-Pyruvic Acid → (P-OH-Phenyl-Pyruvic Acid Oxidase) → Homogentistic Acid

Considering that ascorbic acid catalyzes hydroxylation reactions, it is, consequently, needed in collagen and elastin formation. It also becomes necessary in bone-matrix syntheses. Vitamin C's role in collagen and elastin is in hydroxylation of proline after its incorporation into connective tissue. In the vitamin's deficiency, wounds would not heal, as collagen formation is lacking. In this case, capillaries that require collagen break, and the integrity of blood vessels falls into doubt. Ascorbic acid, as well, aids in hydroxylation of lysine and tryptophan. Of even more importance is its involvement in the hydroxylation of steroids. Hydroxylation of 11-deoxycorticosterone, aldosterone, cortisol, etc., are all thought to have occurred through the aid of ascorbic acid. The same effect of aiding hydroxylation of steroids also results in the making of bile acids.

EFFECT OF ASCORBIC ACID ON METAL IONS

Ascorbic acid has a reducing effect on several metal ions. This is of special importance as these metal ions are cofactors that activate enzymatic systems. For example, ascorbic acid exerts an inhibitory effect on urease as it reduces its metal cofactor from the cupric (Cu^{++}) to the cuprous (Cu^+) ion. It also detoxifies some harmful heavy metals such as mercuric (Hg^{++}) by converting it to mercurous (Hg^+). This reduction of mercuric ions has a beneficial effect as it prevents the inhibitory influence of Hg^{++} on certain enzymes. Ascorbic acid also aids in iron absorption through reduction of the ferric (Fe^{+++}) form to ferrous (Fe^{++}), which is absorbable in the gastrointestinal tract.

Ascorbic acid may cause a relief from cadmium (Cd) toxicity by chelating Cd. It was also reported as increasing fluoride storage in guinea pigs.

ASCORBIC ACID AND ADRENAL FUNCTIONS

The adrenals have a high ascorbic acid content, the uptake of which is energy-dependent. Adrenocorticotropic hormone (ACTH) and the corticosteroids are thought to inhibit this uptake by the glands. ACTH, especially, depletes ascorbic acid from the adrenals.

Some researchers believe ascorbic acid's presence in the adrenal glands is for the purpose of steroid hormones synthesis and release; others believe ascorbic acid's presence in the glands has nothing to do with steroids synthesis. Furthermore, there are those who believe that if a high level of ascorbic acid is introduced to the adrenals, steroid synthesis will be inhibited. Here, the demolase enzyme that causes the cleavage of the cholesterol side chain is inhibited by the high concentration of ascorbic acid. A low level of ascorbic acid is thought to stimulate this cleavage.

This cleavage of the cholesterol side chain is done by demolase before synthesis of steroids takes place.

←Cleavage Point
Demolase

Cholesterol

Thus, an excess of ascorbate inhibits its side chain cleavage, and cholesterol keeps building up and negatively feeds back on itself until further synthesis of cholesterol is inhibited.

Another point in steroidgenesis where ascorbic acid has influence is at the δ^5-3-β-hydroxysteroid dehydrogenase which is responsible for the conversion of pregnenolone to progesterone. The oxidized form of ascorbic acid is believed to have a stimulatory effect on this enzyme.

RELATIONSHIP BETWEEN ASCORBIC ACID AND CHOLESTEROL

Excess ascorbic acid may reduce cholesterol metabolism to its precursors in the body through inactivation of the demolase cleavage enzyme as discussed above. Ascorbic acid may also increase the elimination of cholesterol by forming the ascorbate-3-SO_4 salt of the compound. Sulfation of cholesterol at C-3 renders it water-soluble and increases its excretion.

Ascorbic acid is also involved in the hydroxylation of cholesterol derivatives, thus leading to the synthesis of bile acids, which are the chief factors in getting rid of cholesterol.

There has been reported a highly significant increase in tissue cholesterol in the case of vitamin C deficiency.

This may be largely due to the decrease in the conversion of cholesterol to bile acids (which remove cholesterol from the body). Absence of vitamin C in this case prevents the hydroxylation necessary for making the bile acids.

Smokers, men, and the aged are thought to have lower levels of ascorbic acid than nonsmokers, women, and the young.

Some researchers feel a cholesterol build-up in the arteries is due to vitamin C deficiency. Others feel that the sulfated form of ascorbic acid causes the removal of cholesterol from the body, as it renders it water-soluble.

VARIOUS FUNCTIONS OF VITAMIN C

Ascorbic acid may have a regulatory effect on mitosis, as reports indicate that a high concentration of the oxidized form transforms cells from the nondividing state to the dividing state, probably through the release and activation of the lysozomal hydrolytic enzymes. This is probably achieved as ascorbic acid gives off a free radical that causes oxidation or peroxidation of the phospholipids of the membranes, thus altering their permeability and increasing the hydrolytic enzyme's release, which, in turn, causes cell division.

Ascorbic acid also affects bone calcification. It is necessary for the proper connective tissue and bone-matrix formation. As much as ascorbic acid possesses a role in bone formation, it has also a role in salts mobilization from bone. The vitamin causes an increase in the plasma acid phosphatase enzyme's activity which aids in mobilization of bone calcium. The increase in Ca^{2+} excretion, in this case, is not related to parathyroid hormone but, rather, it is because of Ca-oxalate binding and the effect of ascorbic acid on it.

FACTORS AFFECTING ASCORBIC ACID REQUIREMENTS

A high protein diet increases the requirement for vitamin C. High levels of zinc, copper, and molybdenum also require high levels of ascorbic acid. Males, smokers, and the aged also have a high requirement for the vitamin.

Vitamins A, E, B_1, B_2, and many other B vitamins, if deficient, will result in a decreased ascorbic acid synthesis. A hypervitaminosis of D will increase vitamin C's synthesis, however. The exogenous intake of vitamin C will also determine the body's requirement.

SYMPTOMS OF SCURVY

1. Hyperkeratosis of hair follicles.
2. Loosening of teeth, swelling and bleeding of gums, and swelling of joints.

3. Impairment of collagen synthesis, defective cartilage formation.

4. Improper matrix formation leading to poor bone and teeth mineralization.

5. Poor incorporation of SO_4 into chondriotin-SO_4, leading to a decrease in mucopolysaccharides and defective connective tissue formation.

6. Breaking of capillaries, hemorrhages of tissues.

7. Calcification of tissues.

8. May affect thyroid gland, ovaries, testes, and may cause a decrease in blood clotting.

9. Decline in plasma alkaline phosphatase activity and decrease in iron utilization.

10. Causes an increase in levels of Co, Zn, and Mn in tissues.

11. Functions of the adrenal glands are depressed, and an increase in the excretion of the 17-keto steroids in urine ensues.

12. May result in insulin hormone deficiency and development of ocular lesions.

13. May cause emotional instability and peripheral neuropathy and poor nerve response.

BIOSYNTHESIS OF ASCORBIC ACID

$$\begin{array}{ccc}
\text{H--C=O} & \text{H--C=O} & \text{CH}_2\text{OH} \\
\text{H--COH} & \text{H--COH} & \text{H--COH} \\
\text{HOCH} & \text{HOCH} \xrightarrow{\text{NADH} \quad \text{NAD}} \text{HOCH} \\
\text{H--COH} \xrightarrow{\text{glucurono-}\delta\text{-lactone hydrolase}} & \text{H--COH} \longrightarrow \text{H--COH} \\
\text{H--COH} & \text{H--COH} & \text{H--COH} \\
\text{CH}_2\text{OH} & \text{COOH} & \text{COOH} \\
\text{D-6-glucose} & \text{D-Glucuronic acid} & \text{L-Gulonic acid}
\end{array}$$

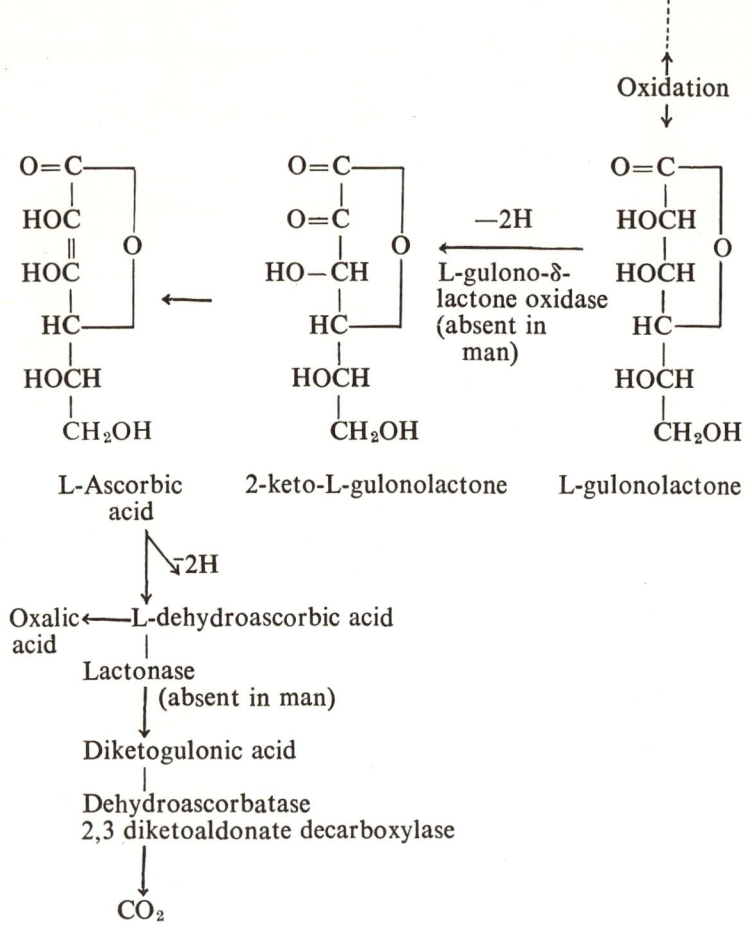

REFERENCES

Abt, A. F., and Farmer, C. J. 1939. Vitamin C. Pharmacology and therapeutics. Chapter XXII of *The Vitamins*. American Medical Association.

Bessey, O. A. 1939. Vitamin C: methods of assay and dietary sources. Chapter XX of *The Vitamins*. American Medical Association.

Dalldorf, G. 1939. The pathology of vitamin C deficiency. Chapter XIX of *The Vitamins*. American Medical Association.

Dunn, F. J., and Dawson, C. R. 1951. On the nature of ascorbic acid oxidase. *J. Biol. Chem.* 189:485.

Ham, A. W., and Elliott, H. C. 1938. Bone and cartilage in scurvy. *Am. J. Path.* 14:323.

Jackel, S. S.; Mosbach, E. M.; Burns, J. J.; and King, C. G. 1950. The synthesis of L-ascorbic acid by the albino rat. *J. Biol. Chem.* 186:569.

Kellie, A. E., and Zilva, S. S. 1939. The vitamin C requirement of man. *Biochem. J.* 33:153.

King, C. G. 1936. Vitamin C, ascorbic acid. *Physiol. Rev.* 16:238.

———. 1950. Vitamin C. *J. Am. Med. Assoc.* 142:563. Chapter IX of *Handbook of Nutrition*. 2nd ed. American Medical Association.

———. and Waugh, W. A. 1932. The chemical nature of vitamin C. *Science* 75:357.

Long, C. N. H. 1947. Relation of colesterol and ascorbic acid to the secretion of the adrenal cortex. *Recent Progress in Hormone Research* 1:99.

Mayer, J., and Krehl, W. A. 1948. The relation of diet composition and vitamin C to vitamin A deficiency. *J. Nutr.* 35:523.

Ralli, E. P., and Sherry, S. 1941. Adult scurvy and the metabolism of vitamin C. *Medicine* 20:251.

Review 1943. Pathologic changes in vitamin C deficiency. *Nutrition Rev.* 1:305.

Review 1944. Metabolism and function of ascorbic acid. *Nutr. Rev.* 2:283.

Review 1950. Ascorbic acid and adrenal function. *Nutr. Rev.* 8:52.

Review 1951. Deficiences of ascorbic acid and pteroylglutamic acid in anemia. *Nutr. Rev.* 9:52.

Samuels, L. T. 1948. The relation of ascorbic acid metabolism in the rat to diets high in protein, carbohydrates or fat. *J. Nutr.* 36:205.

Smith, S. L. 1939. Human requirements of vitamin C, Chapter XXI of *The Vitamins*. American Medical Association.

Weissberger, A., and Lu Valle, J. E. 1944. Oxidation processes. XVII. The autoxidation of ascorbic acid in the presence of copper. *J. Am. Chem. Soc.* 66:700.

Chapter 15

THIAMIN (B$_1$)

INTRODUCTION

Thiamin can be synthesized by the intestinal microorganisms, but whether this is a sufficient source for man becomes doubtful. The vitamin, however, is readily found and is readily utilized. Cereal grains and their by-products are rich sources of thiamin. Meats, generally, are poor sources of this vitamin. Thiamin is not, to any appreciable extent, stored in the body, and it must be phosphorylated in order to be absorbed. For injection, the hydrochloric acid form of the vitamin is usually given.

Thiamin pyrophosphate is the active coenzyme form of the vitamin. The most important reaction it catalyzes is the oxidative decarboxylation of keto acids, a reaction type that bocomes of great significance in carbohydrate metabolism as pyruvate gets oxidatively decarboxylated, through the aid of thiamin, to acetyl-Co-A, the cross-road reaction that leads from the anaerobic glycolysis to the energetic aerobic metabolism.

STRUCTURE

Thiamin Hydrochloride

Thiamin Pyrophosphate (TPP)—"co-enzyme"

FUNCTIONS OF THIAMIN

It catalyzes pyruvic acid conversion to acetyl-Co-A (as mentioned earlier).

Thiamin-pyrophosphate (TPP) is the coenzyme of the transketolase enzymes. The transketolase, in the pentose shunt, activates important reactions in that route of sugar metabolism.

Thiamin-pyrophosphate is also involved in the conversion of α-ketoglutaric acid to succinyl-Co-A, an important reaction in the tricarboxylic acid cycle.

TPP activates transketolases that ultimately are involved in red blood cell synthesis.

TPP also has a role in conversion of tryptophan to nicotinamide.

SPECIFIC REACTIONS CATALYZED BY TPP:

Nonoxidative Decarboxylation

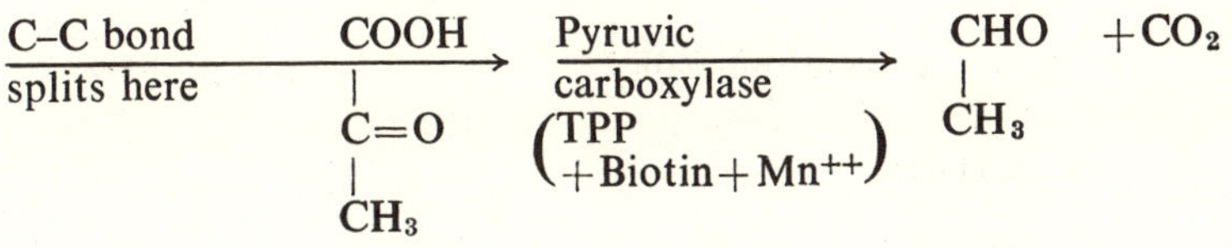

$$\underset{\text{Pyruvic Acid}}{\underset{\text{C–C bond splits here}}{\longrightarrow}\ \begin{array}{c}COOH\\|\\C=O\\|\\CH_3\end{array}} \xrightarrow[\left(\begin{array}{c}TPP\\+Biotin+Mn^{++}\end{array}\right)]{\text{Pyruvic carboxylase}} \underset{\text{Acetaldehyde}}{\begin{array}{c}CHO\\|\\CH_3\end{array}} + CO_2$$

Oxidative Decarboxylation

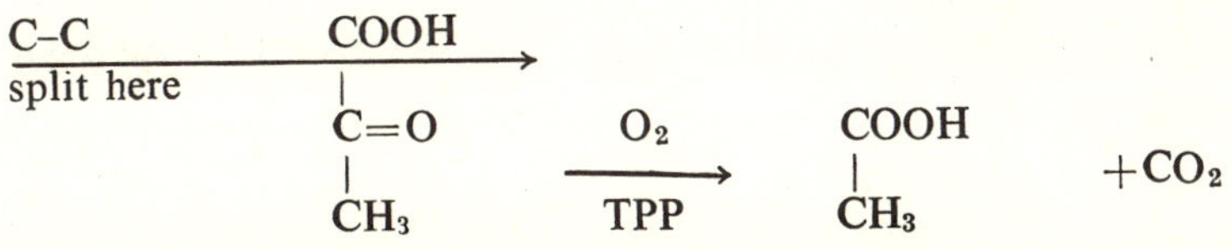

$$\underset{\text{Pyruvic Acid}}{\underset{\text{C–C split here}}{\longrightarrow}\ \begin{array}{c}COOH\\|\\C=O\\|\\CH_3\end{array}} \xrightarrow[\text{TPP}]{O_2} \begin{array}{c}COOH\\|\\CH_3\end{array} + CO_2$$

Carbon Transfer Reactions:

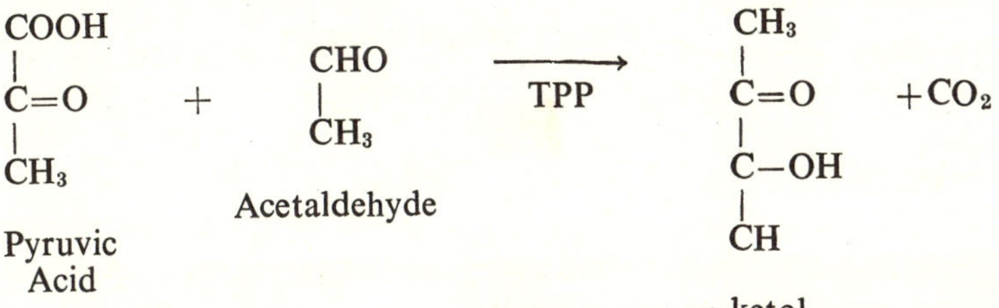

$$\underset{\text{Pyruvic Acid}}{\begin{array}{c}COOH\\|\\C=O\\|\\CH_3\end{array}} + \underset{\text{Acetaldehyde}}{\begin{array}{c}CHO\\|\\CH_3\end{array}} \xrightarrow{TPP} \underset{\alpha\text{-ketol}}{\begin{array}{c}CH_3\\|\\C=O\\|\\C-OH\\|\\CH\end{array}} + CO_2$$

Transketolase Reactions:

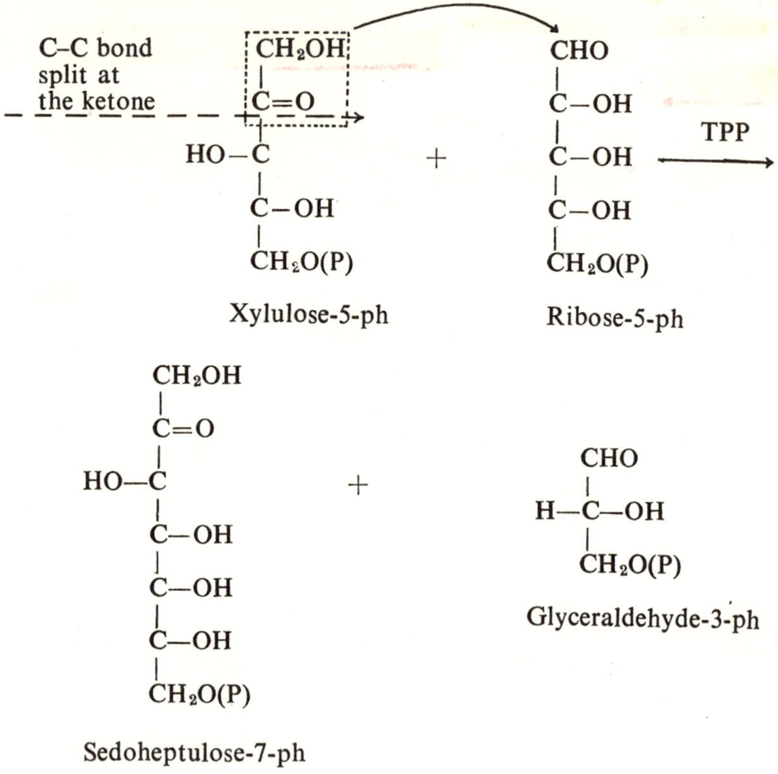

C–C bond split at the ketone

```
  CH₂OH              CHO
   |                  |
   C=O                C–OH
   |          +       |            TPP
HO–C                  C–OH        ⟶
   |                  |
   C–OH               C–OH
   |                  |
   CH₂O(P)            CH₂O(P)
 Xylulose-5-ph      Ribose-5-ph
```

```
  CH₂OH
   |
   C=O
   |                  CHO
HO–C           +       |
   |                H–C–OH
   C–OH               |
   |                  CH₂O(P)
   C–OH
   |               Glyceraldehyde-3-ph
   C–OH
   |
   CH₂O(P)
Sedoheptulose-7-ph
```

Thiamine pyrophosphate is also involved in the large system of the pyruvic dehydrogenase complex.

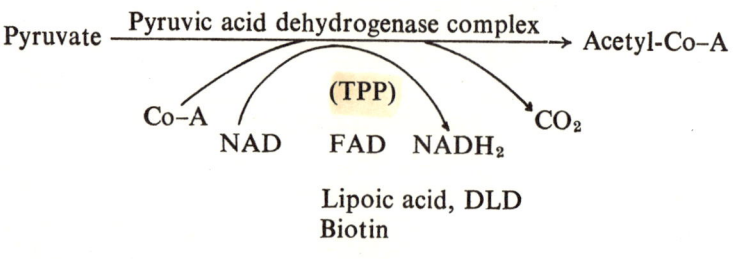

Lipoic acid, DLD
Biotin

POSSIBLE MECHANISM OF NONOXIDATIVE DECARBOXYLATION REACTION:

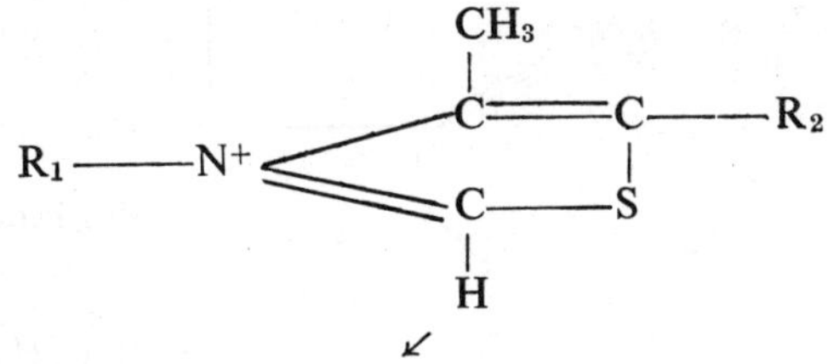

$$R_1-N^+ \diagup \substack{C(CH_3)=C-R_2 \\ \diagdown \\ C-S \\ | \\ H}$$

This hydrogen is very rapidly replaced.

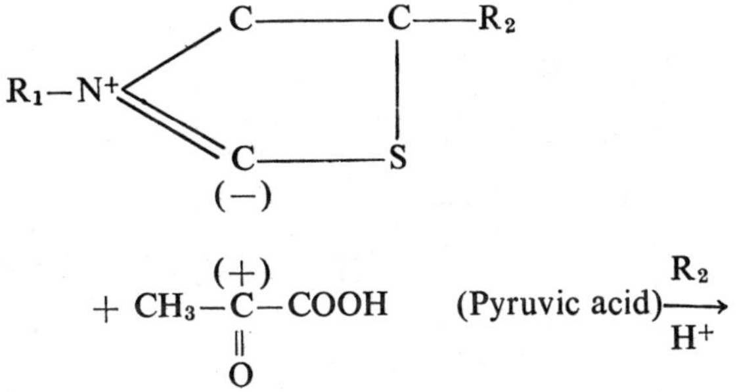

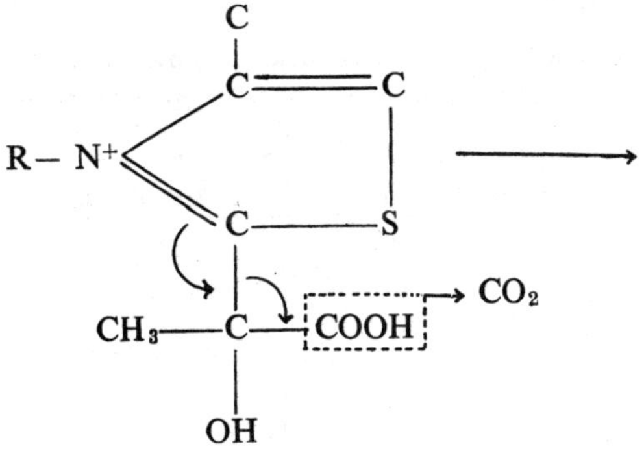

$$R_1-N^+ \diagup \substack{C-C-R_2 \\ \diagdown \\ C-S \\ (-)}$$

$$+\ CH_3-\overset{(+)}{\underset{\underset{O}{\|}}{C}}-COOH \quad \text{(Pyruvic acid)} \xrightarrow[H^+]{R_2}$$

$$R-N^+ \diagup \substack{C(C)=C \\ \diagdown \\ C-S \\ | \\ CH_3-C-[COOH] \to CO_2 \\ | \\ OH}} \longrightarrow$$

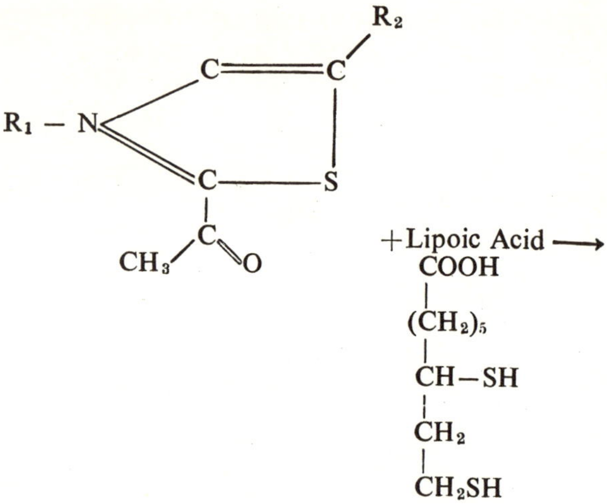

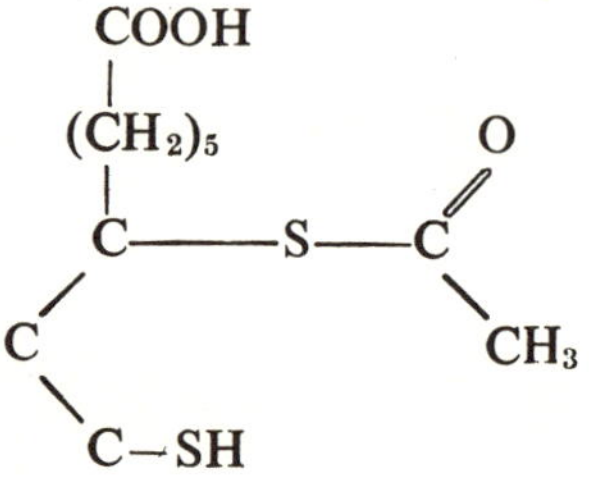

POSSIBLE MECHANISM OF DECARBOXYLATION REACTION

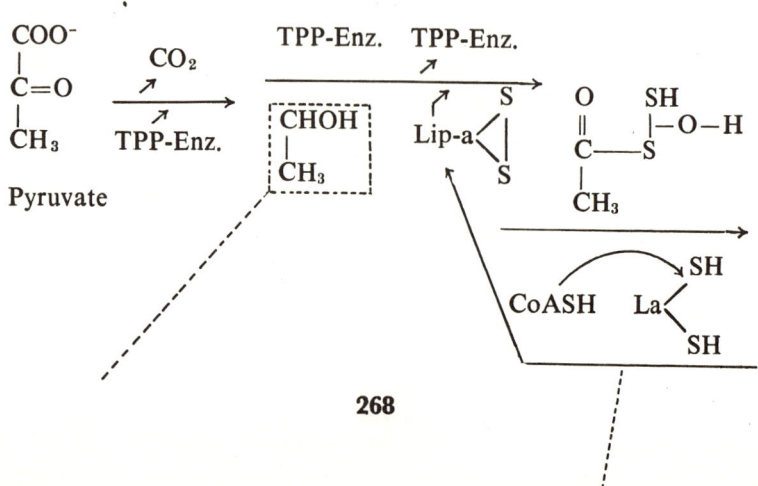

DEFICIENCY SYMPTOMS OF THIAMIN

1. Thiamin deficiency leads to malfunction in carbohydrate metabolism and the accumulation of pyruvic acid.
2. A deficiency in this vitamin results in cerebral dysfunction. This has nothing to do with its coenzyme function, but rather it is a role of thiamin in the nerve membrane transport. It also plays a role in nerve membrane dephosphorylation and rephosphorylation.
3. Besides effects on nerves, a thiamin deficiency may cause alterations in the mitochondria where they become enlarged and tightly packed together.
4. A deficiency in thiamin may cause impairment of absorption of certain amino acids, such as isoleucine and histidine.
5. Glycolysis, pentose sugars shunt, red blood cell metabolism, and thyroid metabolism are all adversely affected in thiamin deficiency.

6. Acetyl-choline synthesis (involved in nerve transmission) is impaired in thiamin deficiency.

7. Beriberi is the disease characteristic of the lack of thiamin. The vitamin's intake in large amounts does not easily produce toxicity.

Chapter 16

RIBOFLAVIN (B₂)

INTRODUCTION

B$_2$ is a fluorescent compound synthesized by yeast, fungus, bacteria, and green plants. It is not synthesized by animal tissue. Although milk is high in riboflavin, it is rapidly destroyed by sunlight. Liver and meat may contain moderate amounts of the vitamin. for its coenzyme action, riboflavin is phosphorylated in the intestinal mucosa. The biosynthesis of the coenzyme forms, flavin adenine dinucleotide (FAD) and flavin mononucleotide (FMN), is highest in liver and kidney.

STRUCTURE AND FORMS:

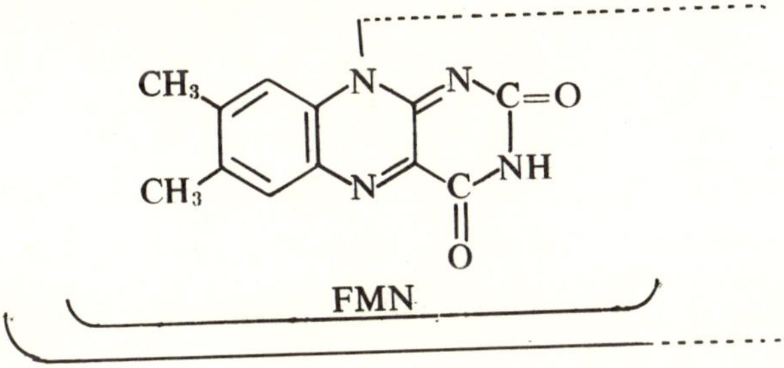

FMN

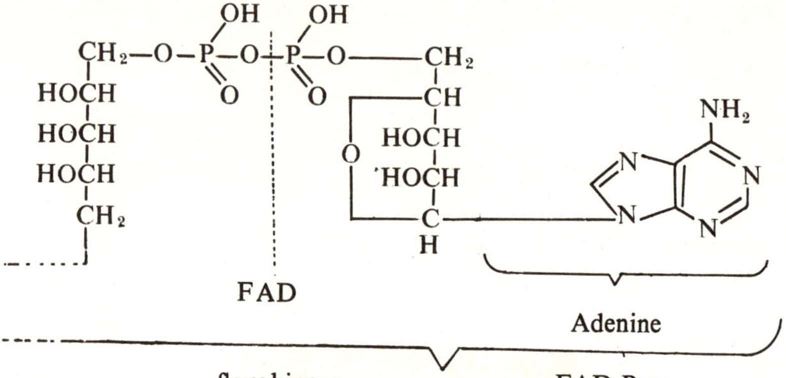

Riboflavin ⇌(flavokinase / FMN phosphatase)⇌ FMN ⇌(FAD Pyrophosphorylase)⇌ FAD

Riboflavin

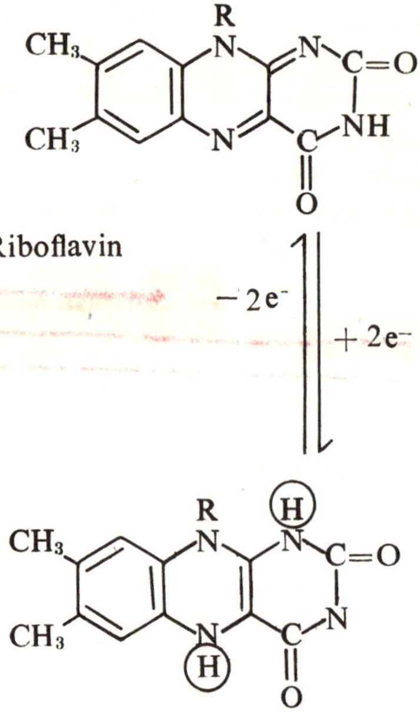

$-2e^-$ $+2e^-$

Dihydroriboflavin

ROLE OF COENZYMES IN ELECTRON TRANSPORT

$NADH_2 \rightarrow FMN \rightarrow Co-QH \rightarrow 2\ Fe^{+++}$ — Cytochrome-b
$NAD^+ \leftarrow FMNH_2 \leftarrow Co-Q \leftarrow 2\ Fe^{++} + 2H^+$

$NADPH_2 \rightarrow FAD \rightarrow 2\ Fe^{++} + 2H^+ \rightarrow \frac{1}{2} O_2$ — Cytochrome-C
$NADP^+ \leftarrow FADH_2 \leftarrow 2\ Fe^{+++} \leftarrow H_2O$

FUNCTIONS OF FAD AND FMN   Redox

The most important function of FAD and FMN is their role in oxidation-reduction reactions in cell respiration, as they aid in hydrogen transfer between FAD and NAD in the electron transport system. They accept electrons from NADH and transfer them to Co-Q.

FMN and FAD are cofactors of specific enzymes such as cytochrome c-reductase, xanthine oxidase, and certain dehydrogenases.

Hypothyroidism causes a decrease in FMN and FAD synthesis, probably through reduction of activity of the enzymes involved. It is believed that thyroid hormones enhance the synthesis of both the protein part of these enzymes and, as well, the FMN and FAD moieties themselves.

The FAD form or coenzyme is the more common one. Most higher animals cannot carry on the reactions that produce FMN and FAD, but microorganisms can.

FMN and FAD's function is similar to NAD's in that they catalyze oxidation-reduction reactions, but they work on the substrate level in the removal of hydrogens, one at a time, in this case. More common of FMN and FAD is that they cause oxidation reduction at the next step.

ENZYMES ACTIVATED BY FMN

1. Cytochrome c-reductase
2. L-amino acid dehydrogenase

ENZYMES ACTIVATED BY FAD

1. D-amino acid dehydrogenase
2. Glycine oxidase
3. Xanthine oxidase
4. Acyl Co-A dehydrogenase
5. Succinate dehydrogenase
6. Butyryl Co-A dehydrogenase
7. Dihydrosphingosine dehydrogenase
8. Aldehyde dehydrogenase
9. Pyruvate dehydrogenase
10. Lipoyl dehydrogenase

METALLO-FLAVOPROTEINS

1. Butyryl Co-A dehydrogenase (Cu)
2. Xanthine oxidase (Fe, Mo)
3. Aldehyde oxidase (Mo)
4. NADH cytochrome reductase (Fe)

DEFICIENCY SYMPTOMS OF RIBOFLAVIN

1. Curled toe paralysis in chickens and degeneration of spinal cord of young chicks.
2. Dermatitis in rats
3. Lip lesions in man, and eye sensitivity to light.

Chapter 17

NIACIN

INTRODUCTION

Niacin is the antipellagara vitamin and is the antiblack-tongue in dogs. Nicotinic acid is present free in plants and is easily converted to the coenzyme form "nicotinamide" in the body.

Lean meat, gravies, liver, yeast, protein supplements, and poultry are all fairly good sources of niacin.

The coenzyme forms of niacin are nicotinamide adenine dinucleotide (NAD)-Co-I, and nicotinamide adenine dinucleotide phosphate (NADP)-Co-II.

Once tryptophan is present in the diet, it is converted by tissues to the vitamin niacin. Certain species, including humans, cannot meet their niacin requirements from tryptophan in the diet, and, thus, their diet must contain the vitamin itself, niacin.

FUNCTION OF THE COENZYMES Redox

Function of NAD and NADP is in transfer of hydrogens on the substrate level. They take these hydrogens on or give them up in a reversible fashion. This takes place nearly everywhere in the metabolism, as NAD and NADP are cofactors for the dehydrogenase type of enzymes.

Alcohol dehydrogenase, for example, has NAD as its prosthetic group, and it causes the reversible oxidation reduction of alcohol to and from acetaldehyde.

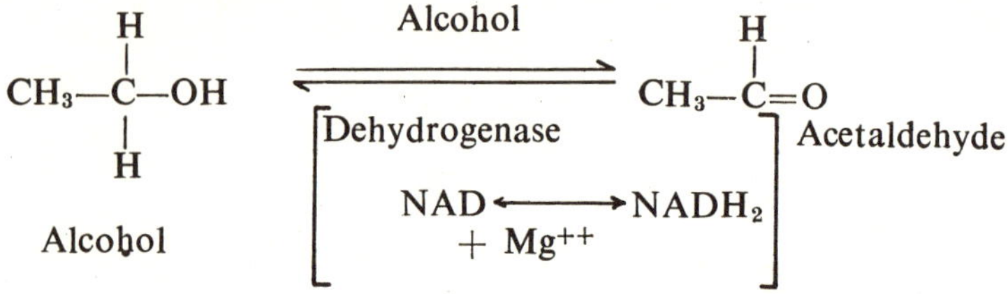

$$CH_3-\underset{H}{\overset{H}{C}}-OH \underset{\substack{[Dehydrogenase \\ NAD \leftrightarrow NADH_2 \\ + Mg^{++}]}}{\overset{Alcohol}{\rightleftharpoons}} CH_3-\overset{H}{C}=O$$

Alcohol — Acetaldehyde

Here, the removal of two hydrogens and electrons is done on the substrate level.

SOURCE OF THE VITAMIN

The body can synthesize the vitamin, and, subsequently, the coenzymes from tryptophan, although synthesis from tryptophan may not provide all the body need of niacin, and intake of the vitamin itself in foods should be practiced.

The synthesis of niacin from tryptophan is influenced by hormones. Estrogen is believed to promote the conversion of tryptophan to niacin, while progesterone and hydrocorticosterone inhibit it.

STRUCTURE AND FORMS OF NIACIN

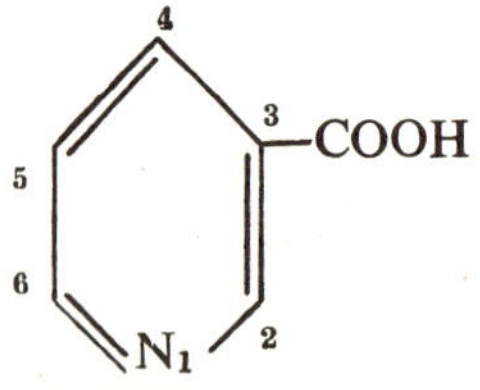

Niacin

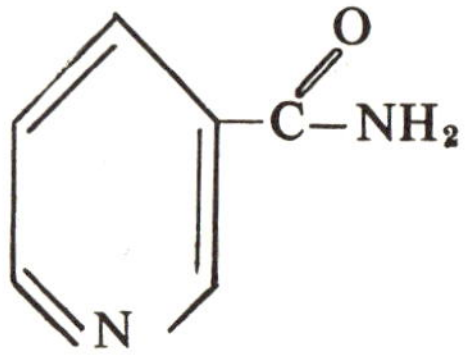

Nicotinamide

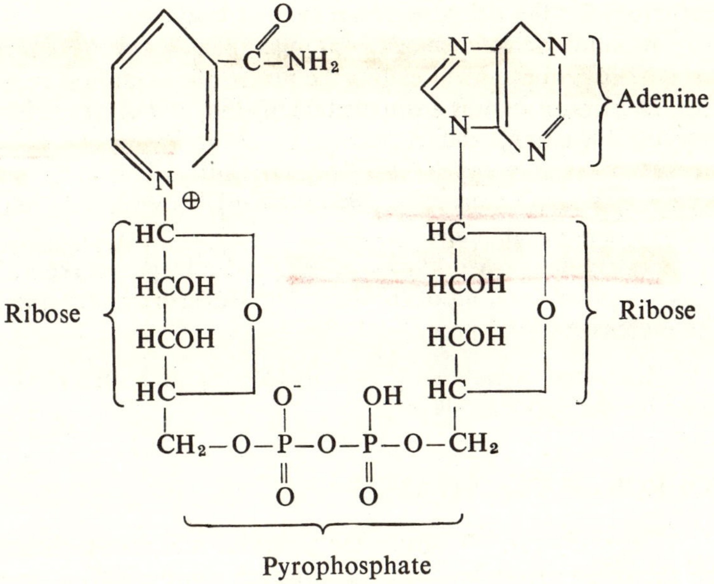

NAD (Co-I)

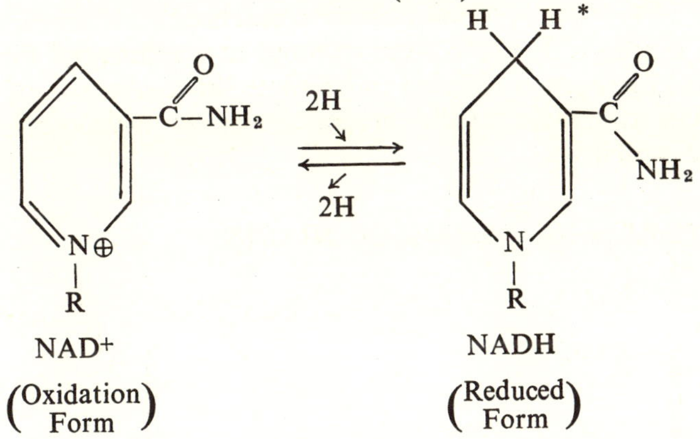

Structure: Nicotinamide – Ribose – Pyrophosphate – Ribose – Adenine

$$NAD^+ \underset{2H}{\overset{2H}{\rightleftharpoons}} NADH$$

NAD⁺ (Oxidation Form) → NADH (Reduced Form)

---

*These two Hs are highly dissociable, which makes NAD very easily shifting from oxidation to reduction and vice versa, the very property that made NAD ⇌ NADH$_2$ a very useful tool in spectrophotometrical methods of analysis.

DEFICIENCY SYMPTOMS OF NIACIN

In man the deficiency symptoms of niacin are: (1) dermatitis; (2) diarrhea; and (3) dimensions.

Dermatitis causes a roughening of skin, and severe itching, burning, swelling, and rupturing of skin. Demensia causes hypertension in people, fear, and irritability. Untreated pellagara may lead to insanity, where it causes irreversible brain tissue and nerve damage.

Niacin has a vasodilation effect, and a large intake of it, thus, causes a flushing of the face, and warmth of skin due to the increased blood flow.

Chapter 18

PANTOTHENIC ACID

INTRODUCTION

The vitamin is found in a bound state in nature, and is present in such form in every living cell.

Gelly bees, liver, yeast, and green leafy plants are rich sources of this vitamin.

The coenzyme factor of pantothenic acid is coenzyme A, which activates acetylation reactions. Both the vitamin and the co-enzyme forms are rather complicated, but the part of the vitamin that man cannot make is the pantoic acid portion, the precursor of which is the amino acid valine. Lower microorganisms and plants can, however, synthesize this moiety from valine. Another part that is essential is the β-alanine, which is derived from asparatic acid.

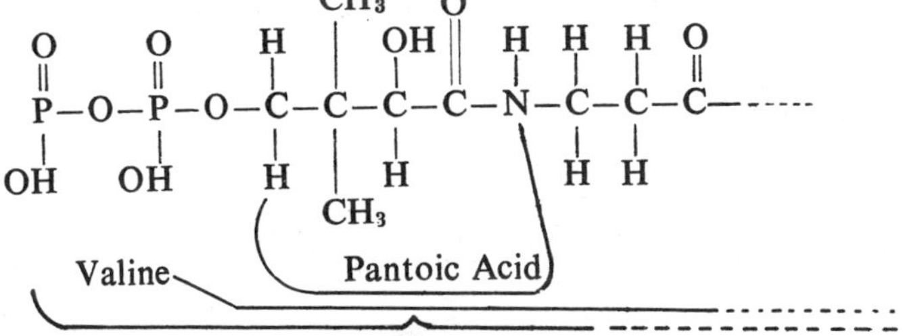

$$\underset{\text{OH}}{\overset{\text{O}}{\underset{\|}{P}}}-O-\underset{\text{OH}}{\overset{\text{O}}{\underset{\|}{P}}}-O-\underset{\text{H}}{\overset{\text{H}}{\underset{|}{C}}}-\underset{\text{H}}{\overset{\text{CH}_3}{\underset{|}{C}}}-\underset{\text{CH}_3}{\overset{\text{OH}}{\underset{|}{C}}}-\overset{\text{O}}{\underset{\|}{C}}-\underset{\text{H}}{\overset{\text{H}}{\underset{|}{N}}}-\underset{\text{H}}{\overset{\text{H}}{\underset{|}{C}}}-\underset{\text{H}}{\overset{\text{H}}{\underset{|}{C}}}-\overset{\text{O}}{\underset{\|}{C}}-\cdots$$

Valine ⎯ Pantoic Acid

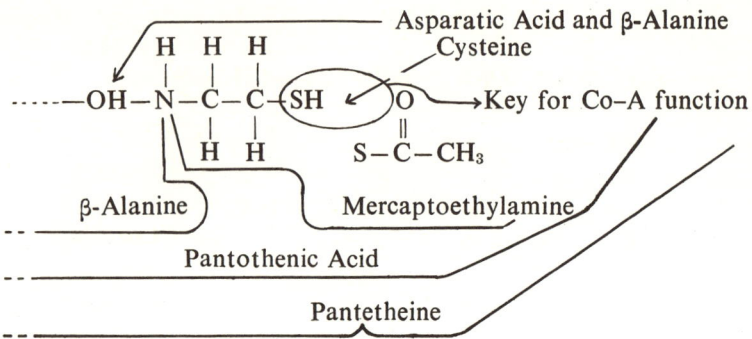

The key group in the co-enzyme reaction is

$$\text{``}S-\overset{O}{\underset{\|}{C}}-CH_3\text{''}$$

the thioester, the acetyl derivative or the activated 2-carbon unit, that is, the active acetate. The metabolism of this 2-carbon unit is the major importance of the coenzyme, since, from acetyl-Co-A, we may go into the metabolism of fatty acids, acetates, carbohydrates, protein, cholesterol, and acetylglucoseamine production.

TYPES OF REACTIONS CATALYZED BY Co-A

Since the thioester (S-$\overset{O}{\underset{\|}{C}}$-$CH_3$) is not a highly stable compound compared, for instance, to oxygen esters, it thus has a high energy of hydrolysis, and it follows that the thioester derivative of Co-A can undergo four types of reactions:

Nucleophillic attacks

The reaction of acetyl-Co-A with choline gives rise to acetyl-choline and Co-ASH.

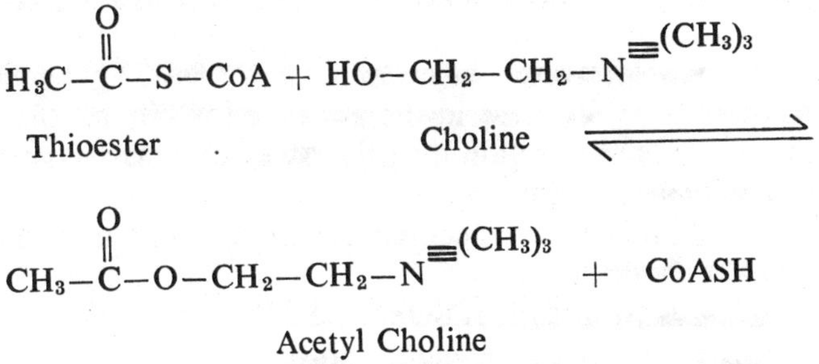

$$\underset{\text{Thioester}}{H_3C-\overset{O}{\underset{\|}{C}}-S-CoA} + \underset{\text{Choline}}{HO-CH_2-CH_2-N\equiv(CH_3)_3} \rightleftharpoons$$

$$\underset{\text{Acetyl Choline}}{CH_3-\overset{O}{\underset{\|}{C}}-O-CH_2-CH_2-N\equiv(CH_3)_3} + CoASH$$

Head-to-tail Condensation

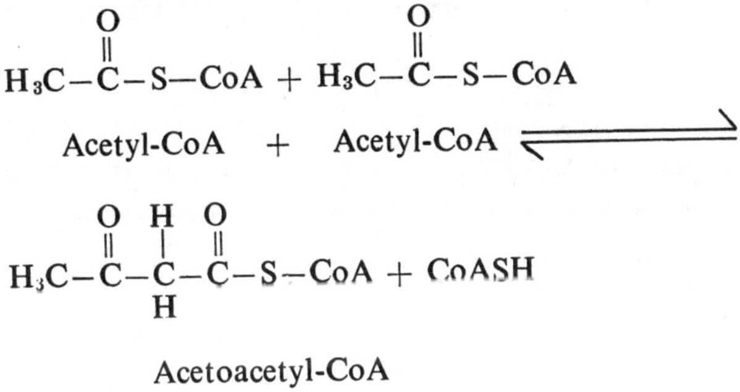

$$\underset{\text{Acetyl-CoA}}{H_3C-\overset{O}{\underset{\|}{C}}-S-CoA} + \underset{\text{Acetyl-CoA}}{H_3C-\overset{O}{\underset{\|}{C}}-S-CoA} \rightleftharpoons$$

$$\underset{\text{Acetoacetyl-CoA}}{H_3C-\overset{O}{\underset{\|}{C}}-\overset{H}{\underset{H}{C}}-\overset{O}{\underset{\|}{C}}-S-CoA} + CoASH$$

SUMMARY OF FUNCTIONS OF PANTOTHENIC ACID

1. Being the prosthetic group of Co-A, it is, thus, necessary for the conversion of pyruvate to acetyl-Co-A, a very important reaction in metabolism leading from the anaerobic glycolytic to the aerobic tricarboxylic acid cycle.

2. Acetyl-Co-A, besides its role in energy metabolism, also gives rise to synthesis of acetylcholine, which is important in nerve transmission.

3. Synthesis of malonyl-Co-A which is necessary for fat synthesis.

4. Acetyl-Co-A is also involved in steroid synthesis.

5. Folic acid and biotin are needed for the proper utilization of pantothenic acid.

DEFICIENCY SYMPTOMS OF PANTOTHENIC ACID

1. Aldosterone is affected in pantothenic acid deficiency, and there is an increase in the appetite for salt as the animal cannot adequately retain his salt, adrenal cortex hypertrophy, and depleted steroids.
2. Animals may become prostrated and may die from internal dehydration.
3. Gastrointestinal defects, as diarrhea.
4. Loss of skin pigment and hair.
5. Increase of sensitivity for insulin.
6. Burning-feet syndrome.
7. Severe dermatitis in chicks, perosis.

Most of pantothenic acid content of some foods is lost during processing. For example, processing of wheat flour results in the loss of half or more of the vitamin content. The same loss occurs upon cooking meats.

Chapter 19

BIOTIN

INTRODUCTION

Biotin is a water-soluble B vitamin. It can occur in both the free and bound states. Half the biotin occurring in feeds is biologically unavailable; however, intestinal microorganisms, including those of man, are capable of providing most of the vitamin requirements. Avians, nevertheless, require biotin in their diet, as their alimentary tract is not long enough and as the vitamin's absorption usually occurs far down in the intestines.

The importance of biotin stems from the fact that, when animals are fed raw eggs, they become biotin deficient, as the protein, avidin, in raw egg white, covalently binds with biotin and prevents its availability.

The most important role of biotin is that it transfers CO_2 from one molecule to another.

Since biotin catalyzes the reactions of volatile fatty acid production in the rumen, ruminants may, consequently, require a certain quantity of it in their diet. Man himself may require the vitamin in his diet to an appreciable quantity if large amounts of raw eggs are consumed.

STRUCTURE

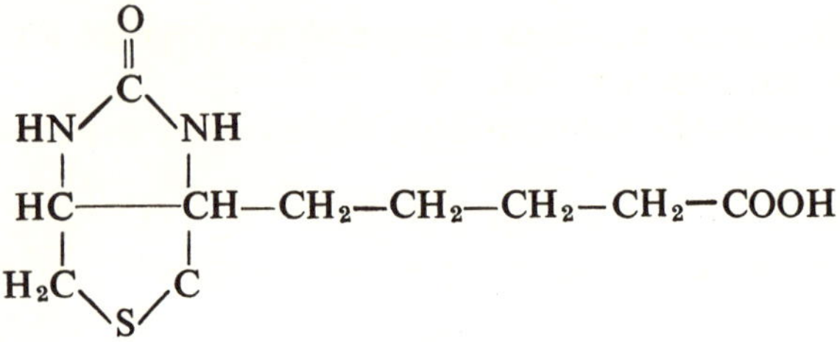

$$\text{Biotin structure: imidazolidone ring fused with tetrahydrothiophene, bearing } -(CH_2)_4-COOH$$

POSSIBLE MECHANISM OF BIOTIN'S BINDING OF CO₂

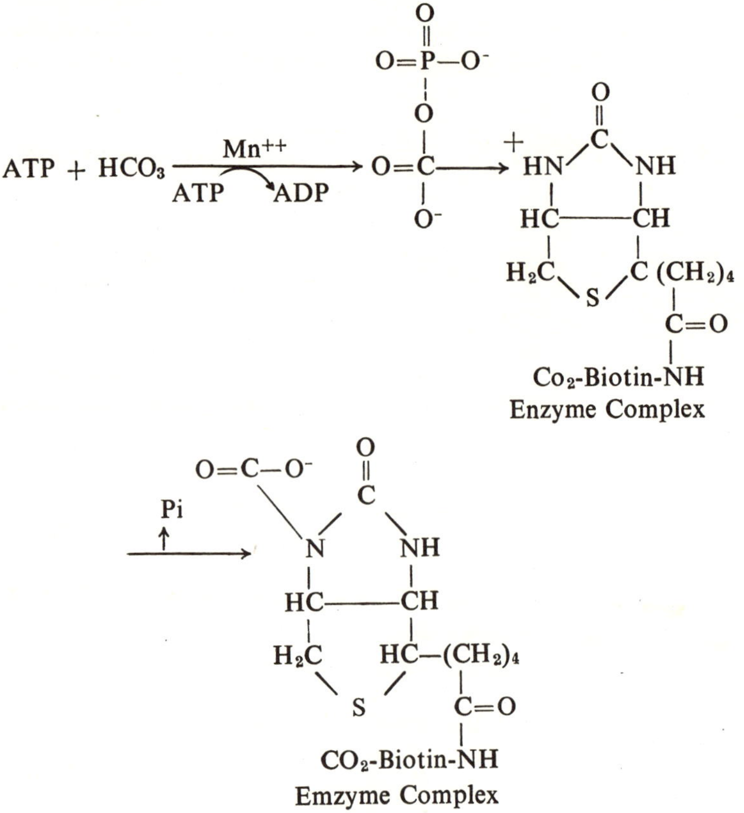

$$ATP + HCO_3 \xrightarrow[\text{ATP} \quad \text{ADP}]{Mn^{++}} \text{carbonyl phosphate} + \text{Biotin-Enzyme}$$

CO₂-Biotin-NH
Enzyme Complex

$$\xrightarrow{\uparrow Pi}$$

CO₂-Biotin-NH
Emzyme Complex

SOURCES OF BIOTIN

Yeast, liver, mollases, eggs, and many green leaf plants are good sources of biotin.

Unlike NAD, biotin is covalently bound to protein.

BIOCHEMICAL FUNCTIONS OF BIOTIN

Role in carbohydrates metabolism

Biotin activates the enzyme pyruvic carboxylase which converts pyruvic to oxaloacetic acid, providing the spark that is needed for tricarboxylic acid cycle and energy production.

$$\text{Pyruvate} + CO_2 + ATP \xrightarrow[\substack{\text{Carboxylase} \\ (\text{Biotin-Mn}^{++})}]{\text{Pyruvic}} \text{Oxaloacetic acid}$$

Role in propionate metabolism

Biotin activates the carboxylation of propionyl-Co-A to methyl malonyl-Co-A. This reaction is important in ruminants as it enables them to use propionate for energy.

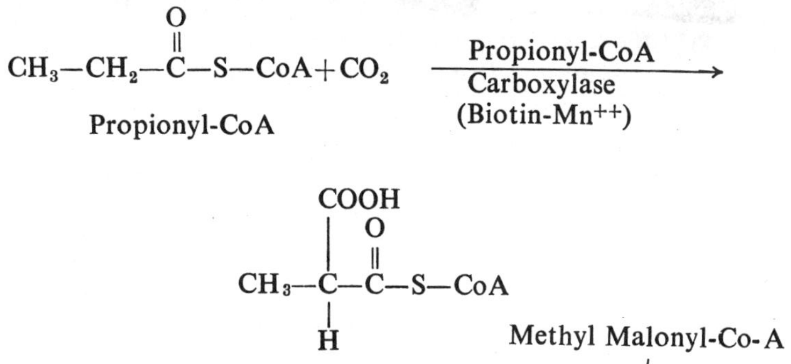

$$CH_3-CH_2-\overset{\overset{O}{\|}}{C}-S-CoA + CO_2 \xrightarrow[\substack{\text{Carboxylase} \\ (\text{Biotin-Mn}^{++})}]{\text{Propionyl-CoA}}$$

$$CH_3-\underset{H}{\overset{COOH}{\underset{|}{C}}}-\overset{O}{\overset{\|}{C}}-S-CoA \quad \text{Methyl Malonyl-Co-A}$$

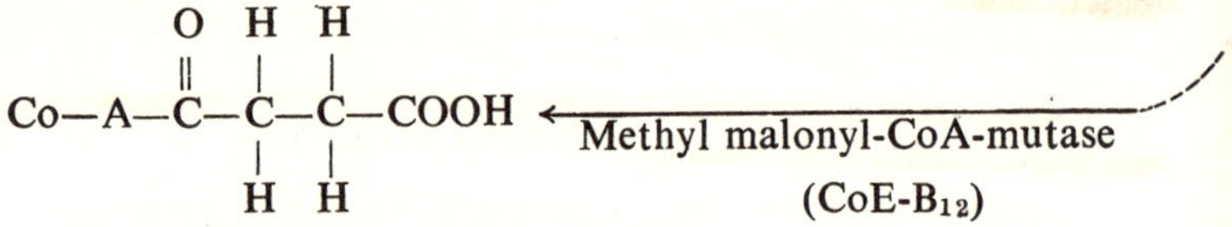

$$\text{Co-A-}\underset{\underset{H}{|}}{\overset{\overset{O}{\|}}{C}}-\underset{\underset{H}{|}}{\overset{\overset{H}{|}}{C}}-\underset{}{\overset{\overset{H}{|}}{C}}-\text{COOH} \xleftarrow{\text{Methyl malonyl-CoA-mutase}}$$
$$(\text{CoE-B}_{12})$$

To TCA Cycle ⟵ Succinyl-CoA

Role in lipid metabolism

Biotin activates the reaction of acetyl-Co-A to malonyl-Co-A, which is a rate-limiting reaction in lipid metabolism, and it is the very first reaction where lipid synthesis begins.

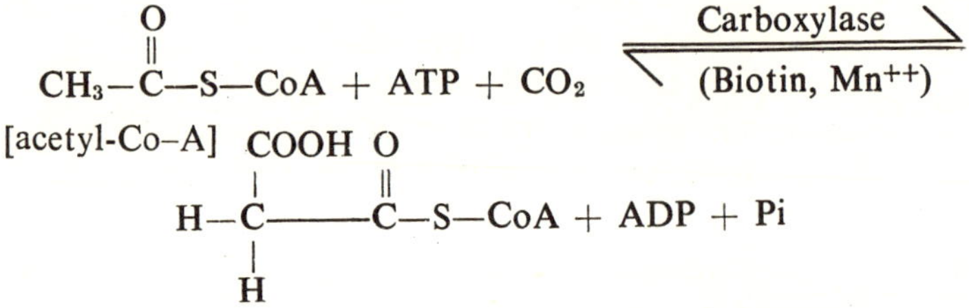

$$CH_3-\overset{\overset{O}{\|}}{C}-S-CoA + ATP + CO_2 \underset{(\text{Biotin, Mn}^{++})}{\overset{\text{Carboxylase}}{\rightleftarrows}}$$

[acetyl-Co-A]

$$H-\underset{\underset{H}{|}}{\overset{\overset{COOH}{|}}{C}}-\overset{\overset{O}{\|}}{C}-S-CoA + ADP + Pi$$

Malonyl-Co-A ⟶ to fatty acid synthesis

Role in Urea Synthesis

Biotin, here also, activates a rate-limiting reaction in urea synthesis, where the activated CO_2 attaches to pyrophosphates and ammonia-forming carbamylphosphate.

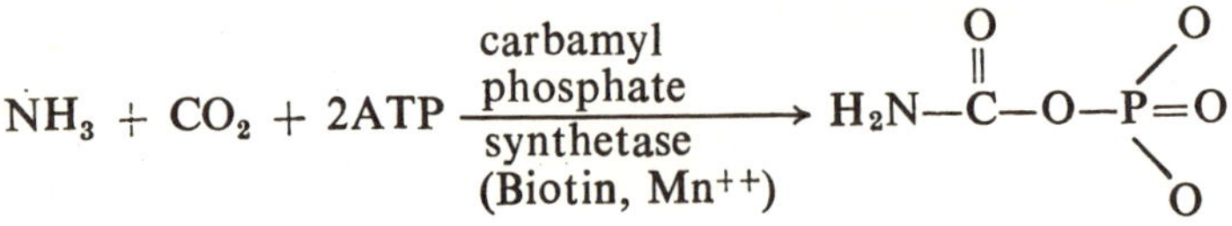

$$NH_3 + CO_2 + 2ATP \xrightarrow[\substack{\text{synthetase} \\ (\text{Biotin, Mn}^{++})}]{\text{carbamyl phosphate}} H_2N-\overset{\overset{O}{\|}}{C}-O-\underset{\underset{O}{\diagdown}}{\overset{\overset{O}{\diagup}}{P}}=O$$

Carbamyl phosphate

Role in Purine Synthesis

Biotin aids in the CO_2 placement on C-6.

$$NH_3 + CO_2 + glycine \xrightarrow[\text{(Biotin, Mn}^{++}\text{)}]{\text{carboxylation}} adenine, guanine$$

SUMMARY OF POSSIBLE BIOTIN ROLES IN BIOCHEMICAL SYSTEMS

Carbohydrate Metabolism

1. Conversion of pyruvate to oxaloacetate
2. Conversion of malate to pyruvate
3. Conversion of propionate to succinate
4. Conversion of oxalosuccinic to α-ketoglutaric acid
5. Phosphorylation of glucose to glucose-6-phosphate

Lipid Metabolism

Conversion of acetyl-Co-A to malonyl-Co-A.

Amino Acid and Nucleic Acid Metabolism

1. Arginine synthesis
2. Urea synthesis
3. Purine metabolism

Enzymes Activated With Biotin

1. Decarboxylases and carboxylases
2. Succinic acid dehydrogenase
3. Asparatic acid deaminase

4. Serine deaminase
  5. Threonine deaminase
  6. Pyruvic and α-ketoglutaric dehydrogenases

Protein Synthesis

Stimulation of RNA and DNA synthesis.

Oxidative phosphorylation

Deamination of asparatic acid, serine and threonine

DEFICIENCY SYMPTOMS OF BIOTIN:

1. Dermatitis of the skin is the most pronounced deficiency symptom of biotin. This is followed by hair loss, skin hemorrhages, and edema. Hair of rats turns grey in biotin deficiency, and there is severe muscle potassium decline.
2. Dermatitis in chicks and turkeys over their entire body.
3. The "bowlegged" condition of cowboys is due to biotin deficiency, where legs become extremely bowed.
4. Hardness of blood vessels, retardation of early sexual development, and impairment of adrenal cortex function in rats and mice.

As microorganism synthesis of biotin is a major source of the vitamin in most animals, drugs that destroy these microorganisms will produce biotin deficiency that will also lead to pantothenic acid deficiency. Biotin intake will cure both deficiencies, however.

Biotin deficiency in pigs causes loss of coordination on the hind quarter, besides hair loss.

Biotin is not toxic if given fairly excessively.

Chapter 20

PYRIDOXINE (B₆)

INTRODUCTION

B_6 occurs in complexes with protein in most foods. It may be present in foods in aldehyde, alcohol, or amine forms complexed with protein. The pyridoxal and pyridoxamine forms are mostly found in animal products, while the alcohol form, pyridoxol, is usually found in plants and seeds.

Liver, red meat, green leafy vegetables, and whole grains are fairly good sources of B_6. Intestinal microorganisms may synthesize the vitamin, but it is doubtful if these amounts are sufficient enough to satisfy man's requirement.

Phosphorylation at the C-5 position produces the coenzyme form of the vitamin. This is done in tissues and, for excretion, dephosphorylation occurs and the compound is ridded of in the acid form.

VARIOUS FORMS OF B_6

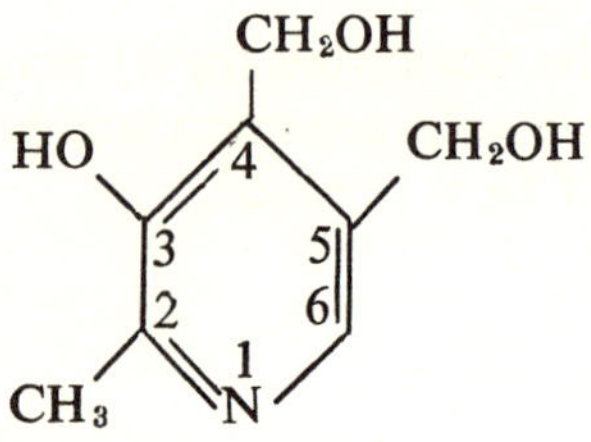

Pyridoxol
(in plants)

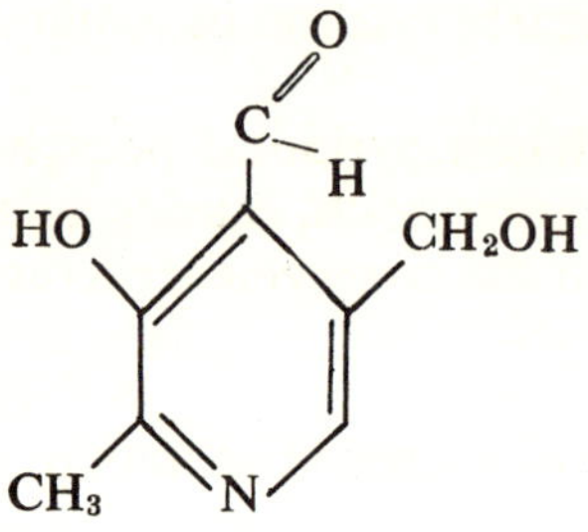

Pyridoxal

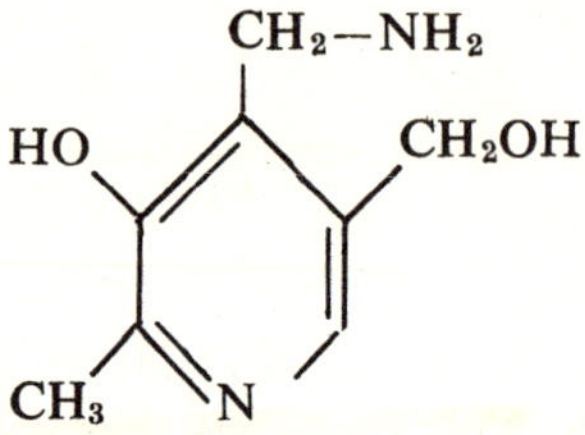

Pyridoxamine
(in animal products)

Coenzyme Form

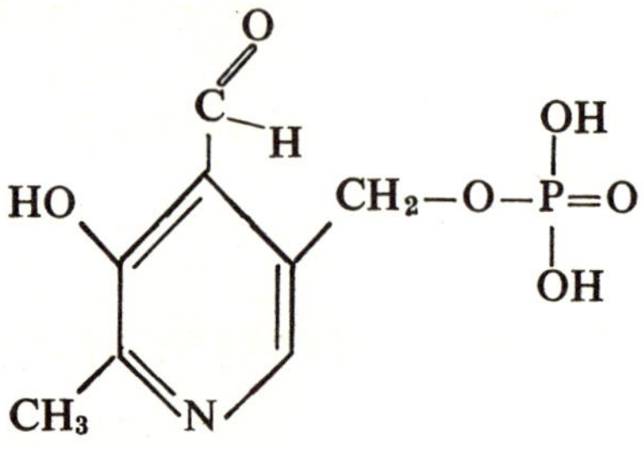

Pyridoxal-5-phosphate (coenzyme form)

ROLES OF B_6

Decarboxylation of amino acids

Where pyridoxal phosphate (the coenzyme form) reacts with amino acids, causing a water split-off, the amino acid gets tied to the coenzyme structure and CO_2 is released.

Transamination of amino acids

Pyridoxal-5-phosphate takes on the amino acid and removes the amino group, forming an α-keto acid.

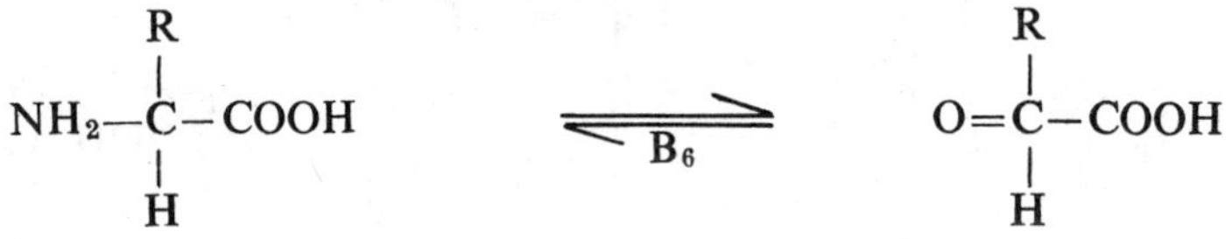

$$NH_2-\underset{H}{\overset{R}{C}}-COOH \underset{B_6}{\rightleftharpoons} O=\underset{H}{\overset{R}{C}}-COOH$$

Racemization reactions (mostly L-amino acids)

It changes the position of the amino group, creating a new isomer of the amino acids.

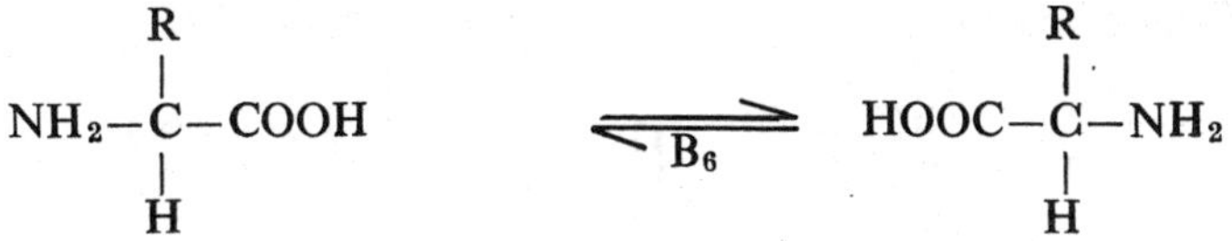

$$NH_2-\underset{H}{\overset{R}{C}}-COOH \underset{B_6}{\rightleftharpoons} HOOC-\underset{H}{\overset{R}{C}}-NH_2$$

L-form　　　　　　　　　　　D-form

POSSIBLE REACTION MECHANISMS OF B_6

1. Decarboxylation
2. Transamination

DECARBOXYLATION

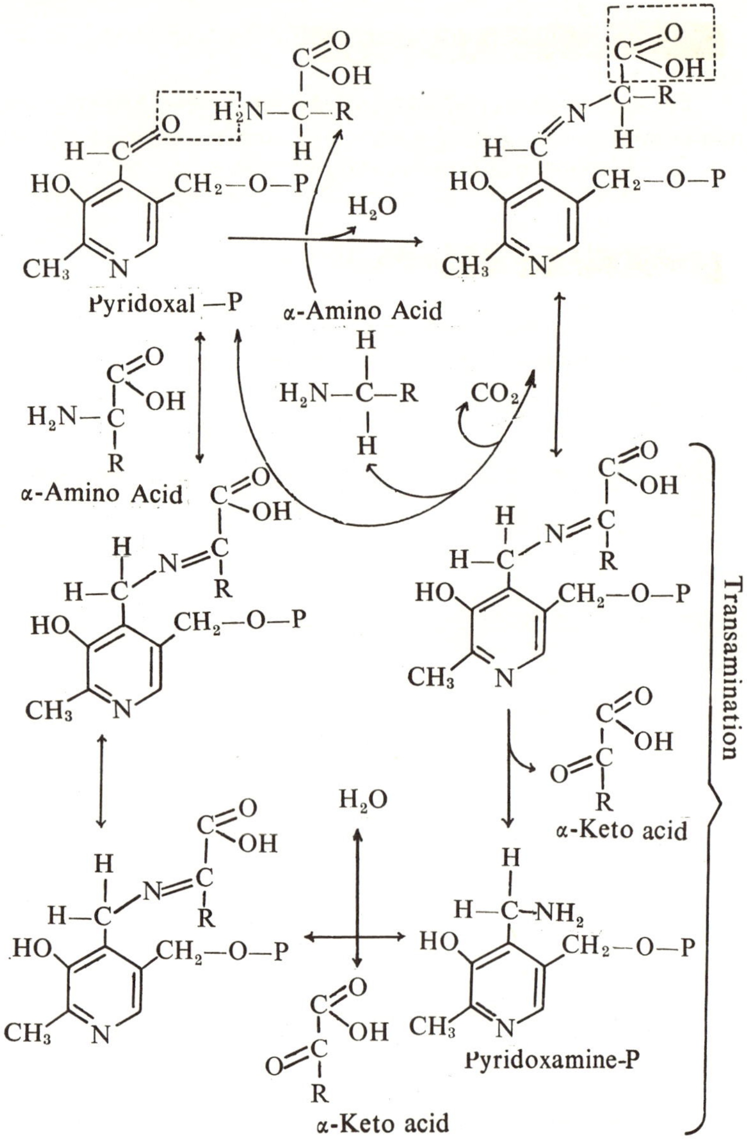

Pyridoxal —P
α-Amino Acid
H_2O
CO_2
α-Amino Acid
Transamination
α-Keto acid
H_2O
α-Keto acid
Pyridoxamine-P

3. Deamination of serine and threonine
4. Tryptophan metabolism
5. Transulfuration, where sulfur is transferred from methionine to serine
6. Desulfation, removal of sulfur from cysteine and homocysteine
7. Transfer of amino acids into cells
8. Oxidation of amines, such as histamine to imidazole acetaldehyde and NH_3
9. In synthesis of γ-aminobutyric acid:

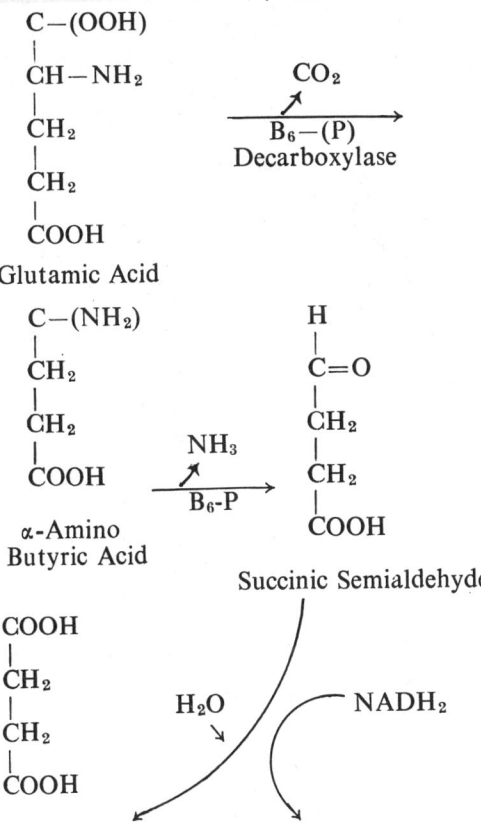

ROLE OF B_6 IN TRYPTOPHAN METABOLISM

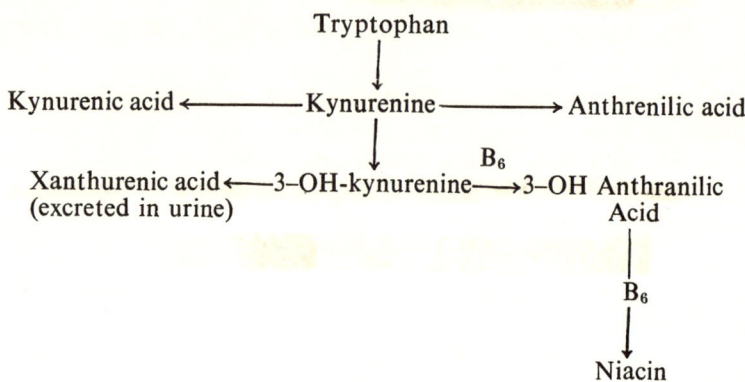

METABOLISM OF B_6

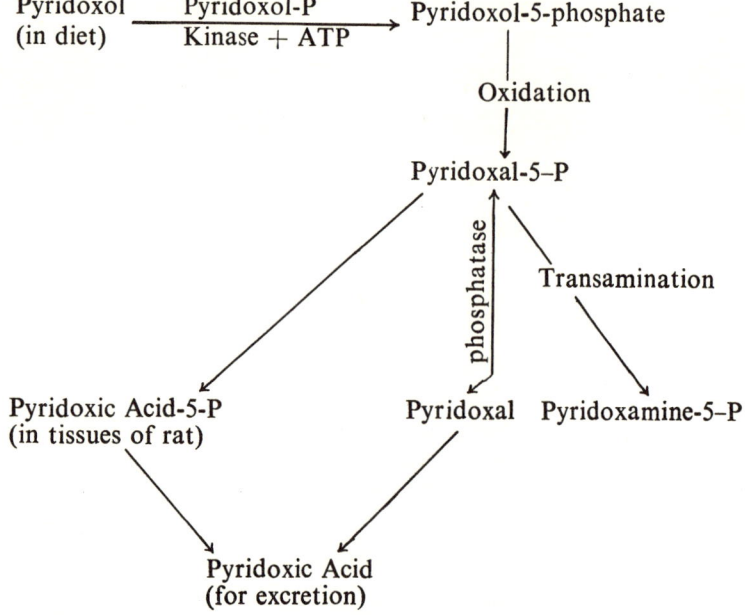

DEFICIENCY SYMPTOMS OF B_6

1. Deficiency of B_6 becomes of special importance in the case of pregnancy, as the fetus's concentration of the vitamin is accumulated at the expense of the mother, causing a deficiency in the mother.

2. Since B_6 is involved in tryptophan conversion to niacin, if B_6 is lacking, the pathway is blocked and there is an accumulation of xanthuronic acid and its excretion in the urine.

3. Intake of birth control pills by women causes a depletion of B_6 and mental depression. The daily requirement for the vitamin in the case of women is higher than in men.

4. A diet high in tryptophan may cause B_6 deficiency in the long run, as B_6 is used up in its conversion to niacin.

5. In a severe B_6 deficiency, demylination of the peripheral nervous system may occur, especially in the young. Hyperirritability and gastrointestinal disturbances also may ensue.

6. B_6 deficiency may aggravate microcytic-hypochromic anemia, and cause an increase in plasma iron, besides impairment of heme synthesis.

7. Certain people may genetically inherit the inability to utilize B_6.

Chapter 21

VITAMIN B_{12}

INTRODUCTION

B_{12} is one of the most recent vitamins chemically synthesized.

The primary source of B_{12} is microbial synthesis, especially by rumens' microorganisms, provided they have cobalt (Co) in the diet.

Meat, liver, kidney, especially of the ruminants, are usually good sources of B_{12} besides eggs, fish, and shellfish.

Cyanocobalamine is the form of B_{12} that is found in nature, the CN group of which is replaceable by OH, Cl, Br, S. etc., the matter that does not impair the vitamin's biological activity. When ribose and adenine replace the cyano group, without influencing the biological activity of the compound, the new form becomes a coenzyme active form.

B_{12} is oxidizable and reduceable by agents such as aldehydes, ascorbic acid, iron salts. This results in reduction of its activity and probably its destruction.

STRUCTURE OF VITAMIN B_{12}

B_{12} is a cyanocobalamine. It is a cobalt chelate where nitrogen is in complex with cobalt in a corrin ring, which is quite similar to the porphyrin ring.

The corrin ring is composed of four pyrrole rings joined by three triangles of single-carbon units. Cobalt, with valency of three, has octahedral geometry and there is coordination of deoxy adenosyl-cobalamide with the nucleotide, the CN forming six ligands and attached to cobalt, granting a high molecular weight to the molecule.

The high molecular weight makes B_{12} not easily absorbed, and this is why it requires the intrinsic factor for absorption in the intestinal tract. The vitamin is stored in the liver in the coenzyme form.

For the coenzyme form to be made, CN is replaced and cobalt is reduced from a valency of three to that of one. Thus, the deoxyadenosyl cobalamide is the active coenzyme form of the vitamin. Here, carbon serves as a ligand to the metal cobalt, a property that is unique in vitamin B_{12}.

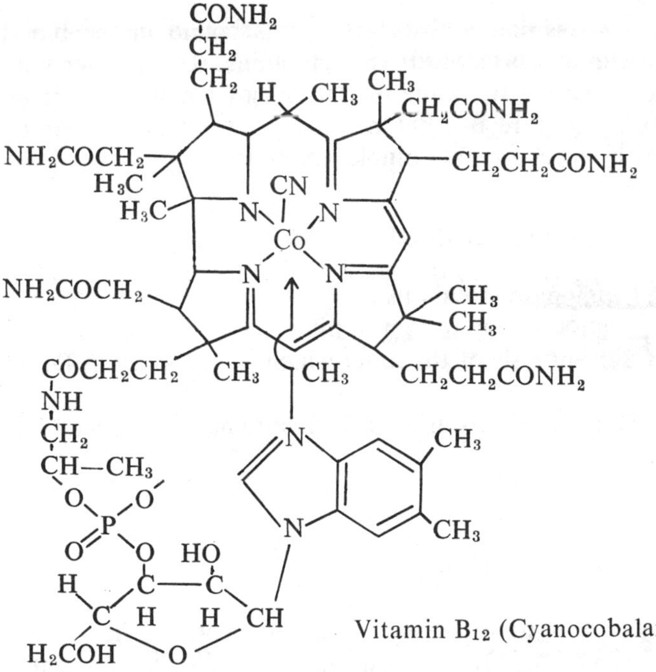

Vitamin B_{12} (Cyanocobalamine)

ABSORPTION OF B_{12}

B_{12} is absorbed from the ileum and it needs two factors for its absorption:

1. An intrinsic factor that is thought to be a glycoprotein produced in the gastrointestinal tract (GI). Recently, this intrinsic factor has been thought of as a transferase enzyme that binds B_{12} and transfers it across the intestinal walls.

2. Hydrochloric acid (HCl) which is produced by the stomach. The corticoid hormones are believed to cause induction of the above transferase enzyme. Recently, there have been reports that tend to favor the idea that there is an exocrine-pancreatic secretion that facilitates B_{12} absorption as well. Hypothyroidism may cause a decrease in B_{12} absorption and utilization, and so does malfunction in fat metabolism, as in Spruce disease.

When B_{12} is finally absorbed, it gets bound in the blood to β-globulin and afterwards to α-globulins. The former may act in transporting B_{12} while the latter may act in its storage. The Liver may store B_{12}, but not to the extent and length of time that it stores the fat-soluble vitamins.

GENERAL FUNCTIONS OF VITAMIN B_{12}

1. Propionate metabolism
2. Homocysteine to methionine
3. Biosynthesis of thymine, which is necessary in DNA synthesis
4. Methylation of uridine to thymidine, which leads to nucleic acid synthesis
5. Synthesis of ribosomal protein

COENZYME FUNCTIONS OF B_{12}

1. The biosynthesis of methyl groups is one of the important functions of B_{12} as a coenzyme in man and higher animals.

Along with folic acid, B_{12} aids in the synthesis, transfer, and in the metabolism of methyl groups.

2. A very important function of B_{12} coenzyme activity is in the ruminants, where it catalyzes the metabolism of propionate, which is used for energy by the ruminants. Here, the need is for cobalt rather than for B_{12}, as rumens' microorganisms, provided with cobalt, will manufacture B_{12} with ease, something man can do either with ease or with difficulty. In the utilization of propionate in the rumen, first, the propionate is activated to the coenzyme A derivative, then it is carboxylated to methyl malonyl-Co-A. The methyl malonyl-Co-A's conversion to succinyl-Co-A is catalyzed by methyl malonyl-Co-A mutase, which requires coenzyme B_{12}. In other words, B_{12}'s role is in activation of the mutase enzyme which produces succinyl-Co-A, a normal metabolite, in the tricarboxylic acid (TCA) cycle, as shown below:

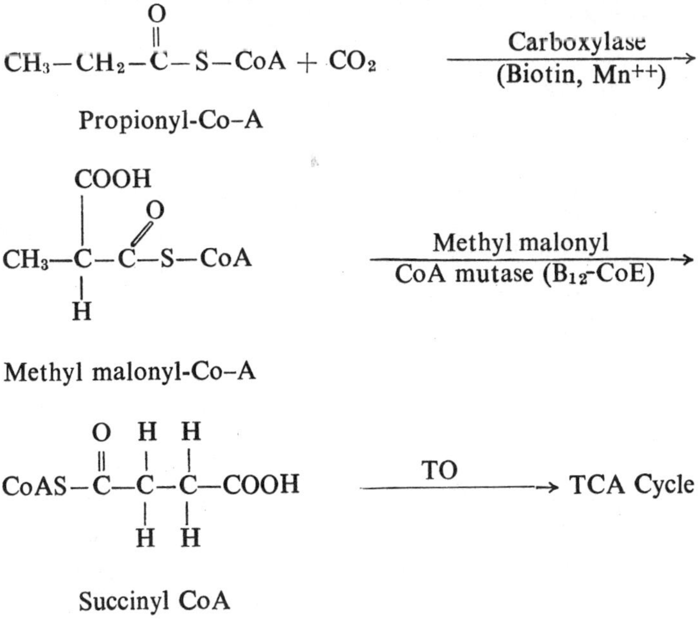

$$CH_3-CH_2-\overset{\overset{O}{\|}}{C}-S-CoA + CO_2 \xrightarrow{\text{Carboxylase} \atop (\text{Biotin, Mn}^{++})}$$

Propionyl-Co-A

$$CH_3-\underset{\underset{H}{|}}{\overset{\overset{COOH}{|}}{C}}-\overset{\overset{O}{\|}}{C}-S-CoA \xrightarrow{\text{Methyl malonyl} \atop \text{CoA mutase (B}_{12}\text{-CoE)}}$$

Methyl malonyl-Co-A

$$CoAS-\overset{\overset{O}{\|}}{C}-\underset{\underset{H}{|}}{\overset{\overset{H}{|}}{C}}-\underset{\underset{H}{|}}{\overset{\overset{H}{|}}{C}}-COOH \xrightarrow{\text{TO}} \text{TCA Cycle}$$

Succinyl CoA

A sign of B_{12} deficiency, thus, may be the presence of methyl malonic acid in the urine.

3. Coenzyme B_{12} is involved in nucleic acid metabolism, specifically in DNA metabolism, where it causes the reduction of ribose to deoxyribose at the number two carbon.

4. B_{12} has synergistic action with N5-methyltetrahydrofolic acid (THF) in amino acids metabolism. Here, THF catalyzes the carbon unit reduction to the methyl group. However, for the removal of the methyl group, B_{12} is needed. Thus, it follows that B_{12} deficiency may induce folic acid deficiency, as folic acid gets trapped in the methyl form and it would require B_{12} to transfer this methyl group to the amino acid.

An illustration of this is the transfer of the methyl group from N5-methyl-THF to homocysteine and the subsequent production of methionine from homocysteine:

$$\text{N5-methyl-THF} + \text{HS-CH}_2\text{-CH}_2\text{-}\underset{\underset{\text{NH}_2}{|}}{\overset{\overset{\text{H}}{|}}{\text{C}}}\text{-COOH}$$

Homocysteine

$$\xrightarrow[\text{S-adenosylmethionine}]{B_{12}} \text{CH}_3\text{-S-CH}_2\text{-CH}_2\text{-}\underset{\underset{\text{NH}_2}{|}}{\overset{\overset{\text{H}}{|}}{\text{C}}}\text{-COOH}$$

Methionine + THF

FACTORS AFFECTING B_{12} REQUIREMENTS

A large intake of vitamin B_{12} does not produce toxicity easily. The requirement for the vitamin is increased in certain circumstances, such as in the rapid growth period and pregnancy, and a decrease in the rate of absorption will increase the requirement for the vitamin. Thus, in old age, where there is a decrease in gastric juice production, and, consequently, the intrinsic factor and HCl are not sufficiently available to cause adequate absorption of the vitamin, the requirement here is expected to rise.

RELATIONSHIP BETWEEN B_{12} AND FOLIC ACID

B_{12} and folic acid are both closely involved in nucleoprotein synthesis. A deficiency in either one results in megaloplastic anemia. This disease is the result of the inability to synthesize DNA, as uracil is not methylated to thymidine.

B_{12} and folic acid are both required for each other's synthesis.

A B_{12} deficiency may cause the derangement of folic acid.

DEFICIENCY SYMPTOMS OF B_{12}

1. Pernicious anemia, which is common only to humans and results from a defect in DNA synthesis. The reduction of ribonucleotides to deoxyribonucleotides in this case is impaired.
2. B_{12} also is believed to play a role in cholesterol metabolism. It is thought to aid in esterifying plasma cholesterol.
3. Magaloplastic red blood cell disease, where huge red blood cells (RBC) with retained nuclei are formed. In this case, the immature RBC are put in bone in their immature form while still retaining their nuclei.
4. A large excretion of methyl malonate in the urine as its metabolism to succinyl-Co-A becomes impaired.

SUMMARY OF B_{12} DEFICIENCY SYMPTOMS

1. Pernicious anemia in man:

    (a) Methyl malonyl in urine

(b) Degeneration of spinal cord
(c) Macrocytic and multinucleated anemia

2. Poor growth in chicks and poor hatchability, perosis, edema, fatty livers, and delayed thyroid development. Reduction in hemoglobin (Hb) and red blood cells (RBC).

3. · In Swine: liver necrosis, fatty liver, reduced growth, and reduced Hb and RBC.

4. In Ruminants: demylination of nerves, poor growth, poor appetite, incoordination, and inability to metabolize propionate.

The half-life of B_{12} is pretty long; for a deficiency of this vitamin to occur requires a prolonged time.

Chapter 22

FOLIC ACID

INTRODUCTION

Folic acid is present in plants and animals. Microorganisms are capable of synthesizing folic acid.

N-5 and N-10 are about the most important groups in the folic acid molecule, as they control the one-carbon metabolism that is the function of folic acid. There may be seven glutamic acids on the molecule, but there is only one free glutamic on the absorbed molecule, since, before absorption, all the extra glutamic acids are cleaved, leaving only one. This hydrolysis probably occurs in the intestinal lumen itself.

To become metabolically active, folic acid has to be reduced by the enzyme dehydrofolic reductase at the C-5 position, as it picks up a formic acid or a methyl hydroxy group at this position or at C-10. The active metabolite after reduction can be either dihydrofolic acid or tetrahydrofolic acid.

Some workers believe that the reduction takes place at the jujenum and then the molecule gets methylated to the 5-methyltetrahydrofolic form, and that the reduction occurring before methylation is important for an effective absorption of the vitamin.

STRUCTURE AND FORMS

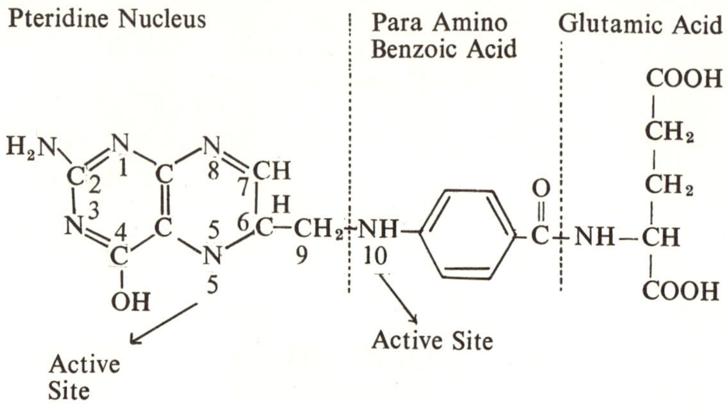

Pteridine Nucleus | **Para Amino Benzoic Acid** | **Glutamic Acid**

Folic Acid (Pteroyl glutamic acid)

Coenzyme forms

$$\text{Folic acid} \xrightarrow[\text{Dihydrofolic reductase}]{NADPH_2 \quad NADP^+} \text{Dihydrofolic acid } (FH_2)$$

$$FH_2 \xrightarrow[\text{Tetrahydrofolic reductase}]{NADPH_2 \quad NADP^+} \text{Tetrahydrofolic acid } (FH_4)$$

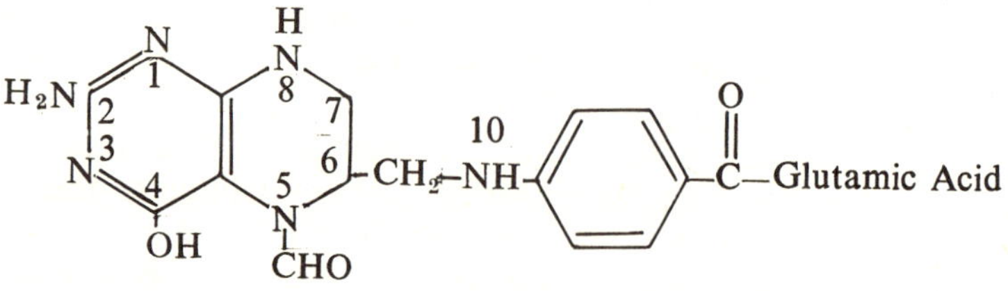

Folinic acid (active coenzyme in metabolism)

Formyl tetrahydrofolic acid isomerase
↓

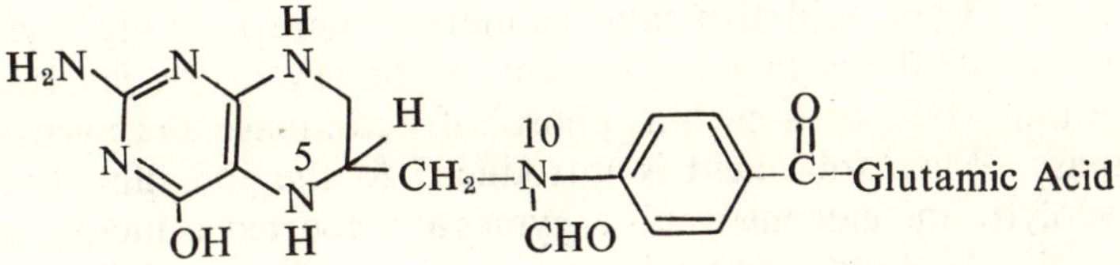

(F-10 FH$_4$) (Active coenzyme in metabolism)

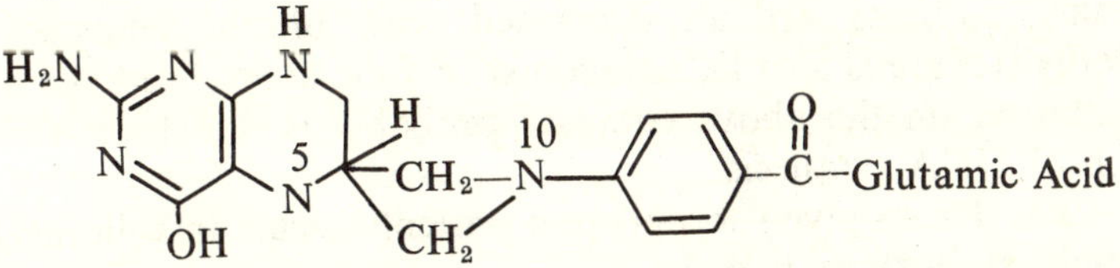

F-5-10 FH$_4$ (Active co-enzyme in metabolism)

The methyl group carried as CH$_2$OH at C-5 and -10 is oxidizable to a formyl group.

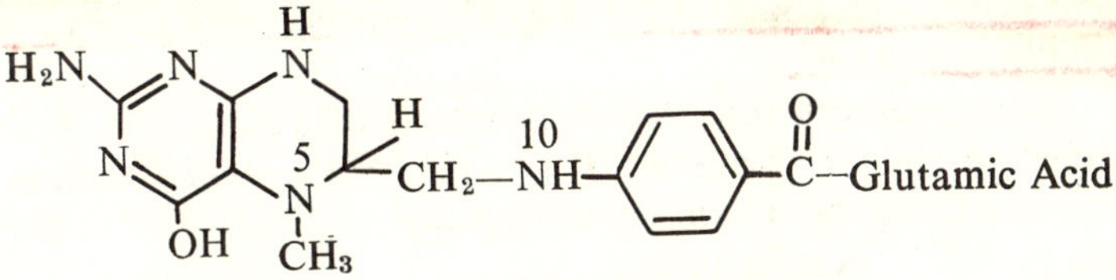

5-methyl-tetra-hydrofolic acid
(form for absorption in circulation)

FUNCTIONS OF TETRAHYDROFOLIC ACID

1. In one-carbon metabolism
2. Catabolism of histidine; in folic acid deficiency, forminoglutamic acid (FIGLU) accumulates in the urine.
3. Purine and thymine synthesis
4. Conversion of glycine to serine
5. Cystine to methionine
6. Uracil to thymine—methylation of pyrimidines
7. Folic acid also causes a marked increase in glycolytic enzyme synthesis in the jujenum, especially in the levels of the important ones such as phosphofructokinase and pyruvic kinase. When folic acid is introduced to the jujenum, both glycolytic and gluconeogenic enzymes are reported as increased.
8. Biotin deficiency has an inhibiting effect on folic acid, as it may affect the enzymes that cause synthesis of the active coenzymatic derivatives of folic acid. Transferases, formyl-tetrahydrofolic acid synthetase, and hydrofolic acid dehydrogenase enzymes are all decreased with biotin deficiency. Methyltrexate is also an antagonist of folic acid. The basis of inhibition on the above enzymes probably is due to inhibition of DNA synthesis.
9. Drugs given to epileptic patients result in folic acid deficiency in those patients.
10. When just folic acid is given, it reduces their B_{12} level.
11. B_{12} and folic acid in serum and cerebrospinal fluids are important in maintaining the mental state of health. Some researchers have reported that giving folate and B_{12} to young children made them brighter than they were, and that folic acid brings about faster desirable mental changes in the young than the old.

Other workers reported that brains of young rats were reduced in size after making the mother folic-acid deficient. They also observed a significant decrease in DNA, RNA, and brain proteins, in such cases.

Sulfanamides are antagonistic to the vitamin, as they replace the paraminobenzoic acid moiety of the molecule.

EXAMPLES OF REACTIONS CATALYZED BY THF

$$\underset{\text{Serine}}{\text{HOC}\underset{\underset{H}{|}}{\overset{\overset{H}{|}}{-}}\text{C}\underset{\underset{NH_2}{|}}{\overset{\overset{H}{|}}{-}}\text{COOH}} + \text{THF} \xrightleftharpoons[\text{methyl transferase}]{\text{Hydroxy}}$$

$$\underset{\text{glycine}}{\text{HC}\underset{\underset{NH_2}{|}}{\overset{\overset{H}{|}}{-}}\text{COOH}} + \text{THF} - \text{CH}_2\text{OH} + 5, 10\text{-methylene-THF}$$

Where a single carbon is transferred from serine to THF at the 5 and 10 nitrogens, serine can give up its methyl group to make 5, 10-methylene-THF. Reactions may go further, as follows:

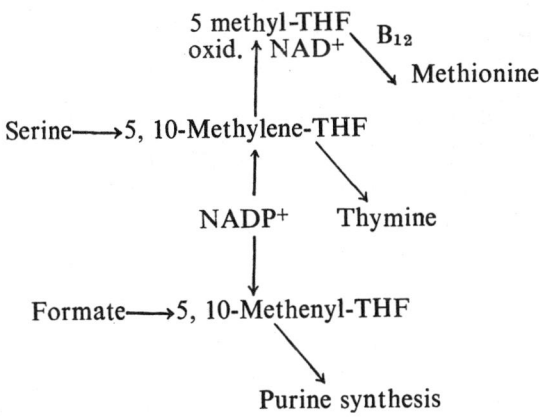

Use of histidine loading as a diagnostic indication of folic acid deficiency in man has been applied as tryptophan is used for B_6 deficiency.

METABOLISM OF HISTIDINE AND THE ROLE OF THF

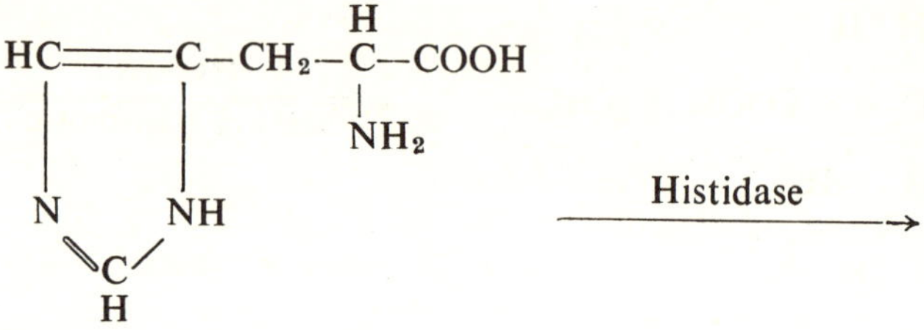

$$HC=C-CH_2-\overset{H}{\underset{NH_2}{C}}-COOH$$
(imidazole ring: N, NH, CH)

Histidine

→ Histidase →

$$HC=C-\overset{H}{C}=\overset{H}{C}-COOH$$
(imidazole ring: N, NH, CH)

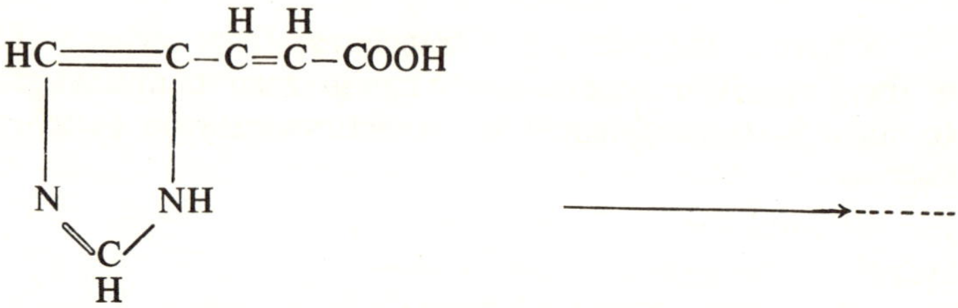

Urocanic Acid

→ - - - - -

← Urocanase —

$$O=C—C-CH_2-CH_2-COOH$$
(imidazole ring: N, NH, CH)

— Hydrolase →

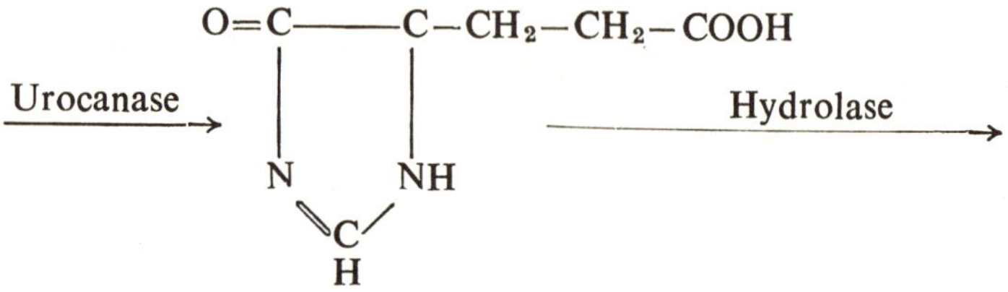

Imidozolone Propionic Acid

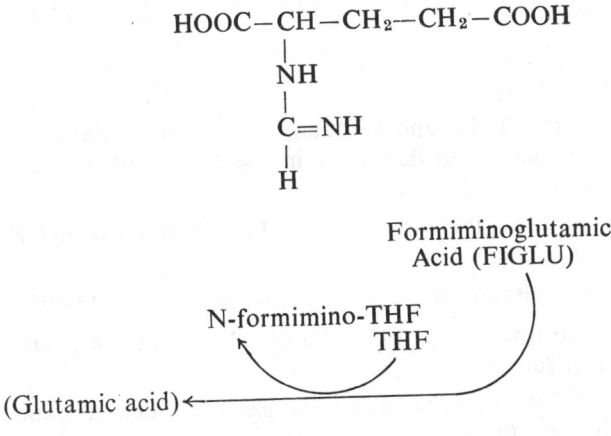

The first step of histidine metabolism is the production of urocanic acid through the action of histidase which deaminates histidine; urocanase causes oxidation of the ring and the production of imidozolone propionic acid, which is hydrolyzed by hydrolase to FIGLU that accumulates and is excreted in the urine in folic acid deficiency, as THF is required for the removal of formiminoimidazolone by its conversion to formimino-THF.

SYMPTOMS OF FOLIC ACID DEFICIENCY

1. Macrocytic anemia, where immature RBC are put in circulation and white blood cell synthesis is also affected
2. Impairment in nucleic acid synthesis
3. Gastrointestinal lesions

REFERENCES

B Vitamins

Chick, H.; Macrae, T. F.; and Worden, A. N. 1940. Relation of skin lesion in the rat to deficiency in the diet of different B_2-vitamins. *Biochem. J.* 34:580.

Elvehjem, C. A. 1944. Present status of the vitamin B complex. *Am. Scientist* 32:25.

———. 1951 The vitamin B complex. *J. Am. Med. Assoc.* 134:960.

Emerson, G., and Folkers, K. 1951. Water-soluble vitamins. *Ann. Rev. Biochem.* 20:559.

Harris, L. J. 1955. *Vitamins in theory and practice.* London: Cambridge University Press.

Harris, R. S., and Thimann, K. V. 1948. *Vitamins and Hormones.* 6 vols. New York: Academic Press.

King, C. G. 1939. The water-soluble vitamins. *Ann. Rev. Biochem.* 8:371.

Lepkovsky, S. 1940. The water-soluble vitamins. *Ann. Rev. Biochem.* 9:383.

Novak, A. F., and Hauge, S. M. 1948. Unidentified growth factor (vitamin B_{13}). *J. Biol. Chem.* 174:235.

Pilgrim, F. J.; Axelrod, A. E.; and Elvehjem, C. A. 1942. The metabolism of pyruvate by liver from pantothenic acid- and biotin-deficient rats. *J. Biol. Chem.* 145:237.

Review 1944. Relationship of choline, betaine, and other compounds in nutrition. *Nutr. Rev.* 2:358.

Rosenberg, H. R. 1945. *Chemistry and physiology of the vitamins.* New York: Interscience.

Sebrell, W. H., and Harris, R. S. 1945. *The vitamins.* Vols. I–III. New York: Academic Press.

Snell, E. E., and Wright, L. D. 1950. Water-soluble vitamins. *Ann. Rev. Biochem.* 19:277.

Stokstad, E. L. R. 1949. The miltiple nature of the animal protein factor. *J. Biol. Chem.* 180:647.

———, and Jukes, T. H. 1949. Water-soluble vitamins. *Ann. Rev. Biochem.* 19:435.

Teply, L. J.; Strong, F. M.; and Elvehjem, C. A. 1942. Nicotinic acid, pantothenic acid and pyridoxine in wheat and wheat products. *J. Nutr.* 24:167.

Voris, L., and Moore, H. P. 1943. Thiamin, riboflavin, pyridoxine, and pantothenic deficiencies as affecting the bodily composition of the albino rat. *J. Nutr.* 25:7.

Williams, R. J. 1942. The approximate vitamin requirement of the human beings. *J. Am. Med. Assoc.* 119:1.

———. 1950. *The biochemistry of B-vitamins.* New York: Reinhold Publishing Co.

———; Eaken, R. E.; Beerstecher, E., Jr.; and Shaive, W. 1950. *The biochemistry of B-vitamins.* Am. Chem. Soc. Monograph No. 110. New York: Reinhold Publishing Co.

Winters, J. C., and Leslie, R.E. 1943. A study of the diet-nutritional status of women in a low income population group. *J. Nutr.* 26:443.

Part III

THE CHEMISTRY OF MINERALS

Chapter 23

MINERALS

INTRODUCTION

Enzymes require a cofactor to make them active, and this can be either a coenzyme or trace element. Thus, the essential elements are required in the diet of animals and for the function of enzymes.

A man of the typical weight of 70 Kg yields 3000 grams of ash, which is 5% on wet-basis weight (depending on temperature of ashing, some carbonate and some oxalate are produced). Calcium constitutes 39% of the ash, and phosphorus 22%. Calcium phosphate is dominant in the skeleton. Potassium is next in order in ash content, composing 5%. Sodium is 2%, chlorine 3%, sulfur 4%, and magnesium is 0.7% of the total ash.

When expressed on a milliequivalent basis, the cations must equal the anions.

MINERALS AND THE BIOLOGICAL SYSTEM

Calcium and phosphorus are important for structure (mainly in bone); magnesium for intra- and extracellular fluids; sodium and chlorine in the extracellular fluid.

The fact that the trace elements are first transitional grants them most of their significance and importance in the biological system. Most the trace elements are transitional with the third electron shell not filled.

Zn^{2+}, in this regard, is not a strictly transitional element as all its orbits are filled. Mn^{2+}, having five unpaired electrons, gives it unique magnetic properties. Some of the uniqueness of the transitional elements is that they may have varying valencies or states of oxidation in losing one or more electrons. Copper, for instance, may be in the cuprous or cupric state, and iron in the ferrous or ferric state.

The transitional elements are also unique in their very effective characteristic of complexing with electron-rich components like proteins. Since the transitional elements are electron deficient in the third orbital, it follows that they have a great affinity for elements that have excess electrons, such as N, O, S, etc., and since proteins and amino acids are rich in O, N, S, etc., thus, proteins do form complexes with the transitional elements. A group that complexes with these elements can have positive or negative charges, and they are called "ligands." CN, for example, is a very strong ligand, as it has N which complexes with iron or other transitional elements.

Fe has a coordination number of six, i.e., it coordinates with six groups, forming an octahedral complex.

$$\begin{array}{c} CN \\ CN \longrightarrow | \longrightarrow CN \\ Fe^{3+} \\ CN \longrightarrow | \longrightarrow CN \\ CN \end{array}$$

Copper has a coordination number of four, and makes a good complex with ammonia (neutral ligand).

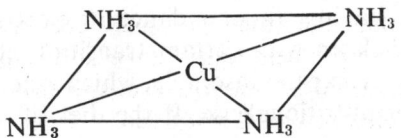

Since manganese has five unpaired electrons in the third orbital, and although it has unique properties of magnetism, still it does not form such strong complexes with electron-rich compounds.

From iron on in the transitional elements table, trace elements generally form strong ligands, especially copper. Some of these elements directly form complexes with proteins, which at times are very hard to dissociate. Normally, strong acids are used to displace the elements from their complexes with the protein or whatever it may be, although this is not true of vitamin B_{12} which is in a very strong complex formation with cobalt.

The best amino acid for complexing with the transitional elements is histidine, which, upon complexing with Cu or Fe, facilitates their absorption.

There are many chelating agents that form a ring-like structure with the transitional element. Some of these formations are strong and do not dissociate readily. Ethylene diamine tetraacetic acid (EDTA) is an example of these chelating agents.

EDTA

Phytates, having an abundance of electron-rich atoms, form strong chelates with certain transition elements. These complexes are strong and insoluble, which renders the element unavailable for nutritional use. If the diet of the young contains huge quantities of vegetable proteins and cereal grains which contain large amounts of phytates, these phytates in the diet may form insoluble complexes, for instance with zinc, and, consequently, render it unavailable for the young's nutrition. This specific matter, that is, the unavailability of zinc for the young, may result in impairment of gonadal development, hair follicules, and, latter on, pubic hair appearance, the properties that availability of Zn in the young, to a good extent, dictates. If, in such an instance, EDTA is given, since EDTA forms a complex with zinc that is soluble (unlike phytates), the EDTA may replace the phytates and the solubility of the EDTA–Zn complex will render zinc available for the young.

Any chelating agent that forms a ring-like structure with other compounds is called a chelate. In this respect, metalloenzymes, hemoglobin, chlorophyll, and vitamin B_{12} are all chelates.

Phytic-Zn complex

Zn^{++} gets bound between the oxygens to satisfy the negative charge and prevent it from being absorbed by the gastrointestinal (GI) tract. There is an abundance of phytic acid in nature. Half of the phosphorus in grains is in the nonsoluble phytate form.

THE ESSENTIAL MINERALS

- Calcium ⎫
- Phosphorus ⎪
- Sodium ⎬ macroelements
- Potassium ⎪
- Chlorine ⎭
- Magnesium } borderline
- Manganese ⎫
- Zinc ⎪
- Iron ⎪
- Copper ⎪
- Cobalt ⎪
- Iodine ⎬ microelements
- Fluorine ⎪
- Selenium ⎪
- Molybdenum ⎪
- Chromium ⎪
- Nickel ⎭

CALCIUM (Ca)

Sources of Calcium

1. Milk
2. Limestone
3. Calcium phosphate

The above are good sources of calcium. Any food source of phosphorus is also a source of calcium as well. Grains and cereals are low in calcium.

Roles of Calcium

1. For bone appetite—formation
2. Affects mitosis
3. Regulates cellular polyferation
4. ATP—hydrolysis requires both Mg^{2+} and Ca^{2+}

5. Inside endothelial blood cells separating blood from bone
6. Plays a role in bone membrane that prevents its solubilization by lactic acid

Deficiency Symptoms of Calcium

1. Increase in basal metabolic rate
2. Decrease in general activity
3. Osteoporesis
4. Tetny
5. Increase in urine volume
6. Increase in tendency of internal hemorrhage

Toxicity Symptoms of Calcium

1. Nephrosis
2. Blood Ca increased, PO_4 decreased
3. May cause deficiency in Mg, Mn, Zn, I, and Cu

Requirement

1% of grown-up diet, on dry weight basis. Toxic level in children is 3% of dry weight of diet, depending on vitamin D, calcitonin, and parathyroid hormone levels.

Human Daily Calcium Requirements

Age		mg Ca/day
0–2	months	400
2–6	″	500
6–12	″	600
1–2	years	700

Age			mg Ca/day
2–6	years		800
6–8	”		900
8–10	”		1,000
10–12	”		1,000
12–18	”		1,200
12–18	”	(females)	1,300
12–18	”	(males)	1,400
18+			800
Pregnancy			1,200
Lactation			1,300

Factors Affecting Calcium Requirement

1. Level of parathyroid hormone. It increases blood calcium and lowers the phosphate level.
2. Calcitonin: which does the opposite from parathyroid hormone, i.e., it decreases blood calcium (takes it to bone).
3. Vitamin D, which increases calcium absorption.
4. Acid foods cause loss of calcium, especially from kidney, as the acid solubilizes it.
5. Level of Ca in diet.
6. Level of protein in diet (calcium is transported protein-bound in blood).
7. Foods that contain Ca chelators, like spinach, with its oxalates that bind Ca.
8. Fluoride may form a complex with Ca, as in bone.
9. Phosphorus complexes with Ca in bone appetite.
10. Mg also complexes with calcium in bone, and in tissues it displaces Ca.
11. Age: as we get older, Ca is depleted from the bones. This starts at about 45 years in women and at 55 years of age in men.
12. Stress causes depletion of calcium.

PHOSPHORUS (P)

Sources of PO_4

Dicalcium phosphate salts, fish meal, bone meal, calcium acid phosphate salts. Many rock phosphates may fit as phosphorus sources for animals, if only their fluorine content is removed. Meats and legumes are high in phosphorus, except that the legume's content is only half available; the other half is in the nonavailable, nonsoluble phytate form.

Biochemical Functions of Phosphorus

1. Phosphorus is a component of the energy compounds of metabolism such as ATP, ADP, AMP, and creatine phosphate.
2. Component of phospholipids.
3. In enzymes.
4. In phosphorylation.

Toxicity Symptoms of Phosphorus

Upset mineral balance, especially of calcium. 0.5% phosphorus in diet may be toxic, depending on calcium.

SODIUM (Na^+)

Biochemical Functions

1. Maintenance of acid-base balance in body.
2. In membrane potential (Na^+ is the extracellular cation).
3. Cation balance for many soluble anions.
4. Influences aldosterone's activity.

Deficiency Symptoms of Na^+

1. Retardation of growth.
2. Deprived appetite.
3. Hypertension, canabolism in some animals (chickens).
4. Gonadal development impairment at early stages.
5. Problems with eyes, bones, and adrenal plasma fluid volume.

Toxicity Symptoms of Na^+

Edema and hypertension. 1.5% of diet may prove a toxic level.

POTASSIUM (K^+)

Potassium is the intracellular cation. It is present in every cell, and it is quite important in the cell membrane potential. Although not many, some enzymes are activated by potassium.

Toxicity Symptoms of K^+

As the requirement of K^+ in man is about 2000 ppm, this may go to as many as 4000 ppm without toxicity. However, 2% K^+ in the diet depresses growth. Toxicity of potassium may cause heart failure, as its accumulation on the outside of cell membrane will poison and kill the membrane potential.

MAGNESIUM (Mg^{2+})

Mg^{2+} is a borderline between the trace and macroelements. It is an important constituent of the skeleton and is found in bone in high concentration. It also occurs in soft tissue and fluids in higher concentrations than most trace minerals, but in lesser concentration when compared to Ca^{++}, $P^=$, or Na^+.

Deficiency Symptoms of Mg^{2+}

1. Grass tetny or hypomagnesemia where the blood level of Mg^{2+} goes down, usually in cattle.
2. Magnesium deficiency in rats causes convulsion or tetny.
3. The most common deficiency symptom of Mg^{++} in animals is metastatism or soft tissue calcification, particularily the kidney tissues. This is because Ca^{++} replaces the absent magnesium in the tissue.

Mg^{++} is thought to be related to fluoride in that, in a magnesium-deficient diet, if fluoride is added, it may relieve the Mg^{++} deficiency and prevent tissue calcification.

Symptoms of Mg^{++} Deficiency

1. Hyperirritability of skin is a common symptom of Mg^{++} deficiency, as calcium occupies Mg^{2+} spaces, thus preventing Na^+ and K^+ movement from outside and inside the cell.
2. Convulsions and irritability.

Role in Enzyme Activation

Magnesium is an activator of numerous enzymes. It activates practically all the kinase enzymes which transfer phosphate groups from one component to another. Mg^{2+} also activates enolase, an enzyme that is inhibited by fluoride.

In Mg^{++} deficiency, the $Na^+ - K^+$ pump becomes upset. The ATP-ase enzymes also, in part, are Mg^{++} dependent, and in part, depend on the $Na^+ - K^+$ ratio, both events which become upset in Mg-deficiency.

Summary of Functions of Mg^{++}

1. Enzyme activator.
2. Structural element in bones (70% of Mg^{2+} in body).
3. In hypernation, serum Mg^{++} goes up.
4. Ca–P–Mg relationships, in bone formation.
5. Depression of central nervous system (CNS), general anaesthesia.
6. Causes decrease in acetyl choline release.
7. Causes a decrease in blood pressure.

The requirement for Mg^{++} is 600 ppm of diet, and the toxic level might be 5000 ppm.

THE TRACE ELEMENTS

Study of trace element metabolism was the tool for predicting their deficiency or toxicity in the past. One trace element, for example, will be left out, and symptoms of observed deficiency are related to it, or one would be added in excessive amounts and toxicity symptoms related to it. This practice is no longer in use as there are interreactions between the various elements. For example, excess iron may induce anemia simply because iron replaces copper. On the same basis, a high level of manganese is antagonistic to iron. High zinc levels may also cause anemia as zinc replaces copper. Selenium toxicity may be relieved by adding arsenic which replaces it.

Even low levels of cadmium are toxic, but this may be overcome by adding zinc since it displaces Cd. Molybdate and tungstate, as well, displace each other, and a relief of either may be achieved by adding the other. Cd and Cu are antagonists to each other, and so is Cu to Ag; and vanadium to chromium.

The similarity between trace elements make many of them act as antagonists to each other. However, in the biological system, this may not be so critical, as specificity of the enzymes is remarkable for having only one metal activator.

IRON (Fe)

Functions

The adult human contains about 4.3 grams of iron, most of which is in the hemoglobin of blood (90% or more). The human body absorbs only whatever it needs of the iron intake and the excess leaks out through the skin, bile, and intestines.

Iron is usually transported through the mucosal walls of the intestines in the ferrous form (Fe^{2+}).

Requirements

About 40 ppm in avians
About 80 ppm in swine
About 25-40 ppm in ruminants

Women can tolerate up to 18 mg/day iron.

Sources of Iron

Blood, liver, and eggs are fairly good sources.

Absorption of Iron

In the mucosal cell walls of the intestines, there is abundance of a protein-bound form of iron, "the apoferretin."

Apoferretin picks up Fe^{3+} that becomes reduced to the ferrous state (Fe^{2+}), which then is transported through the mucosal cells, to plasma where another oxidation process occurs, i.e., to the "transferretin" form which goes to the liver and is stored in this ferretin form.

Biochemical Functions of Iron

It is a component of the hemoglobin, myoglobin, cytochromes, and catalase, and peroxidase enzymes. 70% of the total iron in the body is in hemoglobin, 3% in myoglobin, and rest is in ferretin form, plus 1% in the oxidase enzymes.

The fact that the amount of apoferretin present in the mucosal intestinal cells is large makes iron absorption proceed quickly. An animal may absorb up to 60% of its iron intake if his need requires it.

Iron-Copper Relationship

In copper deficiency, there is a reduction in the red blood cell (RBC) level, as if copper was involved in RBC. Actually, if the animal is extremely copper deficient, his absorption of iron will consequently be impaired. An extreme copper deficiency causes extreme iron deficiency.

COPPER (Cu)

Functions

Cu is involved in tissue metabolism. It causes the cross-linking of the connective tissue proteins, elastin, and collagen, and the maintenance of the integrity of these cross-linkages.

Cu^{2+} activates the enzymes amine-oxidase (involved in synthesis of elastin), cytochrome-oxidase (along with Fe^{2+}), and uricase.

The amine-oxidases cause the oxidative deamination of amines in proteins, elastin, or collagen, using molecular O_2 and H_2O and giving rise to aldehyde, H_2O_2, and NH_3. These enzymes are present in plasma and connective tissue and are metallo-copper-dependent enzymes.

Sources of Cu

Cu sources in foods range from 0.5 to 400 ppm. Organ meat, shellfish, nuts, legumes, and fruits are good food sources of Cu.

The requirement for Cu in man is 4–10 ppm. The toxic level may be 100 times the requirement.

Summary of Functions of Cu

1. Activator of enzymes, such as cytochrome-oxidase.
2. In hemoglobin formation.
3. Regulation of iron metabolism.
4. In respiration.
5. Pigmentation and keratenization of wool.
6. Bone formation.
7. Myelination of spinal cord.
8. Maintenance of red blood cell lifespan.
9. Prevents aorotic rupture (important in turkeys).
10. In ceruloplasm (plasma carrier).
11. In phospholipid synthesis.

Deficiency Symptoms of Cu

1. Swayback in sheep.
2. Bone fracture.
3. Anemia, hypochromic, micro-cytic.
4. Steely wool in sheep, and grey hair in black animals.

Toxicity Symptoms of Cu

1. Leads to molybdenum deficiency.
2. Hemolytic jaundice.

ZINC (Zn^{++})

Requirement

10–50 ppm of dry-diet weight in man.
50 ppm in avians.
Toxic level is 50 times the requirement.

Sources

Sea foods, and animal by-products with bone. Milk is borderline, and so is red meat of chicken. Bone may provide some zinc in storage. Estrogen affects zinc status; and human semen may be as high as 2000 ppm Zn. Testosterone aids in accumulation of Zn in the prostate gland. There is up to 13.8% Zn in the irridescent layer of the eye. Leucocytes are high in zinc. Excretion of Zn is in feces, and some Zn is lost in hair.

Biochemical Functions of Zinc

Zinc, as well as most the transitional trace elements, is an activator cofactor in specific enzymes. Carbonic anhydrase is a zinc-metalloenzyme, where the metal is tenaciously bound to the protein. This enzyme is important in respiration and in acid-base balance, as it catalyzes the production and breakdown of carbonic acid.

$$CO_2 + H_2O \underset{(Zn^{2+})}{\overset{\text{carbonic anhydrase}}{\rightleftharpoons}} H_2CO_3$$

It follows, of course, that zinc deficiency may result in respiration problems. Another metalloenzyme of zinc is carboxypeptidase, which cleaves amino acids at the carboxyl end. Even a more prominent metalloenzyme of zinc is alcohol dehydrogenase, which catalyzes the reversible oxidation reduction of alcohol to aldehyde.

$$R-CH_2-OH + NAD+ \underset{(Zn^{++})}{\overset{\text{alcohol dehydrogenase}}{\rightleftharpoons}} R-\overset{H}{\underset{|}{C}}=O + NADH_2$$

In the metalloenzymes, as mentioned earlier, the metal is tenaciously bound to protein and helps maintain the conformational integrity of the protein. If the metal, in such cases, is removed from the enzyme, the latter will be irreversibly damaged. Alkaline phosphatase and glutamic dehydrogenase also require Zn.

As zinc concentration is increased in the diet, the need for copper increases, since it is replaced by zinc.

Deficiency Symptoms of Zn

1. Parakeratosis (very rough skin).
2. Short, thick bones.
3. Embryonic abnormalities.
4. Testicular degeneration and lack of pubic hair.
5. Failure of muscle relaxation.

6. Cadmium toxicity increases zinc deficiency; so will Cu and Co, as they all replace Zn.

Toxicity Symptoms of Zn^{++}

1. Depressed growth.
2. Hypochromic, microcytic anemia (as it displaces Cu, Zn–Cu–Fe interaction).
3. Defective bone mineralization.

MANGANESE (Mn^{2+})

Functions

Manganese's importance in nutrition came to light only during the past 30 years. A Mn^{++} deficiency results in failure of reproduction in lower animals like mice, rats, and avians, in both the male and female. In the avian species, besides failure in reproduction, Mn^{++} deficiency will also cause infertility of eggs, long-bone-shortening "chondrodystrophy," and perosis or slipped tendon.

Mn-deficiency causes failure of otolith formation, where calcification of the organic matrix and the making of the mucopolysaccharide ground substance of the inner ear becomes defective.

Mn-deficiency may prevent the conversion of glucose to galactose or glucoseamine to galactoseamine in the overall production of mucopolysaccharides.

Biological Roles of Mn^{++}

Of great importance, is Mn^{2+}'s role as a cofactor of certain enzymes. The first clear evidence of Mn^{2}'s being an integral part of a metalloenzyme was observed on pyruvate carboxylase. This metalloenzyme catalyzes the carboxylation reaction of pyruvate to oxaloacetic acid (a very important reaction in metabolism).

$$\text{Pyruvate} + \text{HCO}_3 + \text{ATP} \xleftrightarrow[\substack{\text{Carboxylase} \\ \left(\substack{4\ \text{Biotins} \\ 4\ \text{Mn}^{2+}}\right)}]{\text{Pyruvate}} \text{Oxaloacetic acid}$$

The pyruvate carboxylase is activated by two cofactors: one is biotin, and it also contains one gram atom of Mn^{2+} per one gram atom of biotin. Mn^{2+} in this case is very tightly bound to the enzyme, and dialysis will not cause their separation.

Mn^{2+} also activates the enzyme arginase, only as a stimulant and not as an integral and tight-bound part of the enzyme.

MOLYBDENUM (Mo)

Functions

Mo falls in the second transitional series and has only recently been recognized as an essential element. It is needed in such very minute amounts that it is more often thought of as a toxic mineral than as essential.

There is a Mo–Cu interaction, where Mo is an antagonist to Cu and vice versa. If a pasture is high in Mo to the level of toxicity, $CuSO_4$ is provided where, in this case, the SO_4 mobilizes or moves it faster, and where Mo binds Cu and makes it unavailable in the liver. This becomes important in the case of ruminants where SO_4 is usually fed to move Mo out of the liver.

Xanthine oxidase is a metalloenzyme of Mo. It converts xanthine to uric acid and it requires FAD as another cofactor besides Mo.

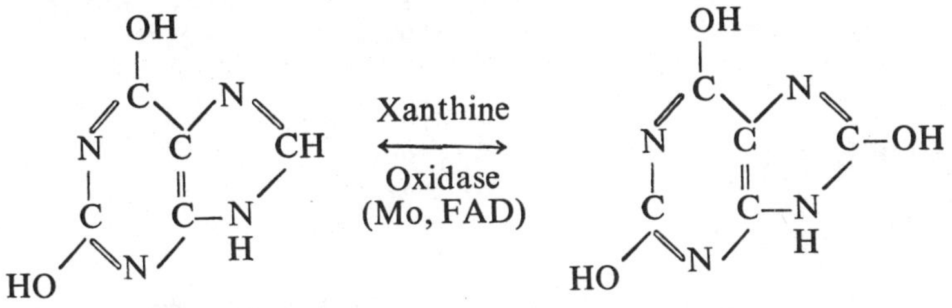

$$\text{Xanthine} \xrightarrow[\text{(Mo, FAD)}]{\text{Xanthine Oxidase}} \text{Uric acid}$$

Aldehyde oxidase and nitrate reductase are also activated by Mo.

Requirement

1 ppm for humans.

Toxic Level

Over 1000 ppm.

Sources

Plants from nondeficient soils are the best source of Mo. Mo is available from alkaline and not acid soils.

Toxicity Symptoms of Mo

1. "Tert," land containing a toxic level of Mo, like England, California, and Nevada, causes severe diarrhea and blood in the urine.

2. A high Mo level results in Cu deficiency.
3. Anemia, and hair color loss due to Cu loss.
4. Testicular degeneration in rats.
5. Decrease in S-oxidase activity, due to Cu loss.
6. Tungstate removes Mo and, thus, reduces its toxicity.

COBALT (Co)

Requirement

It is required for ruminants (50-70 ppb). There is no requirement for monogastrics.

Toxic Level

5 ppm.

Source

From cobalt salts as chlorides and sulfates.

Biochemical Functions of Co

1. A substitute divalent cation for activation of enzymes.
2. Needed for propionate metabolism in ruminants.
3. Metabolism of single-carbon fragments.
4. Stimulates erythroporesis.

Deficiency Symptoms of Co

1. Pernicious anemia.
2. Fatty liver.
3. Hemosidrin in spleen.
4. Breakdown of hemoglobin.

5. Wasty disease in large animals.
6. Vitamin B_{12} deficiency.

Toxicity Symptoms of Co

1. Polycythemia (excess of red blood cells).
2. Interferring with other minerals.

SELENIUM (Se)

Se has also been only recently recognized as an essential trace element. It has been mostly thought of from toxic point of view, rather than as being essential. It may become quite toxic in certain areas, as in the Dakotas, where the soil contains rather high levels of Se and which is transported to plants, namely wheat. Se is considered toxic even in very minute amounts, thus, a few parts per million in wheat is still considered toxic.

When Torula yeast is added to the diet, liver necrosis was prevented in pigs. This is attributed to a component called "Factor 3," present in Torula yeast. Se has been reported as an active component of this "Factor 3."

Se deficiency causes exudative diathesis in the avian species, which is an exudate forming under the skin where vessels' permeability to protein drastically increases. This disease is also characterized by a loss of fluid from plasma into the subcutaneous tissue and the leakage of hemoglobin.

Se deficiency also causes muscle dystrophy in the avain species.

Se and vitamin E are related in their deficiency symptoms, as Se deficiency may cause many symptoms of vitamin E deficiency, and vice versa. However, there are some vitamin E deficiencies that are specific to vitamin E and cannot be prevented by the intake of Se, and the same thing can be said of specific Se deficiencies that are specific of Se and cannot be cured by the intake of vitamin E. For example, resorption sterility in female rats is due to vitamin E deficiency and cannot be prevented by Se; and encephalomalacia can also be prevented only by vitamin E and not by Se. Meanwhile, white muscle disease in ruminants is characteristic of Se and can be prevented only by Se intake and not by vitamin E. On the other hand, there are diseases that can be prevented by the intake of either vitamin E or Se, such as liver necrosis, white muscle disease in chickens, and the prevention of unsaturated fatty acid oxidation.

Animals do require both vitamin E and Se, although ruminants may get enough of the vitamin and may lack Se.

Both vitamin E and Se are important in the protection of the membrane integrity as they prevent membrane breakdown and, consequently, protect lipids from oxidation. Vitamin E is considered a good antioxidant as it easily moves across the membrane and into the cell.

The Se requirement as compared to vitamin E is 1 to 1000 ppm.

FLUORINE (F)

Fluorine has been recently recognized as an essential trace element. It is not essential for life, but is essential for good dental health, as it does prevent dental cavities. F is also thought of as important in bones, especially in women, as it may play a role in prevention of osteoporesis. It is postulated that fluorine in bone may contribute to the stability of the hydroxy-appetite crystals of bones. 1 ppm is required.

Fluorine is also thought to decrease calcification of the aorota.

CHROMIUM (Cr)

Chromium (Cr) has the physiological function of maintaining blood glucose level, i.e., it functions along with insulin in glucose utilization in the cell, probably through activation of the cell membrane surface along with insulin.

REFERENCES

Bessey, O. A.; King, C. G.; Quinn, E. J.; and Sherman, H. C. 1935. The normal distribution of calcium between the skeleton and soft tissues. *J. Biol. Chem.* 3:115.

Cannon, W. B. 1939. *The wisdom of the body*. Rev. ed. New York: W. W. Norton.

Conway, E. J., and Hingerty, D. 1948. Relations between potassium and sodium levels in mammalian muscle and blood plasma. *Biochem. J.* 42:372.

Cunningham, I. J. 1945. The biological function of copper. *N. Z. Sci. Rev.* 3:3.

Denham, H. G., and Gortner, R. A. 1937. Cobalt—an essential element. *Science* 85:382.

Editorial 1944. Fluoride and dental caries. *J. Am. Med. Assoc.* 124:98.

Everson, G. J., and Daniels, A. L. 1934. A study of manganese retention in children. *J. Nutr.* 8:497.

Greenberg, D. M., and Tufts, E. V. 1936. Variation in the magnesium content of the normal white rat with growth and development. *J. Biol. Chem.* 114:135.

Holland, E. B., and Ritchie, W. S. 1941. Trace metals and total nutrients in human and cattle foods. *Mass. Agr. Exptl. Sta. Bull.* No. 379.

Hopkirk, C. S. M., and Grimmett, R. E. R. 1938. Importance of cobalt. Relationship to the health of farm animals. *New Zealand J. Agr.* 56:21.

Hove, E.; Elvehjem, C. A.; and Hart, E. B. 1940. The effect of zinc on alkaline phosphatase. *J. Biol. Chem.* 134:425.

Kehoe, R. A.; Cholak, J.; and Story, R. V. 1940. A spectro-chemical study of the normal ranges of concentration of trace metals in biological materials. *J. Nutr.* 19:579.

Klosterman, E. W.; Dinusson, E. W.; Lasley, E. L.; and Buchnan, M. L. 1950. Effect of trace minerals on growth and fattening of swine. *Science* 112:168.

Lehninger, A. L. 1950. Role of metal ions in enzyme systems. *Physiol. rev.* 30:393.

Lewis, H. B. 1935. The chief sulfur compounds in nutrition. *J. Nutr.* 10:99.

Macy, I. G. 1940. Effect of simple dietary alterations upon retention of positive and negative minerals by children. *J. Nutr.* 19:461.

Maynard, L. A., and Smith, S. E. 1947. Mineral metabolism. *Ann. Rev. Biochem.* 16:273.

McClure, F. J. 1949. 1951. Fluorine and other trace elements in nutrition. *J. Am. Med. Assoc.* 139:711.

McElory, W. D., and Glass, B. 1950. *Copper Metabolism: A Symposium on Animal, Plant, and Soil Relationships.* Baltimore: Johns Hopkins Press.

Review 1947. Manganese metabolism in rats and chicks. *Nutr. Rev.* 2:145.

Review 1947. Significance of trace elements in plants and animals. *Nature* 159:206.

Review 1950. "Trace" minerals requirements. *Nutr. Rev.* 8:178.

Review 1951. Fluoride content of human diets. *Nutr. Rev.* 9:80.

Review 1951b. Ion antagonism and inorganic nutrition. *Nutr. Rev.* 9:135.

Rose, M. S. 1932. The nutritional significance of some mineral elements occurring as traces in the animal body. *Yale J. Biol. Med.* 4:499.

Sadas, V. 1951. The biochemistry of zinc. I. Effect of feeding zinc on the liver and bones of rats. *Biochem. J.* 48:527.

Shohl, A. T. 1939. *Mineral Metabolism.* New York: Reinhold Publishing Co.

Smith, A. H. 1942. Trace elements in nutrition. *J. Am. Dietet. Assoc.* 18:721.

Solomon, R. Z.; Hald, P. M.; and Peters, J. P. 1940 State of the inorganic components of human red blood cells. *J. Biol. Chem.* 132:723.

Stare, F. J., and Elvehjem, C. A. 1933. Cobalt in animal nutrition. *J. Biol. Chem.* 99:473.

Stearns, G. 1939. The mineral metabolism of normal infants. *Physiol. Rev.* 19:415.

Underwood, E. J. 1940. The significance of the "trace elements" in nutrition. *Nutr. Abs. Rev.* 9:515.

Vallee, B. L., and Altschule, M. D. 1949. Zinc in the mammalian organism, with particular reference to carbonic anhydrase. *Physiol. Rev.* 29:370.

Walker, A. R. P.; Fox, F. W. and Irving, J. T. 1948. Studies in human mineral metabolism. *Biochem. J.* 42:452.

Wilder, R. M., and Kendall, E. C. 1937. Intake of potassium, an important consideration in Addison's disease. *Arch. Internal Med.* 59:367.

Part IV

PHYSIOLOGY AND BIOCHEMICAL REACTIONS OF THE HORMONES

Chapter 24

CATECHOLAMINES

HORMONES OF THE ADRENAL MEDULLA

The two hormones are epinephrine and norepinephrine.

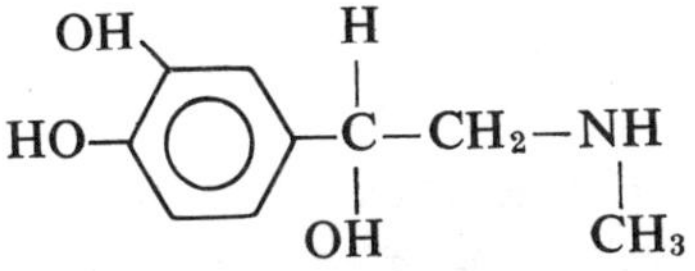

$$HO-C_6H_3(OH)-\underset{OH}{\underset{|}{CH}}-CH_2-\underset{CH_3}{\underset{|}{NH}}$$

Epinephrine

$$HO-C_6H_3(OH)-\underset{OH}{\underset{|}{CH}}-CH_2-NH_2$$

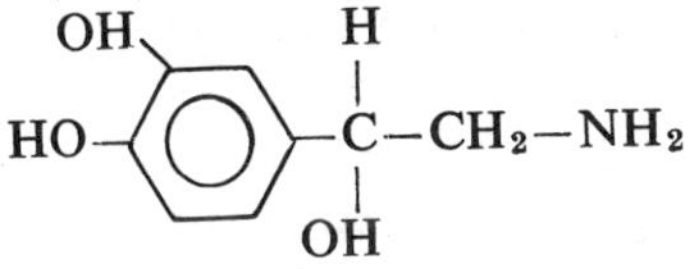

Norepinephrine

The hormones are called catecholamines because they possess a "catechol" nucleus. Norepinephrine comes first in the biosynthetic scheme of catecholamines, and 80% of the medullary secretion is epinephrine and 20% is norepinephrine.

The adrenal medulla is essentially the only body source of epinephrine. Epinephrine was the first hormone isolated and identified.

The adrenal medulla, like the posterior pituitary, is a functional part of the nervous system, or, it is a modified nervous system. It is a specialized sympathetic ganglion regulated by typical splanchnic nerve, upon stimulation of which epinephrine and norepinephrine are secreted, as in the case of an emergency.

An adrenal medullary discharge is one of the first happenings when adaptation to emergency is required. Psychological or physiological stress or excitement will cause an outpouring of epinephrine and norepinephrine to a 1000-fold increase in their release at rest. Thus, the adrenal medulla, being a part of the entire sympathetic system, plays a main role in general adaptation.

Release of epinephrine and norepinephrine occurs under the same circumstances where secretion of ACTH or ADH occurs.

REGULATION OF CATECHOLAMINES SECRETION

Control is entirely nervous via the splanchnic nerves with central nervous system being in the posterior part of the hypothalamus.

Pain, cold, and volume changes in blood due to hemorrhage, emotional excitement, rage, and anxiety—all, through nervous control, will cause secretion of epinephrine and norepinephrine. An oxygen lack acts on certain receptors of the brain or acts on the adrenal medulla itself, thus causing hypoglycemic effects and the subsequent release of the adrenal medulla hormones.

ORIGIN, STORAGE, AND RELEASE OF CATECHOLAMINES

The catecholamines are found within the chromaffin cells. While norepinephrine is synthesized in the granules, epinephrine is synthesized in cytoplasm and then moves into the granules. These granules may contain protein and phospholipids, and are known to be rich in ATP (one ATP for four catecholamines). These granules also have high ATPase activity.

The vesicles are attached to the membrane and they do not leave the cell after secretion of the hormone.

The extract of the adrenal medulla contains both epinephrine and norepinephrine, but for the most part epinephrine is found in the adrenal medulla while norepinephrine is mostly found outside.

SECRETION AND RELEASE OF CATECHOLAMINES

Acetylcholine's reacting with chromaffin cells will cause an increase in the influx of Ca^{2+} in the chromaffin cells. Ca^{2+} then activates the secretion of epinephrine and norepinephrine from the vesicles.

The release of catecholamines is more rapid from the nerve endings.

Epinephrine and norepinephrine are released and metabolized very rapidly by the liver and kidney.

An active, emotional, but aggressive person will release norepinephrine in the urine, while a tense, emotional, but passive individual will release epinephrine instead.

There is also a species difference in the adrenal medullary secretion. A painful stimulus for a cat will cause the secretion of epinephrine.

Epinephrine is known to have more hyperglycemic effects than norepinephrine. Injection of insulin into a subject will reduce the blood sugar, and this will subsequently cause the release of epinephrine. But if the opposite occurs, and the subject is instead hyperglycemic, this would not change the level of norepinephrine much. However, the epinephrine level will go down.

BIOSYNTHESIS OF CATECHOLAMINES

Both epinephrine and norepinephrine have the amino acid tyrosine as a precursor. Tyrosine gets hydroxylated to DOPA in the cytoplasm. DOPA is decarboxylated to dopamine, also in the cytoplasm. Dopamine then moves into the granules and is β-hydroxylated to norepinephrine. Norepinephrine moves out from the granules and back to the cytoplasm for methylation.

The granules are essential for formation of the norepinephrine hormone. A methyl group is needed for making epinephrine from norepinephrine. This methyl group also needs a carrier, which is thought to be S-adenosyl methionine, and this in turn requires an enzyme, the N-methyl transferase. This enzyme in the adrenal medulla is thought to be activated by hydrocortisone. Thus, the adrenal cortex may have an influence on the adrenal medulla.

BIOSYNTHESIS OF CATECHOLAMINES

Phenylalanine —Hydroxylase→ Tyrosine

Tyrosine Hydroxylase → DOPA

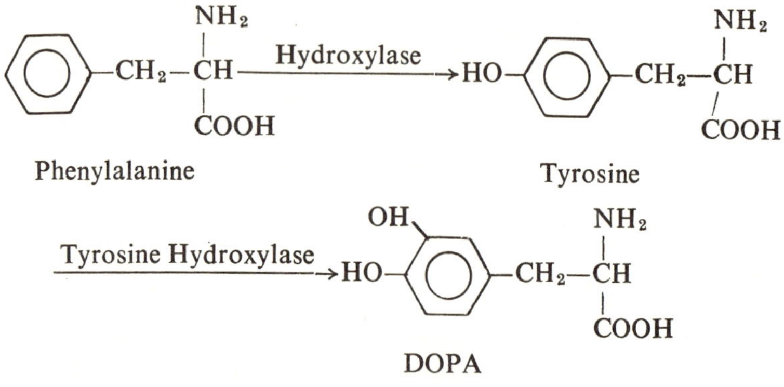

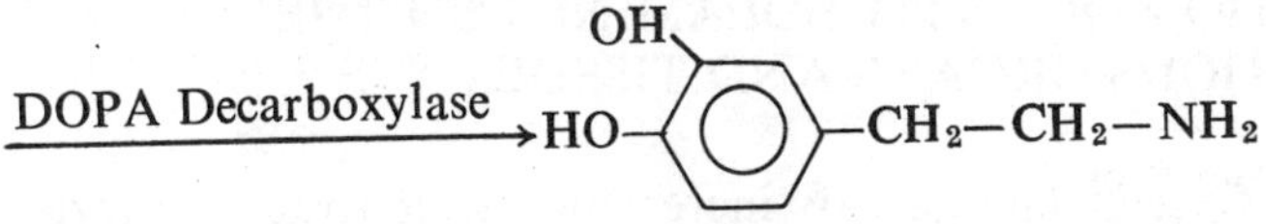

$$\xrightarrow{\text{DOPA Decarboxylase}} \text{HO}-\underset{\text{OH}}{\bigcirc}-CH_2-CH_2-NH_2$$

Dopamine

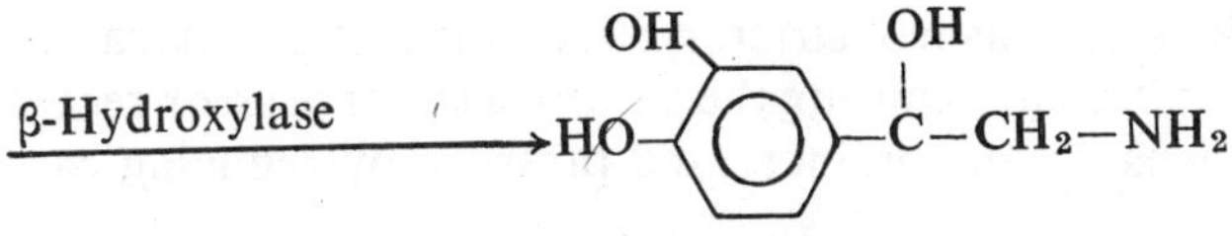

$$\xrightarrow{\beta\text{-Hydroxylase}} \text{HO}-\underset{\text{OH}}{\bigcirc}-\underset{\text{OH}}{C}-CH_2-NH_2$$

Norepinephrine

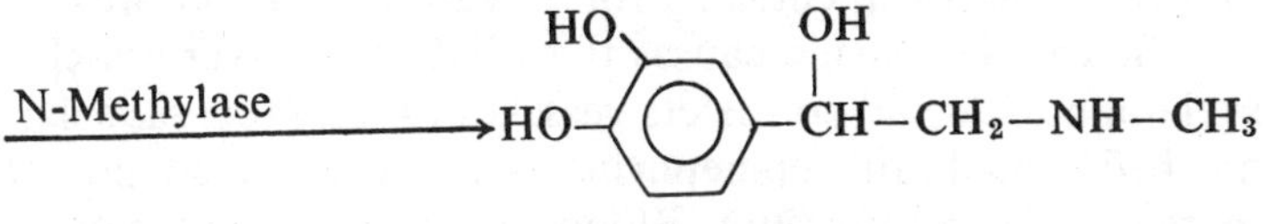

$$\xrightarrow{\text{N-Methylase}} \text{HO}-\underset{\text{HO}}{\bigcirc}-\underset{\text{OH}}{CH}-CH_2-NH-CH_3$$

Epinephrine

PHYSIOLOGICAL EFFECTS OF ADRENAL MEDULLARY HORMONES:

Emergency action, "fight" or "flight" reaction. The results of this reaction are:
1. Increase in cardiac output due to the increases in heart rate and the contractile force.
2. Increase in blood pressure.
3. Increase in the rate and depth of respiration.
4. Increase in blood sugar.
5. Shunting of blood from splanchnic bed to musculature.
6. Lipolysis of fat to FFA which becomes a readily available source of energy.
7. Central nervous system arousal.
8. Decrease in fatigue signal to muscles.

ACTION OF CATECHOLAMINES ON THE VARIOUS ORGANS AND TISSUES

Catecholamines, by increasing the contractility force, increase the stroke volume of the heart. Epinephrine, in this regard, is more effective in stimulating the heart.

Large doses of norepinephrine may decrease cardiac function, since its vasoconstrictor effect may cause elevation of the blood pressure causing the cardiovascular pressor receptors to compensate for the increased pressure by reducing cardiac function.

Although epinephrine increases myocardial excitability, its infusion may lead to extra systoles; norepinephrine, thus, may be given therapeutically for elevation of blood pressure.

Since epinephrine causes the dilation of liver vessels and skeletal muscle and constricts vessels in skin, mucosa, splanchnic bed, and brain, epinephrine is not considered potent in increasing blood pressure. Blood pressure increase caused by epinephrine is usually due to increased cardiac output. The main effect here will be on the systolic pressure.

Norepinephrine is 20% as effective as epinephrine in increasing glycogenolysis in the liver.

Epinephrine causes an increase in glycogenolysis in skeletal muscle, thus increasing the force of contraction in skeletal muscle.

Epinephrine causes constriction of the smooth muscle vessels, except in liver and muscle vasculature where it causes dilation.

Epinephrine causes constriction of the gastrointestinal sphincters while it causes relaxation in the nonsphincter regions.

RECEPTORS

In 1948, Ahlquist theorized that the sympathetic effector cells contain specific receptor substances for the sympathetic neurohumor. He concluded that the sympathetic receptors fall into two classifications, i.e., either "alpha" or "beta," α-receptor being excitatory and most sensitive to epinephrine, as in the following order:

Epinephrine > Norepinephrine > Phenylephrine > Isoproponol
  100   10   1   0

Since β-receptors are inhibitory, their sensitivity to epinephrine is in the following order:

$$ISO > Epi > NE > PE$$
  100 10 1 0

Thus, the β-receptors are most sensitive to Isoproponol.
Epinephrine sometimes may cause stimulation of both α-and β-receptors.

The effect of catecholamines is optimal in the presence of cortisol and the thyroid hormone. In the presence of cortisol and thyroid hormones, the β-adrenergic response is stimulated and c AMP is increased.

BIOCHEMICAL ACTION OF CATECHOLAMINES

Epinephrnie increases blood sugar while norepinephrine does not. This hyperglycemic effect of epinephine is due to the β-adrenergic effect on the liver.

The catecholamines cause the rise of c-AMP which activates the phosphorylase enzyme, which in turn provides glucose from glycogen.

The same thing happens in muscle except that, since the enzyme phosphatase is not present in muscle, the glycogen in this case will give lactic acid as the end product, which goes to the blood.

Epinephrine and norepinephrine also are known to be lipid mobilizing hormones, where they break down fat and cause the release of FFA in plasma.

The oldest known function of epinephrine is its increase of O_2 consumption of the organism.

Epinephrine is also thought to increase the lipase enzyme activity which, in turn, causes the release of FFA into blood.

Epinephrine also may stimulate carbohydrate metabolism by stimulating the release of ACTH, and the subsequent release of glucocorticoids. The glucocorticoids will, in turn, increase blood glucose through gluconeogenesis. The increase in blood glucose level may explain the phenomenon of the increase in O_2 consumption caused by epinephrine since the increase in glucose oxidation will increase the O_2 consumption, and since epinephrine does not increase the glucose output from the liver very much.

The glucose output from the liver, thus, could be stimulated by way of the sympathetic nervous system.

Epinephrine causes the inhibition of insulin secretion through inhibition on the α-receptors.

BIOCHEMICAL MECHANISM OF ACTION OF EPINEPHRINE

Hyperglycemia

This is caused by epinephrine as in the following sequence:

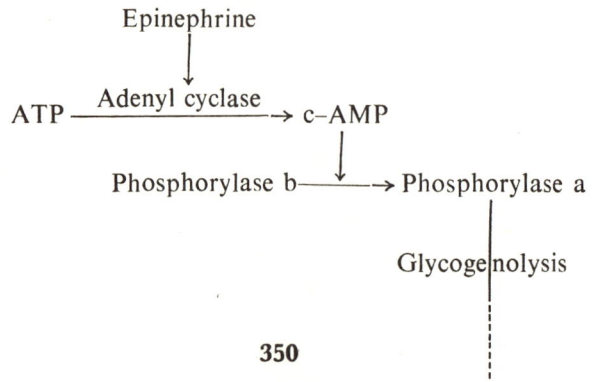

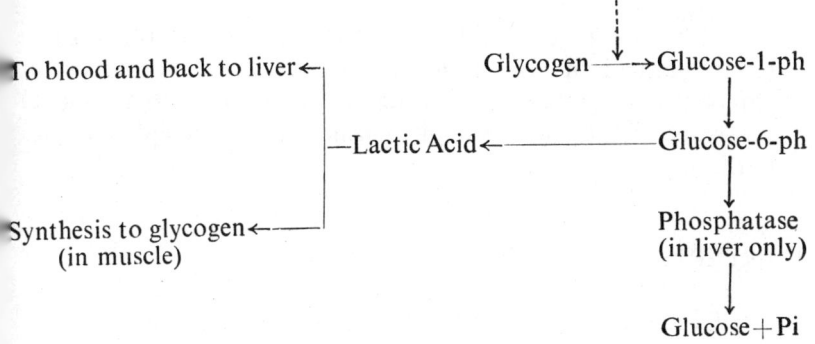

Lipolysis

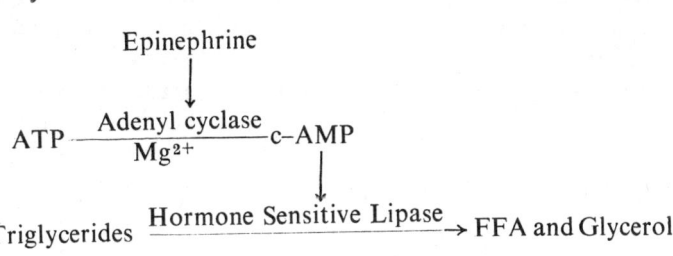

Since epinephrine activates lipase, which in turn causes the release of FFA from depot fat and into blood, oxidation of the FFA will increase the O_2 consumption.

The increase in glucose oxidation caused by epinephrine in skeletal muscle causes the fight or flight reaction in an emergency.

Catecholamines cause an increase in mental awareness, which is related to glucose utilization by the brain.

CONTROL OF SECRETION

There is no negative feedback for epinephrine in the adrenal medulla. Epinephrine and norepinephrine are an intrinsic part of the nervous system and are controlled by nerve impulse.

Response of catecholamines to emergency is transitory; upon removal of the stimulus, output of the hormones is stopped. Any change in the external or internal environment will cause the release of epinephrine or norepinephrine, or both.

INTERNAL NEGATIVE FEEDBACK

Excess epinephrine or norepinephrine inhibits the enzyme tyrosine hydroxylase and stops further synthesis of the catecholamines. Stimulating the sympathetic nervous system will activate this enzyme. In addition, medullary epinephrine will inhibit the enzyme N-methyl transferase, which converts norepinephrine to epinephrine. The build-up of epinephrine in the cell will inhibit this enzyme through a feedback mechanism. This feedback is not like the one in the anterior pituitary. Here, the hormonal output is caused by a stimulus, removal of which will stop this output.

METABOLISM OF CATECHOLAMINES

The biological half-life of catecholamines is very short; the two basic steps in their metabolism being the "O" methylation to metanephrine or normetanephrine, and the oxidation of the amine group. The final end product is vinyl mandelic acid (VMA). Nevertheless, all five products (epinephrine, norepinephrine, metanephrine, normetanephrine, and VMA) are found in urine. The predominant metabolite in urine is VMA, however.

$$HO\text{-}\underset{OH}{}\bigcirc\text{-}\underset{OH}{CH}\text{-}CH_2\text{-}NH\text{-}CH_3 \xrightarrow{\text{Catechol-O-Methyl transferase (COMT)}}$$

Epinephrine or Norepinephrine

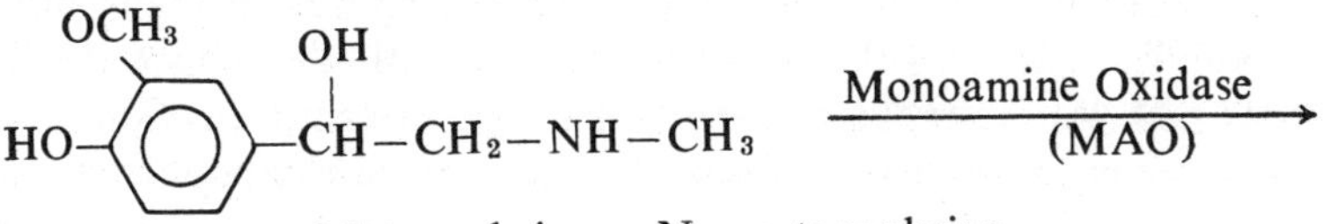

$$\text{HO}-\underset{\text{OCH}_3}{\bigcirc}-\underset{\text{OH}}{\text{CH}}-\text{CH}_2-\text{NH}-\text{CH}_3 \quad \xrightarrow{\text{Monoamine Oxidase (MAO)}}$$

Metanephrine or Normetanephrine

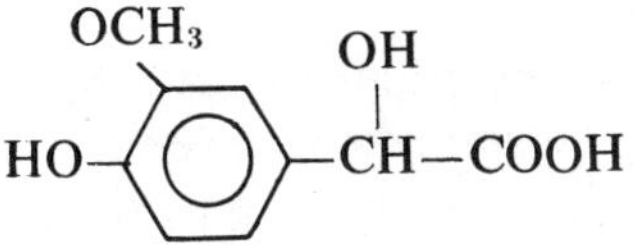

$$\text{HO}-\underset{\text{OCH}_3}{\bigcirc}-\underset{\text{OH}}{\text{CH}}-\text{COOH}$$

Vinyl Mandelic Acid (VMA)

DISTRIBUTION OF URINARY METABOLITES OF EPINEPHRINE

Unchanged Epinephrine—6%
Metanephrine—40%
Vinyl Mandilic Acid (VMA)—41%
4-Hydroxy, 3-Methoxy Phenyl Glycerol—7%
3-4-Dihydroxy Mandilic Acid—2%
Miscellaneous—4%

Epinephrine is conjugated mainly with sulfate rather than glucuronic acid.

In man, the adrenal medulla has four to ten times epinephrine to norepinephrine.

In urine: epinephrine level is ten to 15 μg/day; norepinephrine level is 30 to 50 μg/day.

At birth, epinephrine and norepinephrine are present in equal amounts in the medulla. At three years of age, the epinephrine level rises to 90% and norepinephrine drops to 10%.

ESTIMATION

Epinephrine and norepinephrine are measured in the blood by chemical methods where fluorescent color is developed and read spectrophotometrically.

The biologically active form of both epinephrine and norepinephrine is the L-form. It has been reported as being fifteen times as active as the D-form.

ACTION OF CATECHOLAMINES ON ADIPOSE TISSUE

Epinephrine and norepinephrine are equal in their lipolytic effect, while the central nervous system arousal effect is due entirely to epinephrine and is believed to be due to action on the brain stem.

REFERENCES

Axelrod, J. 1968. The fate of norepinephrine and the effect of drugs. *Physiologist* 11:63.

Burn, J. H. 1965. *The Autonomic nervous system: for students of physiology and of pharmacology.* 2nd ed. Philadelphia: F. A. Davis Co.

Burnstock, G. 1968. The autonomic neuromuscular junction. *Proc. Int. Union of Physiol. Sciences* 6:7.

Burnstock, G., and Holman, M. E. 1963. Smooth muscle: autonomic nerve transmission. *Ann. Rev. Physiol.* 25:61.

Engelman, K., and Portnoy, B. 1970. A sensitive double-isotope derivative assay for norepinephrine and epinephrine: normal resting human plasma levels. *Circ. Res.* 26:53.

Ferry, C. B. 1966. Cholinergic link hypothesis in adrenergic neuroeffector transmission. *Physiol. Rev.* 46:420.

Gillespie, J. S. 1968. Adrenergic mechanisms. *Proc. Int. Union of Physiol Sciences* 6:107.

Guyton, A. C., and Gillespie, W. M., Jr. 1951. Constant infusion of epinephrine. rate of epinephrine secretion and destruction in the body. *Amer. J. Physiol.* 165:319.

Ingram, R. H., Jr.; Szidon, J. P.; and Fishman, A. P. 1970. Response of the main pulmonary artery of dogs to neuronally released versus blood-borne norepinephrine. *Circ. Res.* 26:249.

Malmejac, J. 1964. Activity of the adrenal medulla and its regulation. *Physiol. Rev.* 44:186.

Mauskopf. J. M.; Gray, S. D.; and Renkin, E. M. 1969. Transient and persistent components of sympathetic cholinergic vasodilation. *Amer. J. Physiol.* 216:92.

Nishi, S.; Soeda, H.; and Koketsu, K. 1967. Release of acetylcholine from sympathetic preganglionic nerve terminals. *J. Neurophysiol.* 30:114.

Root, W. S., and Huffman, F. G. (eds.) 1967. *Physiological pharmacology.* Vol. 3, Pts. C and D. New York: Academic Press, Inc.

Von Euler, U.S. 1956. *Noradrenaline.* Springfield, Ill.: Charles C. Thomas, Publisher.

———. 1961. Neurotransmission in the adrenergic nervous system. *Harvey Lect.* 55:43.

———. 1968. Adrenergic transmitter granules. *Proc. Int. Union of Physiol Sciences* 6:95.

Chapter 25

HORMONES OF THE POSTERIOR PITUITARY

OXYTOCIN

Synthesis and Storage

Oxytocin is formed in the paraventricular nucleus and is transported down the axons. It travels conjugated to carrier proteins down the axons and is stored in the axon boutons. Storage of the hormone is in the form of granules in the posterior pituitary.

Release of Oxytocin

Release of oxytocin from its storage area in the posterior pituitary is stimulated by suckling the breast. This is a result of a neurohormonal mechanism that involves spinal nerves' sending afferent nerve fibers to the central nervous system, and causing the release of the hormone from its storage area.

Actions of Oxytocin

Causing milk ejection is the main function of oxytocin. The hormone causes the contraction of the myoepithelial cells of the lactating breast.

Oxytocin inhibits the prolactin inhibitory factor (PIF) and, thus, stimulates prolactin release.

The hormone induces myometrial contraction; thus, an exogenous injection of oxytocin to a term pregnant mother will induce labor. This occurs via the Ferguson Reflex, which causes uterine's smooth-muscle contraction and explusion of the fetus through stretching on the cervix and vagina perceived by stretch receptors. While estrogen increases the oxytocin-contraction effect on the uterus, progesterone decreases this effect.

Oxytocin enhances sperm transport in the oviduct of the female.

The hormone causes naturesis, this loss of sodium through the urine is manifested in rats.

Oxytocin also helps in the ejaculation of sperm from the male's testes.

Assay of Oxytocin

Biological assay is made by measuring rat uterine contractions after an exogenous dose of the hormone, or by measuring milk ejection in guinea pigs after an exogenous dose of oxytocin.

ANTIDIURETIC HORMONE (ADH)

Antidiuretic hormone (ADH) is also called vasopressin. While arginine vasopressin is the form found in humans, lysine vasopressin is found in hogs. The hormone is not species specific, and lysine vasopressin is as effective as arginine vasopressin when given to man.

Synthesis and Storage

Like oxytocin, ADH is a posterior pituitary hormone. It is synthesized in the supraoptic nucleus, transported down the axons, conjugated to protein, and stored in the axon boutons. The hormone is stored in granule form in the posterior pituitary.

Functions of ADH

Antidiuresis is the main function of ADH, where the hormone causes an increase in the permeability of the distal kidney tubule to water, sodium, and urea, thus reabsorbing water from the tubule and producing hypertonic urine.

ADH causes water retention, and the increased blood volume will result in an increased arteriolar blood pressure through constriction of blood vessels.

ADH may increase ACTH release.

A pharmacological dose of ADH may cause naturesis.

Control of ADH Secretion

Increase in plasma osmolarity influences the osmoreceptors in the hypothalamus; thus, impulses to the supraoptic nucleus are increased and more ADH is secreted.

ADH release is also controlled by the volume receptors in the left atrium, where the stretch receptors influence the hypothalamus and increase or decrease pressures, thus causing impulses to be sent via the vagus to inhibit or increase ADH release by lowering or increasing the blood volume, which likewise affects urine volume and hence influences the decrease or increase in the pressure.

While anaesthetics and high temperatures may cause an increase in ADH release, alcohol, coffee, tea, and emotional stress cause a decrease in ADH release.

Assay of ADH

Biological assay induces antidiuresis as a response. Usually the assay for ADH is very difficult as its half-life is extremely short (5-7 minutes, through the liver and kidney).

REFERENCES

Aulsebrook, L.H., and Holland, R. C. 1969. Central regulation of oxytocin release with and without vasopressin release. *Amer. J. Physiol.* 216:818.

Douglas, W. W. 1968. Cellular mechanisms of secretion in the adrenal medulla and posterior pituitary gland: possible methods of stimulus-secretion coupling. *Proc. Int. Union of Physiol. Sciences* 6:63.

Fasciolo, J.C.; Totel, G.L.; and Johnson, R. E. 1969. Antidiuretic hormone and human eccrine sweating. *J. Appl. Physiol.* 27:303.

Orloff, J. 1968. The role of adenyl cyclase in the regulation of antidiuretic hormone action in toad bladder. *Proc. Int. Union of Physiol. Science* 6:163.

Pliska, V.; Rudinger, J.; Dousa, T.; and Cort, J. H. 1968. Oxytocin activity and the integrity of the disulfide bridge. *Amer. J. Physiol.* 215:916.

Chapter 26

HORMONES OF THE ANTERIOR PITUITARY

INTRODUCTION

Since hormones of the hypothalamus or the releasing factors have as their target the anterior pituitary, thus the two organs, i.e., the hypothalamus and the anterior pituitary, do function together.

The hypothalamo-hypophyseal portal system directs the secretion of the hypothalamus to the anterior pituitary. Should the pituitary gland be transplanted to another part of the body, it will revascularize and would secrete its hormones, except at a much lower rate than when it was connected to the hypothalamus via the portal system. In this case, prolactin secretion should be greatly increased because of the lack of any prolactin-inhibiting factor (PIF) which inhibits its secretion.

HISTOLOGY OF PITUITARY GLAND

There are three cell types that stain:

1. Basophils $\begin{array}{l}\nearrow \text{FSH} \\ \rightarrow \text{LH} \\ \searrow \text{TSH}\end{array}$

2. Acidophils — STH, Prolactin

3. Chromophores ⟶ ACTH

The distribution of these cells is not uniform in the pituitary and varies from species to species.

Microscopic examination of cell cross-sections shows (STH) somatotropic hormone cells with an extremely large Golgi apparatus, a large nucleus, and their distinguishing character as large, round secretory granules of the size of 350 to 400 mu.

Lactogenic hormone (LH) also shows large cells with big Golgi apparatus.

The gonadotrophs are usually of smaller cells with abundant rough endoplasmic reticulums, extremely large Golgi apparatus, and their secretory granules are of the size of 200 mu.

The chromophores, or the ACTH-secreting cells, are polymorphic in nature and are difficult to find. They are 8–100 mu in diameter and are present usually in periphery of the cell and close to the basal membrane.

While the anterior pituitary secretes eight hormones, the posterior pituitary secretes only two. Thus, the varied chemical secretion of such a small gland as the pituitary is impressive.

MAJOR GROUPS OF THE PITUITARY HORMONES

The hormones of the pituitary can be arranged into four major groups as follows:
1. α-MSH, β-MSH, Lipotropin, ACTH
2. TSH, LH, FSH, HCG
3. Prolactin, STH
4. Oxytocin, vasopressin

Group 1

Adrenocorticotropic hormone. (ACTH) is a straight-chain polypeptide of 39 amino acids. 1 to 24 amino acids are similar in different species. The molecular weight of ACTH is 4,000.

Lipotrophin is a lipolytic hormone of a straight-chain polypeptide of 90 amino acids. Its molecular weight is 9,500. It is, thus, a large compound and at 40 amino acid, it becomes similar to melanocyte-stimulating hormone (β-MSH).

α-MSH is composed of the first thirteen amino acids of ACTH. It is a straight-chain amino acid polypeptide.

β-MSH is 18 to 22 amino acids of molecular weight of 3,000.

There is a common peptide sequence between MSH, ACTH, and lipotrophin. They seem to have come from a common gene; that is, the three hormones may have a common gene in the nucleus which has gone through mutation.

The similarities between the pituitary hormones make us believe they might have come from a common ancestry.

ACTH and lipotrophin may mimic the action of somatotrophin and prolactin.

All of these compounds have a melanotrophic effect as they stimulate the dispersion and increase the production of melanin.

Group 2

These are glycogenic hormones. Thyroid-stimulating hormone (TSH) contains 15% sugar, such as mannose, fucose, galactose, glucosamine and galactosamine. The hormone is composed of two subunits α, and β. Its molecular weight is 28,000.

LH has 15% sugar, too, and is composed of two subunits, α and β. Its molecular weight is 29,000. The α-subunits of TSH and LH are identical. The two hormones use the same portion of the genome, and probably the partial difference in the β-chains between the two hormones occurred through a mutation that led to the remarkable individual difference in function between them.

Human Choriogonadotropin (HCG) contains 30% sugar, and is composed of two subunits, α and β. The α-subunit of HCG is identical to the α-subunits of TSH and LH.

Follicle-stimulating hormone (FSH) is 20% sugars and is also composed of α and β subunits. Its molecular weight is 30,000.

Although the α-chains of this group of glycoproteins are similar, their β-chains are also related. It is only a small difference in the β-chains, such as a difference in an amino group location or of that of a sugar moiety which gives the hormone its stereochemical specificity for its receptor.

Group 3

STH is a polypeptide hormone of 188 amino acids sequence. Its molecular weight is 22,000. It has two disulfide linkages which may be the key to its biological activity since, as long as these two disulfide linkages are maintained, its biological activity is maintained.

Prolactin is of 180 amino acids sequence. Its molecular weight is 22,500. There is a great overlap between mechanism of its action and that of somatotrophin.

Prolactin has an influence on behavior, reproduction, and on the salt secretion mechanism. It is also involved in osmoregulations in frogs and fish, which drives them from salt to fresh water.

Prolactin is a general metabolic hormone. It mimics the growth hormone's effect, and it can completely replace it in birds.

REFERENCES

Bell, E. T. 1966. Some observations on the assay of anterior pituitary hormones. *Vitamins and Hormones* 24:63–113.

Harris, G. W., and Donovan, B. T. 1966. *The pituitary*. Vol. 3. Berkeley: University of California Press.

Li, C. H. 1961. Some Aspects of the relation of peptide structure to activity in pituitary hormones. *Vitamins and Hormones* 19:313.

Martini, L., and Ganong, W. F. 1966. *Neuroendocrinology*. Vol. 2. New York: Academic Press.

Pincus, G.; Thiman, K.V.; and Astwood, E.B. 1964. *The hormones*. Vol 5. New York: Academic Press.

Chapter 27

ADRENOCORTICOTROPIC HORMONE (ACTH)

CHEMICAL STRUCTURE

It is a single-chain polypeptide of 39 amino acids. Its amino acids segment 1-13 is the minimum part that is essential for biological activity, although, at this level, its activity is low. As amino acids from 14-20 are progressively increased, the activity of ACTH likewise is progressively increased. The amino acids segment 12-39 is where species variation and the immunological properties of ACTH occur. ACTH is not a species-specific hormone. ACTH and melanocyte-stimulating hormone (MSH) have an identical structure of the first thirteen amino acids, thus, ACTH has some MSH activity in humans. MSH is, as well, an anterior pituitary hormone, the secretion of which is controlled by the hypothalamus. It provides protective coloration in lower animals. Although man produces very little MSH, he will still respond to injections of the hormone through a darkening of the skin. In the case of Addison's disease, the destruction of the adrenal cortex causes darkening of the skin due to the large amounts of ACTH present in the body because of the lack of feedback from the corticoid hormones.

BIOLOGICAL ACTIVITY

ACTH causes the stimulation of the adrenal cortex to produce the glucocorticoid hormones, mainly cortisol, in man.

ACTH may also have a direct effect on the isolated tissues such as mobilization of fatty acids from the adipose tissue. In this case, ACTH activates the adenyl cyclase enzyme which in turn catalyzes the production of c-AMP from ATP. c-AMP activates the hormone-sensitive lipase on the fat cell membrane, which in turn breaks up the triglycerides into free fatty acids (FFA) and glycerol as shown in the following scheme:

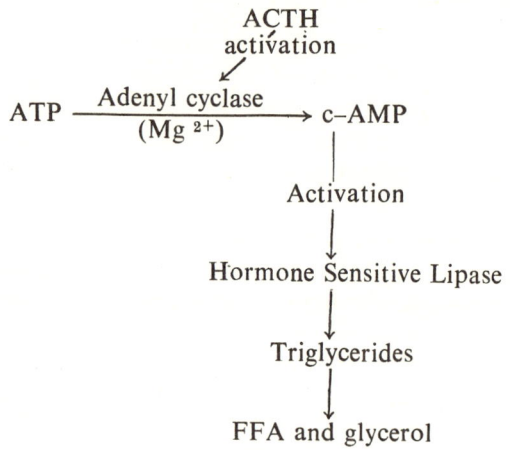

ACTH also may cause an increased deposit of glucose in muscle cells which, in turn, decreases blood glucose level. ACTH may, as well, increase the uptake of amino acids by the muscle cells.

MECHANISM OF ACTION OF ACTH ON CELL MEMBRANE

While the outer layer of the adrenal cortex (zona glomerulosa), which secretes mineralocorticoids, is not under the influence of ACTH, the second layer (zona fasciculata), which secretes glucocorticoids, is under the influence of ACTH.

ACTH causes secretion of the zona fasciculata of glucocorticoids through causing mitosis in those cells.

The specific receptor at the functionality zone of cell membrane is activated by ACTH in a stereochemically specific manner. This will cause release of Ca^{2+} from the outside of the membrane to the interior of the membrane; depolarization of cell occurs where Na^+ moves in and K^+ moves outside the cell. Thus, a wave of excitation moves into the cell, ATP is converted to c-AMP through activation of the adenyl cyclase enzyme, and secretion of glucocorticoids takes place until the effect of the excitation wave is gone. There is no need for RNA or protein to be made. The result of the initial response is secretion and resynthesis.

If the response is sustained for a long time, there may not be enough machinery within the cell to maintain the response. In this case, de novo synthesis of messenger RNA may be imposed on the cell for secretion, resynthesis, and build-up machinery of cell.

Here, as the signal of activation spreads, it speeds up the rate of differentiation of adjacent cells resulting in the movement and expansion of these cells, and there is regeneration and constant removal of hypertrophied cells. Thus, the chemical signal, which in this case is ACTH, causes differentiation of cells, that must be maintained in the proper perspective. The sustained signals cause cells to go through, as with cells activated and pushed, past a point and keep going (but genome reads out stepwise in response to cell differentiation according to physiological demand for the mature glucocorticoid cells).

The ACTH receptor on the plasma membrane is directed toward the outside of the membrane. The interaction between ACTH and its receptor is extremely stereochemically specific. The membrane's phospholipid is very important in maintaining this relationship. If phospholipid is removed, the globular protein of the membrane loses the stereospecificity. Ca^{2+} is a must for ACTH reaction with its receptor.

For every ACTH molecule that hits the receptor, there are 500 c-AMPs produced. The c-AMP moves into the cytoplasm of the cell, where it stereochemically specifically reacts with a second receptor inside the cell which is composed of a stereochemically specific allosteric protein subunit and a specific kinase subunit. Upon the interaction of c-AMP with this inside-the-cell receptor, a change of the conformation of the allosteric protein occurs and the kinase is released. Upon release of the kinase, its active site opens up and the kinase enzyme is thus activated. Otherwise, the inside-the-cell receptor without c-AMP returns to its initial state, where the kinase gets rebound to the allosteric protein and is inactivated.

Any portion of c-AMP not used up in the reaction is broken down by the enzyme phosphodiesterase.

In the adrenal gland, cells, upon activation with ACTH, require an abundancy of acetates.

One particular group of smooth endoplasmic reticulum (SMR) builds up rapidly under the influence of ACTH to cholesterol which is stored in lipid globlets. Synthesis of the corticoid hormones from cholesterol follows, starting with the cleaving of the cholesterol side chain and oxidation at the C-20 position, thus producing pregnenolone as cholesterol is pulled from fat droplets and to the mitochondria for oxidation. Once this point is passed, movement is very rapid and the enzyme's availability is abundant for the subsequent synthesis of the corticoids.

Pregnenolone moves back out to SMR where progesterone is synthesized through oxidation at C-3 and hydroxylation at C-17 and C-21.

The last step in steroidgenesis is when the steroid gets back into the mitochondria and has the C-11 hydroxylated in the final synthesis of corticosterone.

The rate-limiting step is the conversion of cholesterol to pregnenolone.

Steroidgenesis will go under ACTH stimulation for about five minutes without protein synthesis. However, protein synthesis is required after this short initial phase.

SECRETION AND FEEDBACK OF ACTH

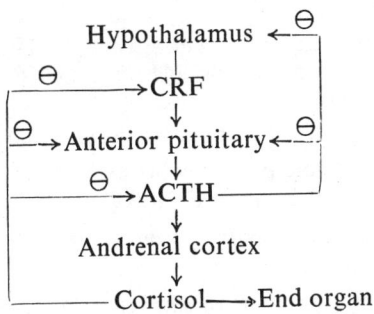

Cortisol may cause negative feedback on the hypothalamus and the anterior pituitary. ACTH, as well, exerts negative feedback on the anterior pituitary, and possibly on the hypothalamus.

ASSAY OF ACTH

1. Bioassay of the ascorbic acid level in the adrenal cortex, where increased levels of ACTH is manifested by a decrease in the ascorbic acid content of the adrenal.
2. Radioimmunoassay, which is the most accurate technique.

REFERENCES

deBodo, R. C., and Sincoff, M. W. 1953. Anterior pituitary and adrenal hormones in the regulation of carbohydrate metabolism. *Recent Progress in Hormone Research* 8:511.

Harris, G. W., and Donovan, B. T. 1966. *The pituitary gland.* 3 volumes. Berkeley: University of California Press.

Harris, G. W.; Reed, M.; and Fawcett, C. P. 1967. Hypothalamic releasing factors and the control of anterior pituitary function. *Brit. Med. Bull.* 22:266.

Munson, P. L., and Briggs, F. N. 1955. The mechanism of stimulation of ACTH secretion. *Recent Progress in Hormone Research.* 11:83.

Sprague, R. C.; Mason, H. L.; and Power, M. H. 1951. Physiologic effects of cortisone and ACTH in man. *Recent Progress in Hormone Research* 6:315.

Thorn, G. W., and Forsham, P. H. 1949. Metabolic changes in man following adrenal and pituitary hormone administration. *Recent Progress in Hormone Research* 4:229.

Venning, E. H. 1965. Adenopypophysis and adrenal cortex. *Ann. Rev. Physiol.* 27:107.

Williams, R. 1968. *Textbook of endocrinology.* 4th ed. Philadelphia: W. B. Saunders Company.

Chapter 28

GROWTH HORMONE (GH)

If the anterior pituitary is removed, changes in the effects of growth hormone will be among the first to be observed.

CHEMICAL STRUCTURE

Growth hormone is a single polypeptide chain of 188 amino acids, which has been synthesized. It is a species specific, and its first observed effect was its influence on bone growth.

BIOLOGICAL EFFECTS OF GROWTH HORMONE

1. Stimulation of growth by increasing amino acids uptake and protein synthesis by the cell. Thus, urea level in blood falls with high GH level.
2. Growth hormone increases fat metabolism which, in turn, causes production of ketone bodies and loss of body fat.
3. Inhibition of glucose phosphorylation within the cell which, in turn, increases the level of glucose in the blood (hyperglycemia), and decreases the utilization of glucose in metabolism. These ketotic and hyperglycemic effects of growth

hormone (both of which are observed in diabetes mellitus) grant growth hormone a "diabetogenic" effect on the body. As a matter of fact, overproduction of growth hormone may burn out the β-cells of the pancreas and, thus, induce diabetes mellitus. Needless to say, this is not normally the cause of the disease, but only as incorporation of amino acids to protein is an energy requiring, and where glucose is mobilized for this purpose.

4. Growth hormone stimulates bone growth through acting at the epiphyseal plates and by increasing the incorporation of sulfate into chondriotin sulfate.

5. Growth hormone would increase potassium retention as a sequence of the increase in cells growth and number, it may also cause an increase in sodium retention. Meanwhile, phosphate excretion goes down with high growth hormone level. This is not a kidney effect, but rather a result of the large production and use of protein.

6. Growth hormone is also considered a lipolytic hormone, where it activates the breakdown of depot fat to free fatty acids (FFA) and glycerol. The production of FFA renders growth hormone ketogenic in its action.

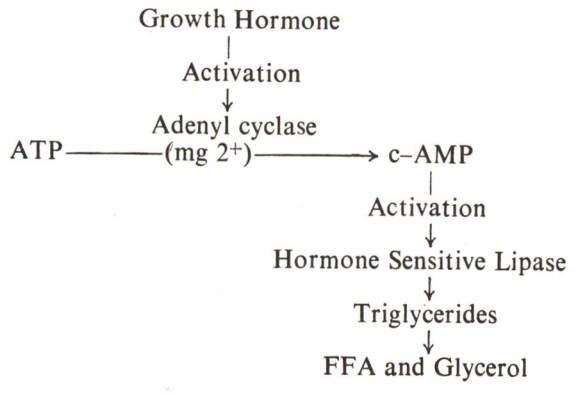

SECRETION AND FEEDBACK CONTROL OF GROWTH HORMONE

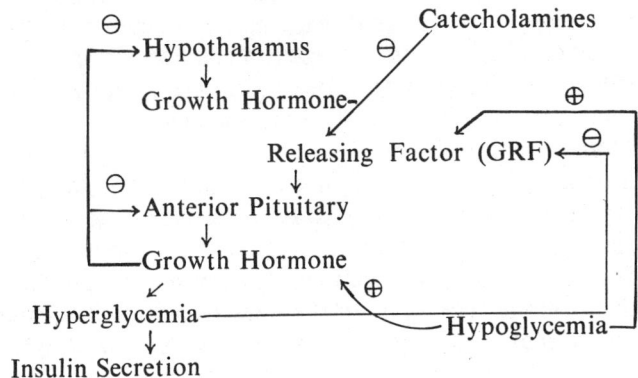

FACTORS INFLUENCING GROWTH HORMONE SECRETION

Growth hormone level in plasma is increased by the increase in amino acids level, a decrease in glucose, and an increase in the free fatty acids (FFA) level.

On the other hand, catecholamines which normally cause an increase in blood glucose level will exert an inhibitory effect on the growth hormone-releasing factor (GRF), though this inhibition effect may not be great. Growth hormone secretion is thought to increase during sleep.

Arginine causes an increase in growth hormone release, and casein intake along with exercise will enhance the hormone's release. Caffeine, alcohol, and other drugs suppress growth hormone's secretion, on the other hand.

Psychic factors and stress may cause the increase in growth hormone-releasing factor (GRF) secretion, and, consequently, growth hormone.

Hypoglycemia, as shown in the diagram, will, to a great extent, cause an increase in growth hormone secretion through positive feedback on both growth hormone and its releasing factor production. Estrogen, epinephrine, and glucagon may cause an increase in growth hormone secretion, while the glucocorticoids are thought to cause a decrease in growth hormone secretion. Thus, estrogen in the female may cause a higher level of growth hormone than in the male. Exercise by itself may also increase growth hormone level.

Thyroxine influences the action of growth hormone as well. For proper functioning of growth hormone, an optimum amount of thyroxine is required. As a matter of fact, thyroxine is necessary for the production of growth hormone itself.

Testosterone in the male may, for a brief period, enhance growth hormone secretion. In the case of a high testosterone secretion, a reduction in growth may result. A similar influence may result from a high secretion of estrogen.

There is a considerable resemblance of growth hormone to prolactin, including function and immunological properties.

ASSAY OF GROWTH HORMONE

1. Radioimmunoassay is the best method for assay of growth hormone.
2. The bioassay method, that measures the uptake of radioactive sodium sulfate into chondriotin sulfate.
3. By measuring the plasma amino acid level, which falls after an injection of growth hormone which incorporates them into tissue protein.
4. Infusion of insulin, causing hypoglycemia, which, in turn, causes an increase in growth hormone secretion.
5. The tibia test, which measures the increased width in epipheseal plates with GH intake. TSH, lactogenic hormone, and testosterone may synergize with GH in increasing the width of epipheseal plates.

REFERENCES

Fraser, T. R. 1963. Human growth hormone. *Scient. Basis Med. Ann. Rev.* 36.

Growth hormone (symposium). *Ann. N. Y. Acad. Sci.* 148:289.

Knobil, E. 1968. Growth hormone. *Proc. Int. Union of Physiol. Sciences* 6:77.

———, and Hotchkiss, J. 1964. Growth hormone. *Ann. Rev. Physiol.* 26:47.

Kupperman, H. S. 1969. *Human endocrinology.* 2nd. ed. Philadelphia; F. A. Davis Co.

Malaisse, W. J.; Malaisse-Lagae, F.; King, S.; and Wright, P. H. 1968. Effect of growth hormone on insulin secretion. *Amer. J. Physiol.* 215:423.

O'Brien, C. P., and Bach, L. M. N. 1970. Observations concerning hypothalamic control of growth. *Amer. J. Physiol.* 218:226.

Chapter 29

PROLACTIN

INTRODUCTION

Secretion of the mammary gland is intermittent and is manifested through integration of the endocrine system with the nervous system. The mammary glands secrete a complex product that is almost complete nutritionally.

Upon the child's suckling of the gland, the sympathetic nerve ending stimulates the reflex going to the spinal cord. This reflex stimulates the neurosecretory neurons of the paraventricular nucleus, which in turn will stimulate the production and release of oxytocin. Oxytocin is released into the bloodstream, reaches the myoepithelial cells, stimulates the contraction of these cells, and, thus, causes the ejection of milk and its release from within. Prolactin production continues on as child muses.

STRUCTURE

A single chain polypeptide (205 amino acids).

BIOLOGICAL EFFECTS

Stimulation of milk production. Other hormones, along with prolactin, are also necessary in milk production, such as estrogen which stimulates mammary duct system growth; progesterone for alveolar structure growth, normal levels of thyroxine, and the adrenal glucocorticoids are also needed.

During puberty, the induction of the mammary gland is under the stimulation of estrogen and progesterone.

Growth and branching of the ducts are highly stimulated by estrogen. Progesterone also causes some ductal growth, but mostly it causes the differentiation of the alveolar ducts.

The full development of the mammary glands is a complex phenomenon depending on the whole complex of the whole healthy, intact individual. The full development of mammary glands would be blocked if there were a lack of any of the other hormones such as growth hormone, thyroxine, insulin, or the corticoid hormones.

The metabolic actions of prolactin are generally similar to those of the growth hormone. In rats, prolactin has a lutealtropic effect, causing the corpus luteum to secrete estrogen and progesterone. In doves and pigeons, it stimulates the crop sac to produce a "milky substance."

ASSAY METHODS

1. Radioimmunoassay.
2. Pigeon Crop Sac Assay (biological assay)

METABOLISM

Little is known about its breakdown. Its estimated half-life is thirty minutes.

CONTROL OF PROLACTIN SECRETION

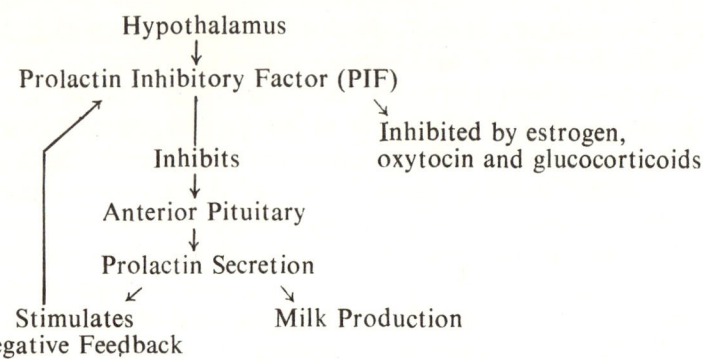

Estrogen inhibits the prolactin inhibitory factor (PIF), thus causing the increase of prolactin production via a double negative feedback loop. Oxytocin and the glucocorticoids also inhibit the prolactin inhibitory factor (PIF), thus increasing prolactin production.

Estrogen and progesterone both induce large number of mitosis in the mammary gland, thus, for it to function, it must be primed with estrogen and progesterone.

After priming with estrogen and progesterone, it is essential at this point that the corticoid hormones exert an effect; then prolactin comes in the picture.

Corticoids seem to induce the receptor site specific to prolactin.

To get prolactin to work, insulin must also be present. The two hormones work synergetically on the mammary gland to cause milk production—this, besides a normal adrenal function, GH, and thyroxine.

At the beginning of pregnancy, the mammary gland is stimulated with estrogen; then progesterone will cause the differentiation of the alveolar sacs.

Prolactin then synergizes with insulin, causing the unfolding, cell differentiation, growth, and production of milk.

REFERENCES

Grosvenor, C. E. 1965. Contraction of lactating rat mammary gland in response to direct mechanical stimulation. *Amer. J. Physiol.* 208:214.

Meites, J. 1968. Inhibition by prolactin of pituitary prolactin secretion and mammary function in rats. *Proc. Int. Union of Physiol Sciences* 6:210.

Meites, J., and Nicoll, C. S. 1966. Adenohypophysis: prolactin. *Ann. Rev. Physiol.* 28:57.

Voogt, J. L.; Chen, C. L.; and Meites, J. 1970. Serum and pituitary prolactin levels before, during and after puberty in female rats. *Amer. J. Physiol* 218:396.

Chapter 30

THYROID HORMONES

INTRODUCTION

The thyroid gland is a shield-shaped, lobulated gland with two lobes connected by an isthmus and is located in the middle third of the neck. Its size and shape are variable and it tends to be larger in women than in men. Its blood supply is very rich, being second only to the adrenal cortex. The gland is innervated through the sympathetic and parasympathetic nervous systems to the vascular system. Its cells, thus, receive no autonomic nervous supply, though the blood supply is regulated depending on the needs of the gland and the body in general. The blood supply, however, does not control the function of the gland.

The thyroid gland is composed of follicles of epithelial cells that are columnar in shape during increased activity of the gland and cuboidal in shape otherwise. The size of the follicular space being inversely proportional to the activity of the gland.

Hypothyroidism is manifested by a larger goiter which is due to large vesicular spaces and low cuboidal epithelium, while hyperthyroidism has some goiter which is due to hyperplasia and hypertrophy. In this case, it is cellular enlargement rather than the previous which is vesicular enlargement. The thyroid can store its hormones, which is a primary difference between thyroid and the other glands which usually do not store any significant quantities.

The thyroid gland is the first endocrine gland to become active in the embryo. In the human, it becomes functional before gestation. The gland is of endodermal origin and becomes an endocrine gland when the enzyme thyroglobulin protease digests the iodinated protein in the gland itself.

The thyroid gland is composed of many glandular follicles.

It is a globe of cells (each with a nucleus) surrounding gelatinous material in the center called a colloid.

The hormone of the thyroid gland is an iodinated amino acid (thyroxine).

$$HO-\underset{I}{\overset{I}{\bigcirc}}-O-\underset{I}{\overset{I}{\bigcirc}}-\underset{H}{\overset{H}{C}}-\underset{H}{\overset{NH_2}{C}}-C\underset{OH}{\overset{O}{\diagup}}$$

It is the derivative of two tyrosine molecules (takes two di-iodo-tyrosines).

THYROXINE (T_4)

The active hormone, contains four iodine atoms.

The dried substance of the thyroid gland has more thyroid physiological activity than could be accounted for on T_4 content, which indicates there are more hormones in the gland than just T_4.

A search revealed the identification of T_3 which is believed to have three times the potency of T_4. However, the rate of T_4 formation is the greatest.

It is suggested that T_4 produced in the gland is reduced to T_3 before it acts at the cellular level.

Usually the iodination of thyroglobulin is not done efficiently, but iodine may be pushed further into it to get more iodination.

Thyroid hormones affect the growth of the thyroid gland itself and its size.

Next to insulin, the thyroid hormone is most studied.

The first radioisotope used was I^{131} on the thyroid gland, and the tracing of thyroid hormone in body came into being.

Thyroglobulin is a protein which remains in the thyroid gland and is released through proteolysis from the cell.

If thyroglobulin is given to a subject, it may function as the hormone, as proteolysis would occur in the digestive tract and the amino acids are broken away to produce T_4 and T_3.

Monoiodo-thyronine (MIT) and diiodo-thyronine (DIT) do not appear in the blood to any great extent, and if given exogeneously they have no response, which rules them out as thyroid hormones.

Triiodothyronine (T_3) and tetraiodothyronine (T_4) are found in the blood but not in the free form. To function, they must be freed from a strong protein binding (thyroxine binding globulin, TBG).

The electrophoeretic pattern of plasma shows TBG as found between a_1 and a_2 globulins. This globulin ties up 50% of T_4 and about 75% of T_3 and has a high affinity constant. The protein binding of the hormone, or the TBG level, goes up along with a high estrogen level in pregnancy. This prevents pregnant women from having a very high level of free thyroxine, but the amount of bound thyroxine becomes abundant.

70% of the free thyroxine in the blood is in the form of T_4; 30% of the circulating level is the T_3 form.

STRUCTURE OF THE THYROID HORMONES

$$HO-\underset{}{\underset{I}{\bigcirc}}-O-\underset{I}{\overset{I}{\bigcirc}}-Alanine$$

3, 5, 3' tri-iodothyronine (T_3)
(mono-plus di-iodothyronine)

$$HO-\underset{I}{\overset{I}{\langle O \rangle}}-O-\underset{I}{\overset{I}{\langle O \rangle}}-\text{Alanine}$$

Thyroxine (T_4)
(di-plus di-iodothyronine)

Other than the above forms, there are no other stable combinations that are known. Thus, T_3 and T_4 (thyroxine) are the main active products found in the colloid in the vesicular space, although T_4 composes 95% over T_3.

The iodination of the tyrosine molecules occurs before coupling.

Since the thyroglobulin, which is the colloid and the polymer of T_3 and T_4 is a very large molecule, it cannot leak out of the vesicular space and into the circulation. TSH activates the protease enzymes, stimulating their production within the colloid, and these proteases break the large thyroglobulin molecule down into the low molecular weight T_3 and T_4 which are the hormones that are released into the circulation. Upon getting into the circulation, T_3 and T_4 are quickly bound to specific plasma proteins (thyroxine-binding globulins). The free forms of T_3 and T_4 are in equilibrium with the protein-bound forms. T_4 is more tightly bound to plasma protein than T_3; however, since it is the free form that hits the cell membrane of the target, T_3 has a considerably faster effect than T_4

FORMATION OF THE HORMONE

Iodine is an essential constituent of the hormone. Ocean spray carries iodine from the sea inland by wind to plants and, subsequently, to humans (land areas separated from the ocean by mountain ranges are iodine deficient).

Ingested iodine is converted to iodide in the intestinal tract and absorbed into the blood.

The thyroid gland has unique affinity for iodide.

The formation and release of the thyroid hormones occur in stages, all of which are sensitive to the thyroid-stimulating hormone (TSH). TSH causes all activities of the thyroid gland to increase. Iodine-trapping by the thyroid gland is the first stage in hormone formation, where the cellular and colloidal concentrations of iodine can be increased up to 350 times that of the plasma, which implies that there is an active uptake of iodide through an iodine pump.

It has a thyroid trap which concentrates iodide by actively transporting it from the circulation to the colloid.

Iodide is picked up by cells of the thyroid and is oxidized to elemental iodine by the enzyme peroxidase inside the thyroid cells.

Cells of the thyroid gland secrete to the lumen a protein called thyroglobulin, the colloid which is in the follicles.

The elemental iodine in cells reacts with tyrosine of the thyroglobulin in the lumen of thyroid follicles that spontaneously undergoes coupling to form thyroxine.

The thyroid cell does not secrete T_4 or T_3. The formation of the hormones occurs in the lumen of the thyroid follicles. This knowledge led to the synthesis of thyroproteins such as iodinated casein (3% thyroxine).

There is no enzymatic activity needed to produce thyroxine.

The thyroxine hormone is released through the proteolysis effect of thyroglobulin protease enzymes on the large thyroglobulin molecule, which hydrolyzes it to T_4 and T_3. T_4 and T_3 then diffuse from the lumen through the cell to the blood.

After release from the gland, T_4 and T_3 are carried chiefly in a loose combination with protein. 99.9% of the hormones circulate as protein-bound in the blood, compared to about 0.1% in the free form of T_4 or T_3.

The kidney picks up iodine and can recycle 90% of the iodine in the body. Thus, it takes a long time to become iodine deficient.

A lower iodine intake may cause more T_3 formation because of the lack of sufficient iodine to form T_4. T_3 being more active, it circulates faster than T_4.

As TSH reacts with specific receptors on the membrane, a sequence of reactions occurs in the membrane which include depolarization, lowering the energy charge in the cell, release of Ca^{2+}, activation of adenyl cyclase, and production of c-AMP.

When the membrane is depolarized, there is a rapid influx of substrates, such as amino acids and sugars, into the cell by facilitative diffusion.

The function of the thyroid cells is to produce thyroglobulin in large amounts. TSH aids in incorporation of amino acids and the intercore sugar which is put immediately onto the polypeptide chain.

The thyroglobulin molecules move to the vesicles, fold, and fuse to the Golgi apparatus, taking a tertiary structure at this point.

Molecular iodine accumulating in the intercellular spaces is actively pumped into the cell under the influence of TSH. Iodine is oxidized at the point of being added to the tyrosine molecule.

The thyroglobulin molecule has an abundance of tyrosines on its surface.

As molecular iodine is oxidized by the thyroid oxidase enzyme (which is present in large amounts in the thyroid gland), at the same time of thyroglobulin synthesis, the peroxidases are being synthesized.

Peroxidases get incorporated on the inside of the membrane.

When iodine is pumped into the cell, if one iodine is attached to the tyrosine molecule on the thyroglobulin at the apex of the cell, monoiodotyrosine is produced. If two iodines get attached to tyrosine, diiodotyrosine is produced. Alanine is lost and the two rings become linked together through oxygen, thus forming the hormone thyroxine.

$$\text{HO}-\langle\bigcirc\rangle_I^I-\text{CH}_2-\underset{\text{NH}_2}{\overset{\text{H}}{\text{C}}}-\text{COOH} + \text{HO}-\langle\bigcirc\rangle_I^I-\text{CH}_2-\underset{\text{NH}_2}{\overset{\text{COO}^-}{\text{CH}}} \longrightarrow$$

Alanine

$$\text{HO}-\langle\bigcirc\rangle_I^I-\text{O}-\langle\bigcirc\rangle_I^I\text{CH}_2-\underset{\text{NH}_2}{\overset{\text{H}}{\text{C}}}-\text{COOH}$$

Thyroxine turnover time is about sixty days, since, if molecular iodine is not readily available to the cell, a mechanism to hold it for I_2 availability is developed.

The smaller the follicles in size, the greater the synthesis; the larger the follicles in size, the more the release of the hormone.

CONTROL OF THE RELEASE OF THE THYROID HORMONES

The hypothalamus produces the thyrotropic-releasing factor (TRF) which stimulates specific cells in the anterior pituitary to release TSH which, in turn, activates the protease enzymes. These digestive enzymes cause the breakdown of the large thyroglobulin molecules and, thus, the release of T_3 and T_4. Both T_3 and T_4 have a negative feedback effect, mainly on the anterior pituitary and secondly on the hypothalamus, the mechanism through which a constant plasma level of the hormones is maintained.

Regulation of TSH from anterior pituitary is influenced by direct feedback on the pituitary (thyroxine acts directly on the pituitary and also through afferent input of the hormone on the thyrotropic area of the hypothalamus).

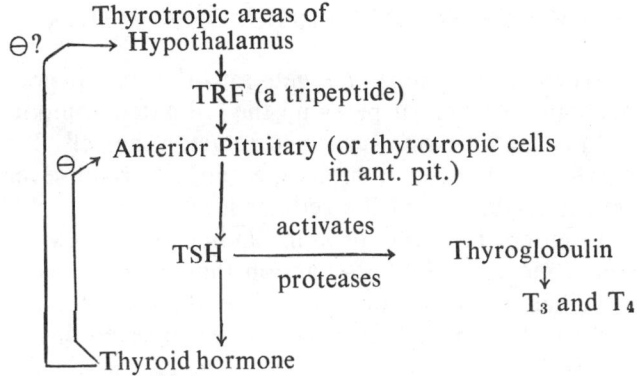

There is very good evidence of feedback of thyroxine on the anterior pituitary, put this is not clear with regard to the brain.

The effect of hormones on feedback on the hypothalamus is a sustained effect with a nonmajor role. The effect on the pituitary, however, is direct.

FUNCTIONS OF THE THYROID HORMONES

Increase in the metabolic activity on every cellular level, a system designed for protection of the organism against cold. Cold stimulates thyroid hormone production through the autonomic nervous system stimulation of the hypothalamus, causing an increase in TRF release and, thus, the whole system is turned on. In cold climates, there is developed a new, increased steady state of plasma levels of T_3 and T_4 by balancing the cold stimulation and the feedback inhibition.

The free level (which exerts the action) is small and is estimated in ng quantity.

THEORIES OF THYROID HORMONE ACTION

When thyroid hormone gets into the cell, it goes to the mitochondria where it picks up the iodinated compound.

T_3 is the hormone suspected to act in the cell. T_4 may get converted to T_3 after it enters the cell, (is about seven times more active than T_4 at the cellular level).

T_3 enters less into protein-binding than does T_4, which makes it more available for action than T_4.

T_3 has a shorter turnover time than T_4. Its half-life is thought to be 1.5 to 3 days, compared to seven days for T_4.

ACTION OF THYROXINE DEFICIENCY IN THE YOUNG

1. Decrease of O_2 consumption
2. Decrease in heat production and metabolism
3. Impairment of growth
4. Reproduction failure and deficiency of mental development

Hypothyroid adults are mentally sluggish. Hyperthyroid individuals show higher intelligence, have thin hair, and the mitochondria of certain of their tissues undergo swelling as a result of a high turnover of ATP and uncoupling of oxidative phosphorylation and the production of heat. This may account for the increase in O_2 consumption.

Giving a small quantity of the hormone will cause an increase in protein synthesis, an increase in the number of mitochondria, an increase in oxidative phosphorylation, and more heat production.

A large amount of the hormone, on the other hand, will definitely cause catabolism.

Euthyroid = histologically of normal appearance.

Hypothyroid = not producing enough of the hormone; may occur early or late in life. If early, it may cause irreversible brain damage and lead to cretinism. If it occurs later, it causes myxedemia (reversible with the hormone in any form).

SPECIFIC ACTIONS OF T_3 AND T_4

This is the calorigenic effect, where they increase the metabolic activity at almost all cellular levels. This results in:

1. Increased heat production.
2. Increase of fat, protein and carbohydrate utilization.
3. Increase in respiration rate and CO_2 production.
4. Increased heart rate.

All deficiencies are dependent on a lower metabolic rate.

Some tissues are affected by the thyroid hormones apart from metabolic rate, such as the reproductive system.

Thyroxine induces changes in membrane permeability and in enzyme synthesis. It also affects nucleic acid metabolism.

If thyroxine is exogeneously injected, after it circulates some of it is deiodinated, and some is secreted in bile partially as conjugated with glucuronic acid and partially as free thyroxine.

The free form gets chelated in the small intestines and stays in the lumen to be secreted.

Artificial hyperthyroidism can produce thyroprotein by iodination of protein, as an example—casein, which will make an animal hyperthyroidic and enhances its productivity and milk production. Milk secretion in this case may be increased by 20%. Iodinated casein, however, does not selectively increase metabolism. The increase in metabolism here is a general increase in the metabolic rate or more conversion of feed to milk as the animal eats more as well.

Hyperthyroidism is not usual in man; Grave's Disease is a manifestation of hyperthyroidism in man which is due to excess TSH activity.

RELATION OF THYROXINE TO OTHER HORMONES

1. Will increase requirement for insulin due to the increase in the metabolic activity.

2. Vitamin requirements will increase due to the fast usage of foodstuffs.

3. Since there is an increase in the metabolic activity in the muscle cells and not in nerve cells, there is a decrease in synaptic resistance and, thus, the central nervous system becomes hyperexcitable.

4. Increase in muscular activity.

5. Hyperthyroidism will result in a high heart rate, strong heartbeat, exhaustion, irritability, free perspiring, an in the normal built or thin, a great increase in skeletal growth rate of the young person unless plasma levels of T_3 and T_4 go so high that catabolic processes exceed metabolic processes, and results in great appetite.

Relationship between the Adrenal Cortex and Thyroid Hormone

In a moderately cold environment, ACTH and TSH production follow the same direction. However, in other circumstances their relationship is inversely related.

Effect of Thyroid Hormone on the Gonads

Generally, thyroxine stimulates an increase in the development of the gonads. Too much thyroxine, however, may cause sperm and hormone production derangement. Excess or deficiency of the hormone will influence menses in the female.

Relation to Parathyroid Glands

Thyroxine tends to increase osteoblastic activity more in the young, while it increases osteoclastic activity more in the old.

CHARACTERISTICS OF THYROTOXICOSIS

Generally, this is due to excessive activity of the anterior pituitary's cells that produce TSH rather than an excessive activity of the thyroid gland itself. This case is more common in females than in males. The gland's size may increase 2-3 times and its output of the hormone may increase up to fifteen times. Treatment may be through the removal of the thyroid gland itself, or its destruction with radioactive iodine.

GOITROGENIC COMPOUNDS

These are compounds that:
1. Block the uptake of iodine by the thyroid.
2. Block the incorporation of iodine through the membrane of epithelial cells.
3. Block iodine incorporation into thyroglobulin.

EXAMPLES OF GOITROGENIC COMPOUNDS

Thiouracil: Which prevents the oxidative coupling of tyrosines.

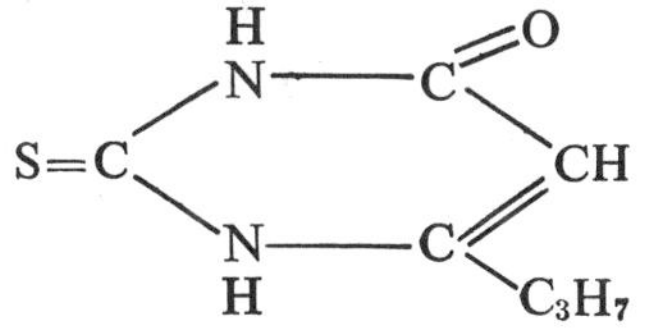

$$\begin{array}{c} \quad H O \\ \quad N\text{———}C \\ S=C CH \\ \quad N\text{———}C \\ \quad H C_3H_7 \end{array}$$

This becomes more effective if a side group is added to it, thus becoming methylthiouracil, or propylthiouracil (as above).

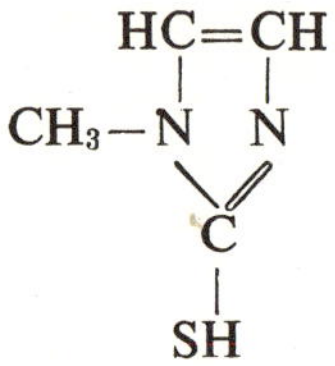

$$\begin{array}{c} HC=CH \\ | \quad | \\ CH_3-N \quad N \\ \diagdown \!\!/\!\!/ \\ C \\ | \\ SH \end{array}$$

Or, perchlorates $KClO_4$ and thiocyanates $KCNS$

Here chlorine replaces iodine and is taken up by the thyroid. If enough iodine is given in this case, it competitively inhibits chlorine. These compounds decrease the uptake of iodine by the cell, and production of T_3 and T_4 is stopped but the colloid formation continues and TSH level increases, resulting in a goiter.

Paraminobenzoic acid

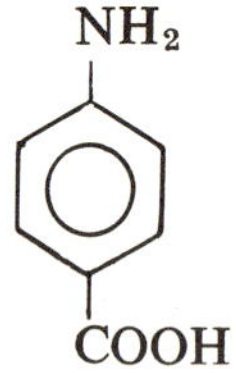

$$\underset{COOH}{\overset{NH_2}{\bigcirc}}$$

has some goitrogenic activity, but is not used as much as the previous ones.

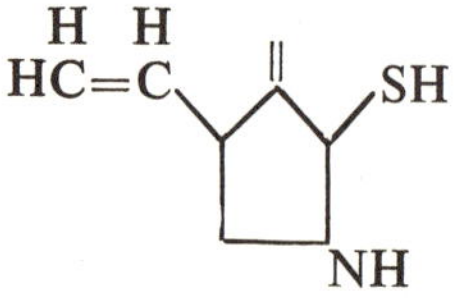

Goitrin, from cabbage.

ASSOCIATED DISEASES AND DYSFUNCTIONS

Graves' Disease of Hyperthyroidism

This is thyrotoxicosis associated with tumors.

Plumner's Disease

This is where a moderate hyperthyroidism is the case.

Hashimoto thyroiditis

An autoimmune disease where the thyroglobulin molecule leaks out to the blood and leads to the development of antibodies.

Myexedemia

A disease having symptoms of hypothyroidism, characterized by dullness, lack of vigor, low O_2 consumption, puffing of skin, high serum cholesterol, low protein-bound iodine, low I^{131} uptake by the thyroid, and low metabolic rate.

Cretinism

Cretinism is caused by a lack of production of thyroid hormone in childhood. Individuals with this condition are small, mentally deficient, and cannot reproduce. The disease may be related to iodine deficiency in soil. Exogenous iodine intake in salt does not reverse cretinism as the disease occurs at an early stage where the fetus does not receive the right amount of iodine at the right stage.

Cretinism will not respond to iodine.

In this case, there is an absence of the thyroid hormone. The cretin is short in stature, and his gastrointestinal tract grows relatively larger than does his skeletal area resulting in the cretin's having a very large abdomen. His skull does not grow as quickly as it should, but his tongue does, resulting in his tongue's filling his mouth. The development of his central nervous system will be impaired unless he is provided treatment within a few months after birth.

Hyperthyroidism is generally hypersecretion in the pituitary of TSH which leads to hyperproduction of T_3 and T_4, or an abnormality of the thyroid gland. To treat this, a goitrogenic compound is given. If this does not work, giving a high level of iodine will suppress the production of TSH.

The thyroid gland may be taken out partially or totally, and the patient may be given thyroid hormone exogeneously to bring a normal level of thyroxine in blood.

LATS

Long-Acting Thyroid Stimulating, this is found only in humans, not in animals (because humans have a longer lifespan). It acts in stimulating thyroid as TSH does, except that it comes from blood and is produced in immune cells. LATS is related to γ-globulin and has a longer half-life of eight hours, compared to thirty minutes for TSH.

Graves' Disease (hyperthyroidism)

Edema is formed behind the orbit of the eye and fatty degeneration of the extraoccular muscles occurs. When the eye is pushed forward, the condition is called exopthalmus. Graves' disease is caused by long-acting thyroid-stimulating compound (LATS), and its production is not limited by thyroxine levels.

Thyroiditis

This infection and damaging of the thyroid gland would lead to Hashimoto's disease. The damage to the thyroid gland's cells will cause thyroglobulin release into the circulation, which, in effect, will give rise to production of antibodies against thyroglobulin. The autoimmune response, thus, inactivates the thyroglobulin with the hormone attached, and hence the thyroxine is lost.

If hypothyroidism is caused by an iodine lack, and if checked early, it can be cured with iodine. If not checked early, the gland may get exhausted and iodine treatment may become ineffective.

Idiopatic colloid goiter may be caused by the intake of huge quantities of goitrogenic foods, such as cabbage, turnips, etc., which depress iodine intake or inhibit the oxidative coupling of the tyrosines.

Hypothyroid subjects are usually sensitive to cold, sluggish, just love to sleep, have scaly skin, develop edema, and their vocal cords are swollen causing their voices to be deep. When edema involves the heart of a hypothyroid individual, it may cause bradycardia.

The L-form of thyroxine is usually the biologically active form. This may be given for the purpose of lowering blood cholesterol at times. The form of tetraiodothyroacetic acid is 250% as effective in lowering blood cholesterol as thyroxine, but has small ability to stimulate BMR and O_2 consumption.

REFERENCES

Challoner, D. R. 1969. A direct effect of triiodothyronine on the oxygen consumption of rat fat cells. *Amer. J. Physiol.* 16:905

Danowski, T. S. 1962. *Clinical Endocrinology.* Vol. II: Thyroid. Baltimore: The Williams & Wilkins Co.

DeGroot, L. J. 1965. Current views on formation of thyroid hormone. *New England J. Med* 272:243, 297, and 355.

Deiss, W. P., Jr. 1963. Transport of thyroid hormones. *Fed. Proc.* 21:630.

Greer, M. A. 1962. The natural occurrence of goitrogenic agents. *Recent Progress in Hormone Research* 18:187.

McKenzie, J. M. 1968. Humoral factors in the pathogenesis of Grave's disease. *Physiol. Rev.* 48:252.

Means, J. H.; DeGroot, L. J.; and Stanbury, J. B. 1963. *The thyroid and its diseases.* 3rd ed. New York: McGraw-Hill Book Co.

Pitt-Rivers, R., and Trotter, W. R. 1964. *The thyroid gland.* Vols. 1 and 2. London and Washington: Butterworth Scientific Publications.

Rosenberg, I. S., and Bastomsky, C. H. 1965. The thyroid. *Ann. Rev. Physiol.* 27:71.

Straw, J. A., and Fregly, M. J. 1967. Evaluation of thyroid and adrenal-pituitary function during cold acclimation. *J. Appl. Physiol.* 23:825.

Szepesi, B., and Freedland, R. A. 1969. Effect of thyroid hormones on metabolism. IV comparative aspects of enzyme responses. *Amer. J. Physiol.* 216:1054.

Werner, S. C., and Nauman, J. A. 1968. The thyroid. *Ann. Rev. Physiol.* 20:213.

Wolff, J. 1964. Transport of iodide and other anions in the thyroid gland. *Physiol. Rev.* 44:45.

Chapter 31

PARATHYROID HORMONE

INTRODUCTION

The parathyroid glands belong mostly to higher mammals since animals with large bones that serve as sink for calcium required a well-developed system of parathyroid gland. Fish have not acquired a parathyroid gland, while birds have a parathyroid complex.

The parathyroid hormone is produced in specialized cells in the parathyroid gland (the chief cells) and which secrete about 1 mg per day. The gland contains the "chief cells" and the "oxyphil cells."

The chief cells are more abundant in the secretion and excretion of the hormone. Oxyphil cells become more predominant at old age. Rats, cats, dogs, and infant humans have no oxyphil cells in their parathyroid gland. The parathyroid glands in the neck vary in number in the various species. Reptiles have four, birds two to four, chickens two and there are four in pigeons. In mammals, there are two or four, and their position changes.

In humans, the parathyroid gland weighs about 16 mg of tissue, of which the parathyroid hormone weighs about 0.01

percent. The hormone is a polypeptide of 83 to 85 amino acid residues.

FUNCTION OF THE PARATHYROID GLAND

The primary role of the parathyroid gland is the regulation of calcium metabolism. Complete removal of the gland results in tetny and death. If a low calcium diet is used, the parathyroid gland's weight will markedly increase, while calcium and vitamin D given in excess amount will suppress the weight of parathyroid.

The parathyroid hormone itself relies on or must have, a small amount of vitamin D in order to cause calcium absorption. Vitamin D enhances calcium movement in the gut, and the movement of calcium and phosphorous in bone.

The thyroid gland in most mammals contains follicular cells and parafollicular cells which stain differently. The parafollicular cells are the "c-cells" and produce calcitonin.

As the parathyroid is the coarse adjuster in removing calcium from bone to blood, calcitonin is the fine adjuster and removes calcium from serum back to bone. Calcitonin or thyrocalcitonin is composed of ten amino acid residues, and has a molecular weight of 9500.

Parathyroid has one sulfhydryl bridge, which is essential for the hormone's activity.

CONTROL OF CALCIUM METABOLISM

There are three endocrine substances involved in the control of calcium metabolism. These are: parathyroid hormone, thyrocalcitonin (from the thyroid gland), and vitamin D from the skin.

Vitamin D is a hormone produced in the skin from a steroid precursor present in the diet and regulates calcium metabolism. The hormone form or the cellular active form of vitamin D is 25-OH-cholehydrocalciferol, which has a great influence on parathyroid hormone and on calcium metabolism.

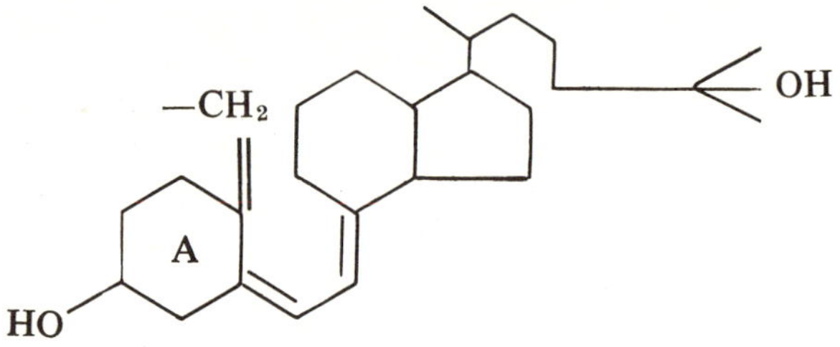

25-OH-cholehydrocalciferol

This form of vitamin D renders the cells of the intestines permeable to calcium, and after the calcium is absorbed by the intestines, it acts on bone.

Thus, the precursor is generated in the skin through the action of ultraviolet light, and is hydroxylated at carbon-25 in the liver. This substance in the nuclei of the brush border of the intestines is involved in the synthesis of RNA from DNA and the production of a substance that is the "transferase" which transfers calcium across the membrane.

INTERACTION OF PARATHYROID-CALCITONIN-VITAMIN D

The parathyroid hormone removes calcium from bone into blood, and also removes it from bone to put it back into bone; i.e., it remodels bone.

Parathyroid hormone renders the plasma membrane more permeable to calcium but is not involved in the formation of the carrier protein. Thus, the hormone changes the membrane permeability for calcium, getting it to the carrier protein. The hormone cannot work without the aid of vitamin D, while vitamin D, on the other hand, can do its work without the aid of parathyroid hormone.

Without parathyroid hormone, the bone does have some ability to remodel itself based just on calcium solubility alone.

The synthesis of carrier protein for calcium transport is done through the aid of vitamin D. Parathyroid hormone speeds up calcium transport. Some transport of calcium may occur without the aid of vitamin D, but this is done rather slowly. On the other hand, parathyroid hormone is not active without the aid of vitamin D. Thyrocalcitonin, meanwhile, is active without vitamin D, but it does require protein synthesis.

Parathyroid hormone increases phosphate excretion by the kidney and increases blood calcium.

Thyrocalcitonin feeds back in just the opposite direction, where a high level of blood calcium activates thyrocalcitonin which in turn moves blood calcium into the bones.

Vitamin D, meanwhile, is not related to the blood calcium level. Instead, it controls its own level through synthesis of the 25-OH-cholecalciferol form in the liver, the level of which is controlled through feedback mechanisms in the liver.

REFERENCES

Arnaud, C. D., Jr.,; Tenenhouse, A. M.; and Rasmussen, H. 1967. Parathyroid hormone. *Ann. Rev. Physiol.* 29:349.

Bell, N. H., and Stern, P. H. 1970. Effects of changes in serum calcium on hypocalcemic response to thyrocalcitonin in the rat. *Amer. J. Physiol.* 218:63.

Bronner, F. 1968. Studies in calcium homeostasis. *Proc. Int. Union of Physiol. Sciences* 6:271.

———, and Aubert, J. P. 1965. Bone metabolism and regulation of the blood calcium level in rats. *Amer. J. Physiol.* 209:887.

Copp, D. H. 1970. Endocrine regulation of calcium metabolism. *Ann. Rev. Physiol.* 32:61.

Danowski, T. S. 1962. *Clinical Endocrinology*. Vol. III: Calcium, phosphorus, parathyroids and bone. Baltimore: The Williams and Wilkins Co.

Fourman, P. 1968. *Calcium metabolism and the bone*. 2nd ed. Philadelphia: F. A. Davis Co.

Gaillard, P,; Talmage, R. V.; and Budy, A. M. 1965. *The parathyroid glands*. Chicago: University of Chicago Press.

Gitelman, H. J.; Kukolj, S.; and Welt, L. G. 1968. Inhibition of parathyroid gland function of hypermagnesemia. *Amer. J. Physiol* 215:483.

Hirsch, P. F., and Munson, P. L. 1969. Thyrocalcitonin. *Physiol. Rev.* 49:548.

Hurwitz, S.; Stacey, R. E.; and Bronner, F. 1969. Role of vitamin D in plasma calcium regulation. *Amer. J. Physiol.* 216:254

Jackson, W. P. U. 1967. *Calcium metabolism and bone disease* Baltimore: The Williams & Wilkins Co.

Krawitt, E. L., and Schedl, H. P. 1968. *In vivo* calcium transport by rat small intestine. *Amer. J. Physiol.* 214:232.

Morii, H., and Deluca, H. F. 1967. Relationship between vitamin D deficiency, thyrocalcitonin, and parathyroid hormone. *Amer. J. Physiol* 213:358.

Rasmussen, H. 1961. The parathyroid hormone. *Sci. Amer.* 204 (4):56.

Richelle, L. J. 1968. Calcification in biological systems: bone, mineral. *Proc. Int. Union of Physiol. Sciences* 6:265.

Sammon, P. J.; Stacey, R. E.; and Bronner, F. 1970. Role of parathyroid hormone in calcium homeostasis and metabolism. *Amer. J. Physiol.* 218:479.

Schedl, H. P.; Osbaldiston, G. W.; and Mills, I. H. 1968. Absorption, secretion and precipitation of calcium in the small intestine of the dog. *Amer. J. Physiol.* 214:814.

Snapper, I., and Bloom, M. 1964. The parathyroid glands. In Duncan, G. G. (ed.) *Diseases of Metabolism*. 5th ed. Philadelphia: W. B. Saunders Company, p. 1280.

Urist, M. R. 1969. Rarefying diseases of bone (with a comment on translocation of remodelled bone in osteoporosis). In Bittar, E. E., and Bittar, N. (eds.) *The biological basis of medicine.* New York: Academic Press, Inc., Vol 3, p. 425.

Vaes, G. 1968. Biochemical aspects of bone resorption. *Proc. Int. Union of Physiol Sciences* 6:274.

Chapter 32

INSULIN

INTRODUCTION

The endocrine secretion of the pancreas is less than 1% of its total secretion.

All vertebrates have small aggregates of 20-100 various cell types which compose the pancreas islets.

They have the alpha$_1$ cells and the delta cells which sercrete gastrin and alpha$_2$ cells which secrete glucagon. The beta cells secrete insulin.

Insulin is long known in human medicine because its functional absence is recognized by the disease diabetes mellitus. This is a widespread disease and was recognized by the Egyptians before a 1000 years B.C.

In 1960 the first human protein completely synthesized was insulin.

Insulin is composed of two amino acid chains with three disulfide linkages which are necessary for its biological activity.

Insulin is the "anabolic hormone." it has different effects on different cells. It basically regulates carbohydrate and fat metabolism.

Insulin and chymotripsin have many similarities. They have similar structural configuration, proteolytic activity, and disulfide groups that came from the same source.

Insulin is a double amino acid chain which is formed as a single chain of 84 amino acids. In this form, it is called proinsulin which is secreted into the bloodstream as well as insulin, and also has action on carbohydrate and fat metabolism.

SYNTHESIS

The precursor of insulin is the single-chain polypeptide "proinsulin" with the following structure:

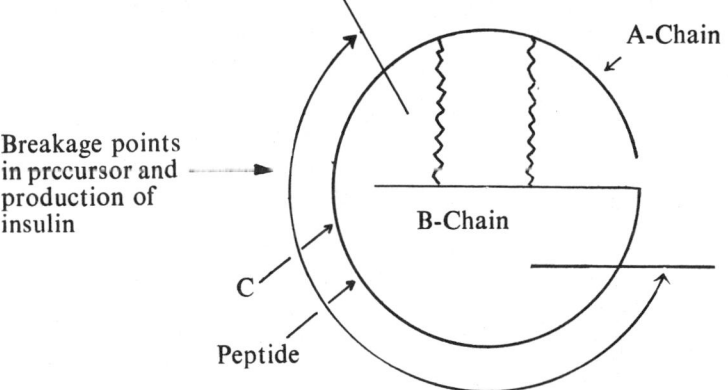

Proinsulin has little biological activity, but is measurable similarly in the blood, and does respond to glucose changes similarly to insulin.

There are two splits in proinsulin: one between the site 30-31 amino acids; the other between 63-64. The part removed from proinsulin is the C-peptide. Upon the removal of the C-peptide, the molecule becomes insulin which, by far, is a more active form.

Insulin has cross linkages between 19-20, 6-11, and 7-7 amino acids. These cross linkages give the molecule its curling property. These disulfide cross linkages also are required for insulin's activity.

The human pancreas contains about one million of the islets of Langerhans in one gram weight of mostly β-cells. These β-cells are continually secreting insulin at an output of 1–2 mg per day.

Half of the insulin secreted goes to the liver.

Pancreatic secretion of insulin is primarily determined by the circulating level of blood glucose.

INSULIN'S ACTION ON MEMBRANES

Insulin's action on membranes is proteolytic in nature. Upon its release in the blood, it will cause a deep cleft on the membrane surface.

FACTORS THAT INFLUENCE INSULIN RELEASE

1. Digestion metabolic products such as glucose, amino acids, and ketones stimulate insulin release.
2. Gastrointestinal factors, such as digestive enzymes and gut-glucagon.
3. Glucagon.
4. Orinase, which stimulates insulin secretion provided there are some functional beta cells present. Thus, orinase has no effect on a young patient with juvenile diabetes since there are no functional beta cells.

BIOLOGICAL ACTION

1. Influence of insulin on the permeability of glucose to various tissues.
2. Insulin's role in the inhibition or stimulation of enzyme synthesis involved in intermediate metabolism.

3. Changes in intermediate metabolism due to the build-up of substrates.

INSULIN AND CARBOHYDRATE METABOLISM

Due to the varying permeability of glucose in tissues, insulin's role becomes important in this aspect. The brain and liver are the only tissues that do not require insulin for glucose entrance into the cell.

The mechanism of insulin's role in the entry of glucose into the cell is not an active process, but is a facilitative diffusion process. Insulin helps glucose couple with some unknown membrane component which then passively diffuses across the membrane's concentration gradient. The presence of the gradient is due to glucose's not existing in the free glucose form inside the cell. Thus, insulin causes the increase in glucose movement instead of transport into cell. This action of insulin is a true example of facilitative diffusion and does not require energy. Insulin simply binds with the cell membrane to let glucose diffuse through. In other words, insulin enhances the permeability of glucose through the cell membrane.

An absolute requirement for insulin secretion is Ca^{2+}, and glucose is needed in an optimal amount. Na^+ is also required for the movement of glucose into the cell through facilitation.

There are many factors, other than blood glucose, that increase insulin release; these are: ACTH, glucagon, amino acids, other monosaccharides, and medium-and short-chain fatty acids. Epinephrine is the only compound that causes inhibition of insulin release. Insulin also will influence carbohydrate metabolism as it stimulates or inhibits various enzymes, mainly in the liver. Phosphorylation of glucose with the presence of insulin and the stimulation of glucose-6-phosphatase which dephosphorylates glucose-6-phosphate with the lack of insulin are the two most important systems influenced by insulin or its absence. The latter case will result in shifting the metabolism of the liver to glucose secretion instead of glucose utilization.

Up to the age of 8-9 days in humans, glucose moves freely in all tissues without the aid of insulin. After this age, however, insulin is required.

Both insulin and proinsulin are active in the body. Proinsulin exerts a greater effect on fat cells than other cells. It is, however, 10% as active as insulin, although its activity lasts longer than insulin.

INFLUENCE OF INSULIN ON SKELETAL MUSCLE CELLS

Besides insulin's facilitating the movement of glucose into muscle cells, it also causes an increase in the glucokinase enzyme activity which converts glucose to glucose-6-phosphate.

This enzyme, consequently, increases the conversion of glucose into glycogen. Thus, insulin stimulates glycogen synthesis and deposition.

Insulin may also accomplish this directly by increasing the synthesis of the glycogen synthetase enzyme and by decreasing the phosphorylase enzyme activity. Insulin is the only hormone that decreases the activity of adenyl cyclase, causing the net decrease of c-AMP in cell, and this may be the reason of the decrease in phosphorylase activity. Insulin also causes an increase in the phosphodiesterase enzyme activity.

By increasing glucose movement into cell, insulin causes movement of K^+ into the cell and the increase of intracellular space.

Insulin increases the activity of phosphofructokinase and pyruvic kinase, thus increasing the rate of glycolysis.

Insulin, thus, increases glucose conversion to CO_2 by increasing glucose oxidation.

Exercise may increase glucose movement inside the cell without the need for insulin. In this case, exercise probably causes a change in the membrane's electrical potential which changes the membrane permeability and lets glucose move in.

INSULIN AND FAT METABOLISM

Insulin binds to the fat cell surface and facilitates the movement of glucose in the fat cell as well. Also, through activation of glucokinase and glycogen synthetase in the fat cell, glycogen synthesis inside the fat cell will be increased.

Insulin's lack in fat metabolism will very highly complicate diabetes. With its lack, no glucose enters the adipose tissues which normally convert glucose to triglycerides for storage, and there is a breakdown of the triglycerides to fatty acids causing an increase of fatty acids in the blood. These fatty acids become the major energy source for most tissues. The liver, too, becomes overburdened handling the fatty acids without insulin. The huge influx of fatty acids will cause a buildup of acetyl-Co-A, which, in turn, blocks pyruvate formation. As a consequence, the liver produces ketone bodies such as acetoacetate, acetone, and β-hydroxy butyrate which, all being metabolic acids, will result in metabolic acidosis.

Insulin will increase the α-glycerol phosphate level leading to increased fat synthesis.

It also activates the hexose monophosphate pathway, thus increasing the availability of NADPH which is necessary for fat synthesis.

Insulin increases the conversion of acetyl-Co-A to fatty acids, thus increasing synthesis of fat from glucose, which is species-dependent. It also increases the citrate cleavage enzyme activity. As acetyl-Co-A inside the mitochondria combines with oxaloacetic acid forming citrate, citrate diffuses out of the mitochondria and is cleaved by the citrate cleavage enzyme to oxaloacetic acid and acetyl-Co-A. Getting acetyl-Co-A outside the mitochondria will cause an increase in fat synthesis.

Insulin, thus, is an anabolic hormone which increases fat synthesis.

There is an increase in the lipoprotein lipase enzyme which acts on the triglycerides in the blood-stream that are bound to protein, thus releasing free fatty acids (FFA) inside the cell and leaving protein outside.

In the fat cell, most of the deposition of lipid caused by insulin is due to the activation of the lipoprotein lipase which causes the deposition of FFA inside the fat cell.

Insulin causes a decrease in the cytoplasmic lipase enzyme activity which breaks down the triglycerides into FFA and glycerol. In this case, however, the FFA are released outside the cell, and glycerol goes back to the liver for metabolism.

INSULIN AND PROTEIN METABOLISM

Insulin is thought to increase the permeability of cells to amino acids.

It may activate RNA synthesis and, thus, stimulate protein synthesis within the cell and in specific locations. Insulin, thus, is essential for normal growth, and here its role is similar to that of growth hormone, or it might work in synergism with growth hormone. Insulin increases incorporation of amino acids into protein, increases glutathione synthesis, decreases the release of amino acids from the cell, and increases the incorporation of precursors into nucleic acids.

MISCELLANEOUS ACTION OF INSULIN

1. It alters membrane potential.
2. Increases incorporation of sulfate into chondriotin sulfate.
3. Decreases the free sulfhydryl content of muscle.

4. Increases chondriotin sulfate in:
   a. Skeletal muscle
   b. Cardiac muscle
   c. Adipose tissue
   d. Leucosites
   e. Crystalline lens
   f. Aqueous humor
   g. Pituitary.

Insulin does *not* affect the following: brain cells, kidney tubules, intestinals mucosa, red blood cells and liver cells.

SYMPTOMS AND CAUSES OF DIABETES MELLITUS

1. Increase of blood glucose.
2. Great increase of ketone excretion.
3. Increased gluconeogenesis and increased nitrogen excretion.
4. Loss of Na^+, K^+, and Cl^- as there is increased glucose in the urine which pulls water and electrolytes out with it.
5. Polyuria, as the reabsorptive capacity of the kidney is exceeded.
6. Polydipsia, due to extracellular and cellular dehydration.
7. Polyphagia.
8. Acetone breath, as the ketone bodies' production rate exceeds the ability of the muscle and heart cells to metabolize them.
9. Asthenia, due to the generally poor metabolic state.
10. Cardiovascular abnormalities, due to the shift to fatty acid metabolism, and the large increases in cholesterol resulting in arteriosclerosis; poor peripheral circulation, hypertension, and strokes may occur.
11. Coma, which is an acid-base problem due to the increase in ketone bodies resulting in metabolic acidosis. When the pH falls to 6.9-7.0, coma occurs.

BIOASSAY FOR INSULIN

1. Rate of glucose utilization by rat diaphram (related to radioimmunoassay).
2. Glucose Tolerance Test.

Through glucose utilization by rat diaphram, overproduction of adrenal corticoid hormones leads to deamination of protein to a-ketoacids which are in turn converted to glucose. This raises the blood glucose and leads to more secretion of insulin and subsequently the exhaustion of the β-cells.

ACTH also will raise blood glucose by stimulation of glucocorticoid production which increases insulin secretion.

Growth and lactogenic hormones both are hyperglycemic and are antagonistic to insulin.

In the growing person, it is advantageous to have high secretion of insulin and growth hormone, where there will be a sparing action of insulin on protein, as glucose will be used for energy and amino acids are spared for build-up of tissues.

To eliminate insulin production in an intact animal, alloxan is given, which differentially destroys the β-cells of the islets of Langerhans, resulting in zero insulin secretion.

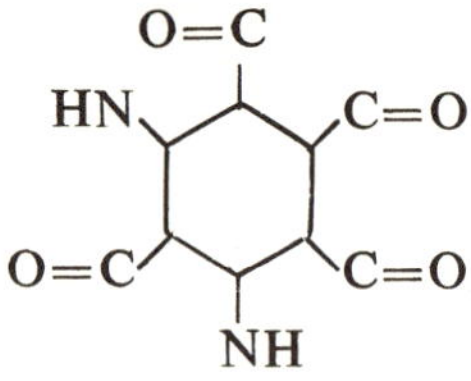

Alloxan

ORALLY EFFECTIVE AGENTS FOR CONTROL OF DIABETES MELLITUS

Tolbutamide

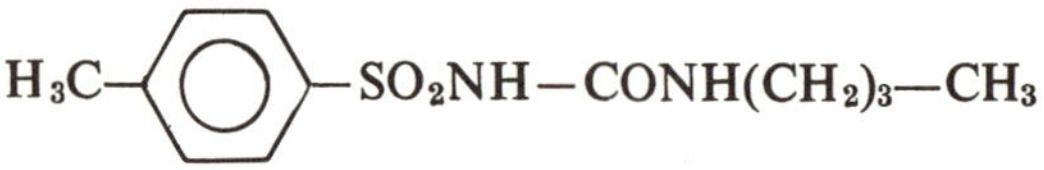

$H_3C-\langle \bigcirc \rangle-SO_2NH-CONH(CH_2)_3-CH_3$

Cortutamide

$$H_2N-\text{C}_6H_4-SO_2NH-CONH(CH_2)_3-CH_3$$

Chlorpropamide

$$Cl-\text{C}_6H_4-SO_2NH-CONH(CH_2)_2-CH_3$$

All of the above stimulate the release of insulin from β-cells of islets of Langerhans.

Phenformin

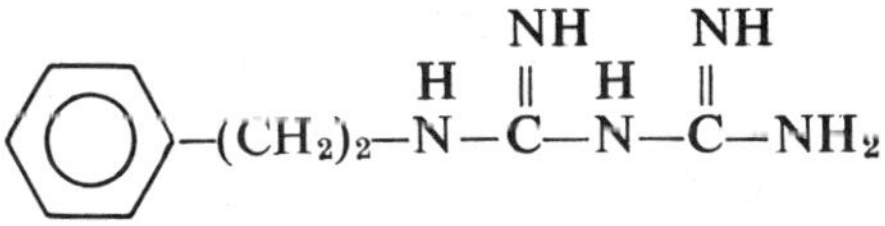

$$\text{C}_6H_5-(CH_2)_2-\underset{H}{N}-\underset{\parallel}{\overset{NH}{C}}-\underset{H}{N}-\underset{\parallel}{\overset{NH}{C}}-NH_2$$

It has a direct effect on the increase of glucose uptake by cell.
The third way to treat diabetes is by injection of insulin.

TYPES OF INSULIN

1. Zn-Insulin: Half-life is 25–30 minutes, slows down absorption.
2. Protamine zinc-insulin: further slows down its absorption.
3. Lente-insulin: utilized slowly, highly effective.
Secretion rate of insulin in the human is 50 I.U./day.

REFERENCES

Bishop, J. S.; Steele, R.; Altszuler, N.; Dunn, A.; Bjerknes, C.; and deBodo, R. C. 1965. Effects of insulin on liver glycogen synthesis and breakdown in the dog. *Amer. J. Physiol.* 208:307.

Cahill, G. F., Jr.; Owen, O. E.; and Felig, P. 1968. Insulin and fuel homeostasis. *Physiologist* 11:97.

Diabetes mellitus and obesity. 1968. *Ann. N.Y. Acad. Sci.* 148:573.

Duncan, G. G. 1964. Diabetes mellitus. In Duncan, G. G. (ed.) *Diseases of metabolism.* 5th ed. Philadelphia: W. B. Saunders Company.

Frohman, L. A. 1969. The endocrine function of the pancreas. *Ann. Rev. Physiol.* 31:353.

Grodsky, G. M., and Forsham, P. H. 1966. Insulin and the pancreas. *Ann. Rev. Physiol.* 28:347.

Hales, C. N. 1968. The glucose fatty acid cycle and diabetes mellitus. In Bittar, E. E., and Bittar, N. (eds.) *The biological basis of medicine.* New York: Academic Press, Inc., Vol. I, p. 309.

Henderson, M. J.; Morgan, H. E.; and Park, C. R. 1961. Regulation of glucose uptake in muscle. IV. The effect of hypophysectomy on glucose transport, phosphorylation, and insulin sensitivity in the isolated, perfused heart. *J. Biol. Chem.* 236:273.

Issekutz, B., Jr., and Paul, P. 1968. Intramuscular energy sources in exercising normal and pancreatectomized dogs. *Amer J. Physiol.* 215:197.

Issekutz, B., Jr.; Paul, P.; and Miller, H. I. 1967. Metabolism in normal and pancreatectomized dogs during steady-state exercise. *Amer. J. Physiol.* 213:857.

Kanazawa, Y.; Kuzuya, T.; and Ide, T. 1968. Insulin output via the pancreatic vein and plasma insulin response to glucose in dogs. *Amer. J. Physiol.* 215:620.

Levine, R. 1961. Concerning the mechanisms of insulin action. *Diabetes* 10:421.

Levine, R. A., and Lewis, S. E. 1967. Hepatic glycogenolytic activity of cyclic 3′, 5′-AMP and its monobutyryl derivative. *Amer. J. Physiol.* 213:768.

Mallaisse, W. J.; Mallaisse-Lagae, F.; and Wright, P. H. 1967. Effect of fasting upon insulin secretion in the rat. *Amer. J. Physiol.* 213:843.

Morgan, H. E.; Henderson, M. J.; Regan, D. M.; and Park, C. R. 1961. Regulation of glucose uptake in muscle. I. The effects of insulin and anoxia on glucose transport and phosphorylation in the isolated perfused heart of normal rats *J. Biol. Chem.* 236:253.

Park, C. R.; Morgan, H. E.; Henderson, M. J.; Regen, D. M.; Cadenase, E.; and Post, R. L. 1961. The regulation of glucose uptake in muscle as studied in the perfused rat heart. *Recent Progr. Hormone Res.* 17:493.

Park, C. R.; Reinwein, D.; Henderson, M. J.; Cadenas, E.; and Morgan, H. E. 1959. The action of insulin on the transport of glucose through the cell membrane. *Amer. J. Med.* 26:674.

Post, R. L.; Morgan, H. E.; and Park, C. R. 1961. Regulation of glucose uptake in muscle. III. The interaction of membrane transport and phosphorylation in the control of glucose uptake. *J. Biol. Chem.* 236:269.

Pruett, E. D. R. 1970. Glucose and insulin during prolonged work stress in men living on different diets. *J. Appl. Physiol.* 28:199.

Rieser, P. 1968. *Insulin, membranes and metabolism*. Baltimore: The Williams & Wilkins Co.

Role of insulin in membrane transport (symposium). 1965. *Fed. Proc.* 24:1039.

Sutherland, E. W., and Robinson, G. A. 1969. The role of cyclic AMP in the control of carbohydrate metabolism. *Diabetes* 18:797.

Weber, G. 1968. Hormonal control of gluconeogenesis. In Bittar, E. E., and Bittar, N. (eds.) *The biological basis of medicine*. New York: Academic Press, Inc., Vol. 2, p. 262.

Chapter 33

GLUCAGON

INTRODUCTION

Glucagon is a polypeptide hormone of 29 amino acid residues. The portion between 19-22 amino acids is the part necessary for lipolysis. The 24-29 amino acids portion is necessary for glycogenolysis. Thus, glucagon has two different immunogenic sites.

FUNCTION

There are two types of glucagon, the gut or gastrointestinal and the pancreatic glucagons, the latter being secreted by the a_1-cells of the pancreas. A decrease in blood sugar will cause the release of pancreatic glucagon which activates adenyl cyclase enzyme, thus making c-AMP from ATP. c-AMP activates phosphorylase-b to phosphorylase-a which, in turn, breaks down glycogen to glucose-1-ph leading to the release of glucose in blood.

ACTION VS. INSULIN

Glucagon, in contrast to insulin, may decrease glycogen synthetase activity, thus causing release of glucose and a more prolonged activation of gluconeogenesis. This may be done at the expense of protein deamination which is activated by glucocorticoids. The activation of phosphorylase enzyme by glucagon is a fast process, and glucose, consequently, is released in the blood in a short time.

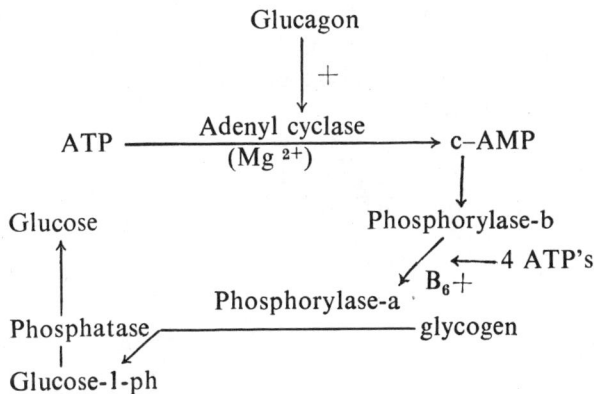

In this case, if glucagon ensures a continuous energy supply, it may enhance insulin's release to an extent. Although pancreatic glucagon may indirectly cause some release of insulin, the gastrointestinal glucagon causes a major release of insulin and the flow of nutrients. After a rich meal, for example, release of gut-glucagon will be followed by a release of insulin.

Glucagon's action is similar to that of epinephrine. Perhaps there are two different receptor sites with two different adenyl cyclases on the hepatic cell membrane, one of which is sensitive to epinephrine, the other to glucagon. The release of epinephrine may activate glucagon's release from the pancreas. Although glucagon is justifiably taken for granted as an antagonist to insulin, yet glucagon may also be thought of as a synergist of insulin. Simply, when insulin's release causes low blood sugar, this in itself causes the release of glucagon to raise back the level of blood sugar, and vice versa.

GLUCAGON AND LIPOLYSIS

Glucagon is considered a lipolytic hormone. It activates adenyl cyclase on the cell membrane which produces c–AMP from ATP. C–AMP activates a hormone-sensitive lipase on the fat cell membrane by further phosphorylating it. The phosphorylated lipase mobilizes the depot fat by breaking down triglycerides to free fatty acids (FFA) and glycerol. Glycerol goes back to liver for metabolism, while FFA go to the blood and are used for energy, as shown in the following scheme:

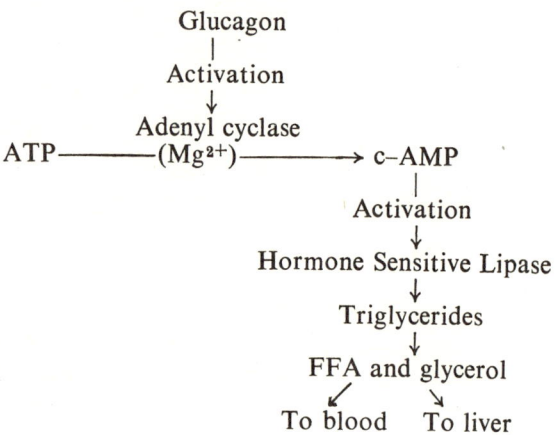

REFERENCES

Foa, P. P., and Galansion, G. 1962. *Glucagon: chemistry and function in health and disease.* Springfield, Il.: Charles C. Thomas, Publisher.

LaRaia, P. J.; Craig, M.; and Reddy, W. J. 1968. Glucagon: effect on adenosine 3′–5′ monophosphate in the rat heart. *Amer. J. Physiol.* 215:968.

Mayer, S. E.; Namm, D. H.; and Rice, L. 1970. Effect of glucagon on 3′, 5′–AMP, phosphorylase activity and contractility of heart muscle of the rat. *Circ. Res.* 26:225.

Moir, T. W., and Nayler, W. G. 1970. Coronary vascular effects of glucagon in the isolated dog heart. *Circ. Res.* 26:29.

Chapter 34

GONADOTROPIC AND SEX HORMONES

STEROID STRUCTURES

The steroid molecule is a small one. Its molecular weight is in the area of 300 compared to 300,000 for protein.

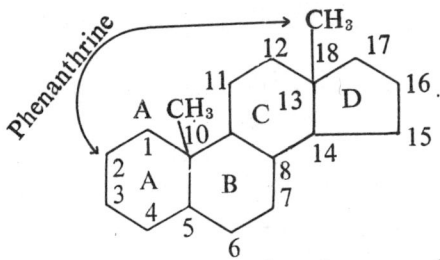

A, B, and C rings are 6-carbon member rings, and are saturated. They are a perhydrophenanthrine.

The D ring is a cyclopentane and is a fully saturated 5-carbon ring. The steroid nucleus is thus a cyclopentanoperhydrophenanthrine. It is a hydrocarbon by itself and is fully saturated.

The steroid itself is composed of the first three basic rings with number five carbon a key point in the configuration.

If a steroid is of eighteen carbons, it is called estrane.

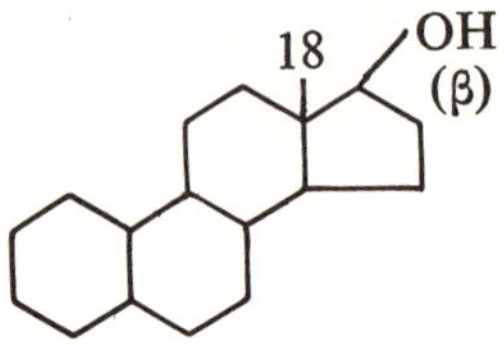

Estrane

If it is of nineteen carbons, it is called androstane (all-male type)

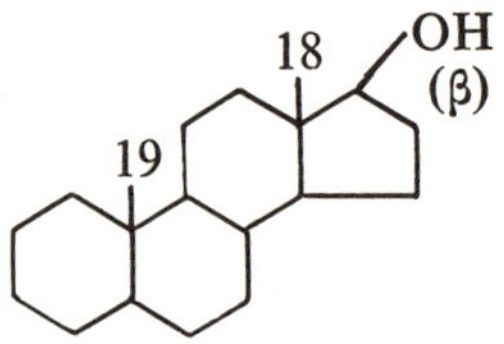

Androstane

Estrane includes all the female types or the estrogen type.

C-21 Pregnane

---

*This OH being an α type (dotted lines), it points away from the methyl group and is on the other or opposite plane. Where this OH points, i. e., if it is α or β, makes a big difference to the receptor site. Its relation to where the methyl group points causes all the configuration changes on the molecule.

This OH is usually involved in binding the steroid to protein. If its configuration is β, it is pointing toward the onlooker along with the methyl group and is drawn in solid line.

The α form is drawn in dotted lines and is pointing away from the onlooker.

NAMING

E_1 is estrone.
E_2 is estradiol.
An androstane would be as follows:

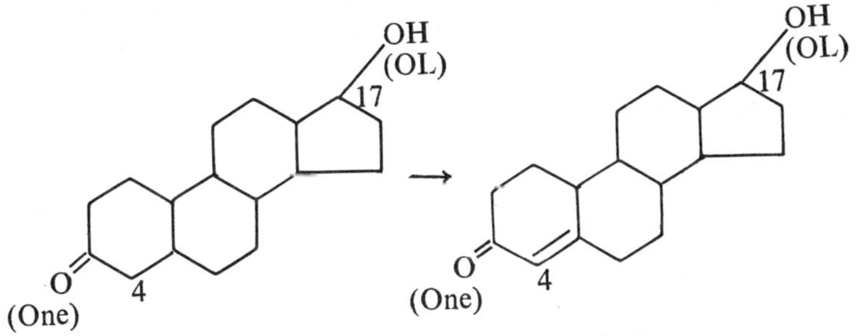

Androstane → Androstene

If it is OH on C-17, spelling will be (OL); if it is O group, its spelling is (one).

The double bond at C-4 will make the name δ^4-androstene.

Naming for the OH or O group comes first. Thus testosterone will be named δ^4-androstene-17β-OL-3-one.

If estrane has three double bonds, it will be called estriene and will have $\delta^{1,3,5}$ meaning that the double bonds are at C-1, 3, and 5.

All estrogens have conjugation in the A ring and have phenolic hydroxyl groups. This sets them apart from all other steroids. The unsaturation of their A ring makes them more polar than the rest of the steroids.

STEROID BIOSYNTHESIS

The cleavage enzyme desmolase cleaves the side chain of cholesterol at carbon 17, thus producing pregnenolone which is the precursor to progesterone. A reductase enzyme converts δ^5 to δ^4, i.e., the unsaturation comes at C-4. An isomerase converts the OH group at C-3 into a keto group, and thus progesterone is synthesized.

All steroid biosynthesis begin with a C-21 compound.

A 17 a-OH-hydroxylase enzyme puts a 17a-OH group at the C-17 position, thus converting pregnenolone into dehydroepiandrosteine. Another desmolase splits away the $CH_3-C=O$ side chain.

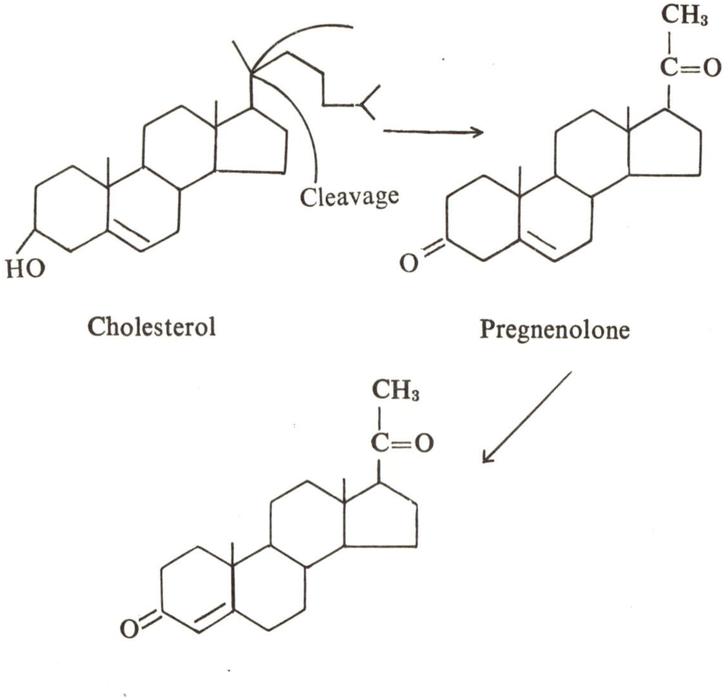

BIOSYNTHESIS AND METABOLISM OF SEX HORMONES

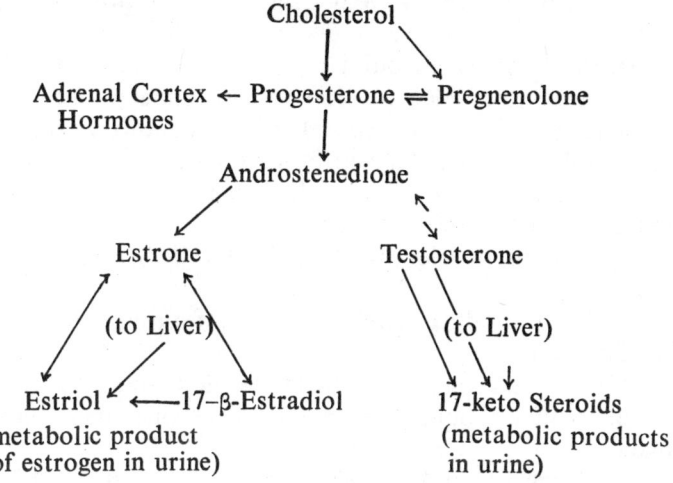

As shown in the scheme, progesterone is the intermediate for both the gonadal hormones and the adrenal steroids, and androstenedione, present in both males and females, is the immediate precursor to testosterone and estradiol.

FOLLICLE-STIMULATING HORMONE (FSH)

FSH stimulates both male and female gametes.

The primitive cell, spermatogonia, does undergo some differentiation into primary spermatocyte. Through the aid of FSH, this primary spermatocyte is converted to spermatid which, in turn, becomes the mature spermatozoa through the action of testosterone.

Spermatogonia $\xrightarrow{\text{Differentiation}}$ primary spermatocyte

$\xrightarrow{\text{FSH}}$ Spermatid $\xrightarrow{\text{Testosterone}}$ Mature spermatozoa

As this process goes on, testosterone substitutes for FSH.

LH acts on the Leydig cells for secretion of testosterone.

In the ovary, cells build around the ovum to make follicles. Primary follicles develop into secondary and tertiary follicles. The primary stage develops under pituitary influence. Certain follicles reach a maturity state, but some do not. Very few of the thousands of ova available reach the maturity stage to be fertilized by spermatozoa.

FSH helps develop the follicles, while LH causes ovulation.

Follicles produce small amounts of progesterone along with estrogen.

After ovulation, the follicle develops into the corpus leuteum which produces mainly progesterone also some estrogen.

Although LH and FSH seem to act independently of each other, they are difficult to obtain when not contaminated with each other.

The placenta also produces gonadotrophic hormones. Woman and the horse are the best examples in this regard.

HORMONES OF THE OVARY

Estrogens

This is a category of hormones that causes vaginal cornification. Estrogen is secreted by the ovarian follicle cells and the corpus leuteum. The placenta and testes also secrete it.

Progesterone is also another category of hormone and is defined as a substance that produces secretory changes in the already estrogen-primed uterus. Unlike estrogen, progesterone produces very little external observable effect.

Action of Estrogen

One of estrogen's major effects is on the reproductive organs. It acts on the female genitalia to cause their development and puts it in a functional state. As the level of estrogen increases at puberty, it causes the vagina to increase in mass and in size and in the cornification of its lining. It also causes a change in the glycogen content and in the pH. Estrogen will also cause the endometrium to undergo hyperplasia, hypertrophy, and an increased vascularity, thus the endometrium gets primed for implantation of the ovum. Estrogen causes a lowering in the membrane potential cells of the myometrium resulting in an increase in the contractility, size, and number of its cells. It increases the mobility for the sperm, and in the egg transports in the oviduct.

Estrogen helps initiate the development of the breasts, controls distribution of fat and hair, causes sodium retention that is dependent of aldosterone, stimulates bone growth at puberty, and enhances the closure of the epiphyseal plates, thus limiting bone growth after puberty.

Estrogen prepares the female's tubular genitalia for fertilization and it also affects the female's nervous system.

It is responsible for the female sex drive and preparation for copulation.

Estrogen causes the duct system of the mammary gland to polyferate.

It is also responsible for the female secondary sex characteristics and body contours.

Estrogens are somewhat anabolic and they cause some acceleration of growth, deposition of nitrogen, and formation of protein.

As it causes rapid growth at the time of puberty in the female this growth of the skeleton is related to Ca^{2+} metabolism, which becomes obvious in birds where there is a several-fold increase in estrogen by the increase in the concentration of blood Ca^{2+}.

Estrogen, through its effect on Ca^{2+} metabolism, may play a part in osteoporesis, a case that becomes a problem when the ovary is no longer producing estrogen.

There are three major estrogens:

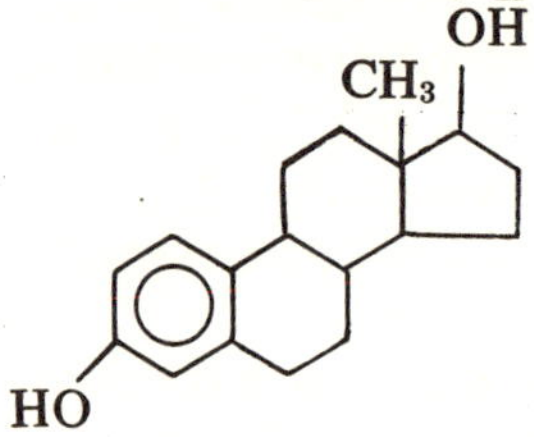

Estradiol 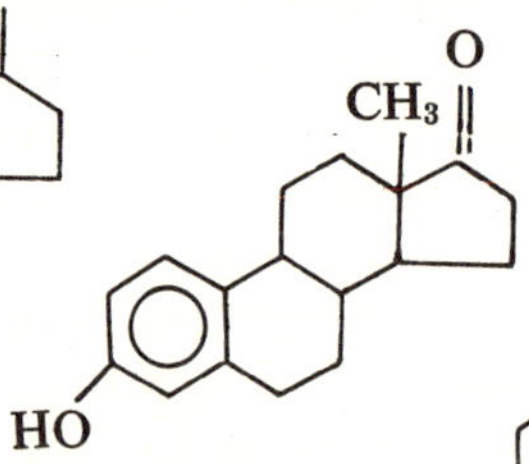   Estrone 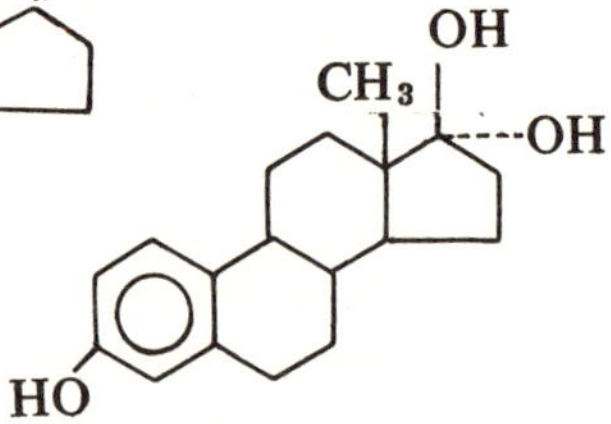   Estriol

Types of estrogens

1. Naturally occurring (the above three most common)
2. Synthetic estrogens
3. Plant estrogens

Sources of the naturally occurring estrogens in the body

1. Ovary
2. Testes
3. Placenta
4. Adrenal Cortex

If the major source is the ovary, estrogen is released into the blood and exists in the free or protein-bound forms.

Estrogens are conjugated in the liver with sulfuric acid for ease in excretion.

Solubility

Free estrogens are lipid-soluble, while the conjugated estrogens are water-soluble. They pass out through the kidney to the urine.
Estradiol and estrone are produced in the ovary.
Estradiole 17-β is the most biologically active estrogen.

Biological Effects of Estrogens

1. They cause edemia (increased water retention).
2. They cause hyperemia (increase in blood flow).
3. They increase cell size and cell number.
4. They cause hyperplasia.

Bioassay for Estrogens

1. Allan and Doisy Test: where estrogen will cause vaginal cornification.
2. Estrogen will cause an increase in uterine weight in immature or castrated rats.
3. Estrogen will cause vaginal opening in immature rats.
4. Kober Test (chemical): where sulfuric acid plus estrogen will give an orange color, plus water which will have a green fluorescence and develop a red color that will absorb at the wavelength of 282 mu in the ultraviolet region of the spectrum.
5. Estrogen may be measured through the use of gas chromatography.
6. Estrogen is also measurable through fluorometric methods.
7. The best method, at present, for measuring very minute amounts of estrogen would be radioimmunoassay.

Synthetic Estrogens

These do not have the same chemical structure but do mimic the action of estrogen in the body. The most famous synthetic estrogen is diethylstilbestrol (DES).

DES

Diethylstilbestrol acts as estrogen because the distance between the OHs is the same in the two cases, i.e., both of them have the same key to fit into the proper slot.

Until very recently, DES was used in beef cattle for fattening.

DES, however, is not conjugated or broken down in the liver. It, thus, circulates around for a long time and is slowly eliminated from the liver.

DES may activate the pituitary for the release of growth hormone in ruminants where efficiency of rate of gain and food utilization is enhanced.

In nonruminants, however, DES will have the estrogenic effect but not the growth-stimulating effect.

In lower animals, a high level of DES can override progesterone and abortion may result.

MGA

MGA is another synthetic estrogen produced by the Upjohn Company of Michigan. It is more progesterone-like in structure than estrogen. It is a 17 α-acetoxy-6-methyl, 16 methylene pregnen-4, 6-diene, 3, 20-dione.

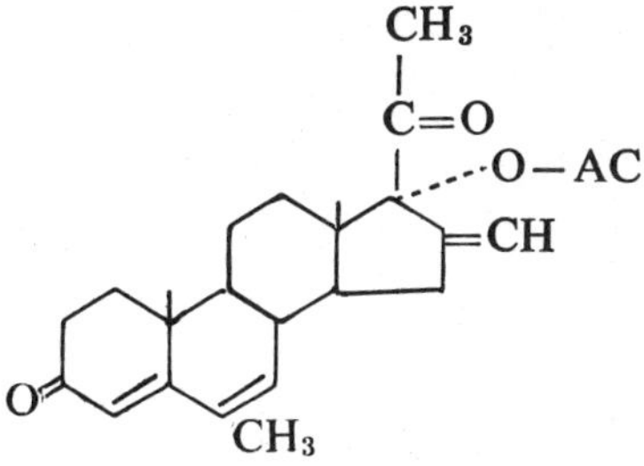

MGA

It has a growth-promoting effect similar to progesterone.
Other synthetic estrogens which are used in birth control pills are:

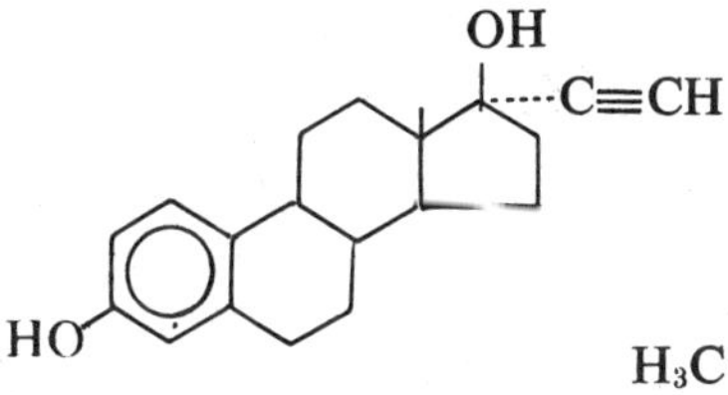

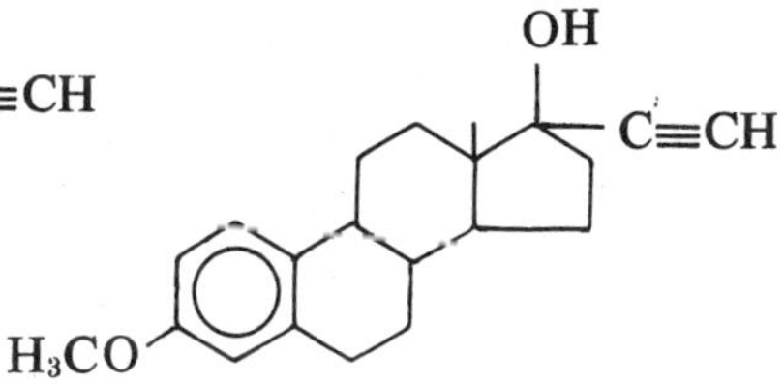

Ethinyl estradiol   Mestranol

Plant Estrogens

A most common plant estrogen with isoflavone structure is shown below, and which source is subtranean and red clover.

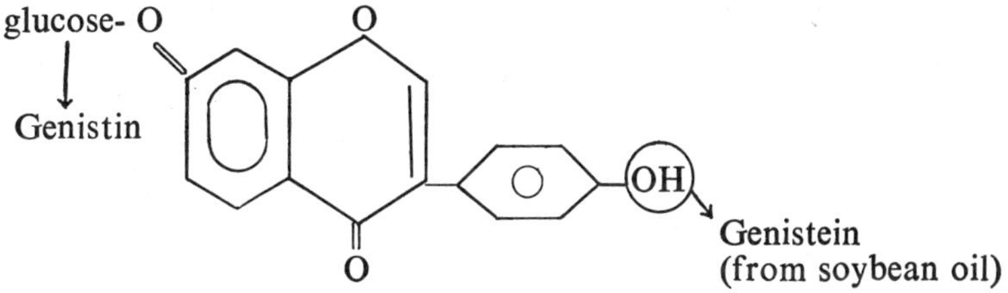

Genistein
(from soybean oil)

Another similar compound to the isoflavone above is biochanin A (from red clover).

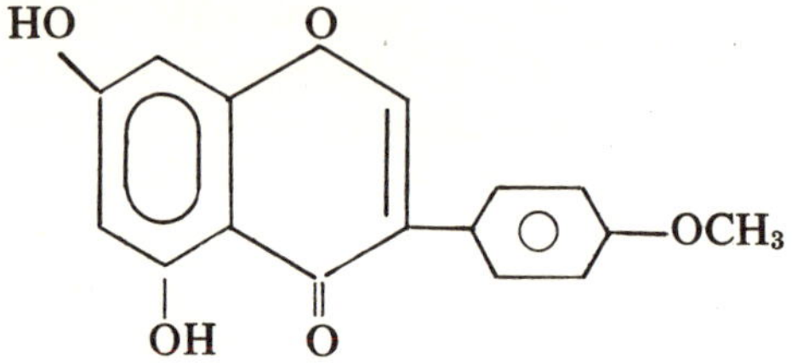

[Structure of biochanin A]

Isoflavone derivatives of plant origin are similar in structure to the isoflavones. Colimestrol, which is biologically active because the distance between its hydroxyl groups, is similar to that of estrogen.

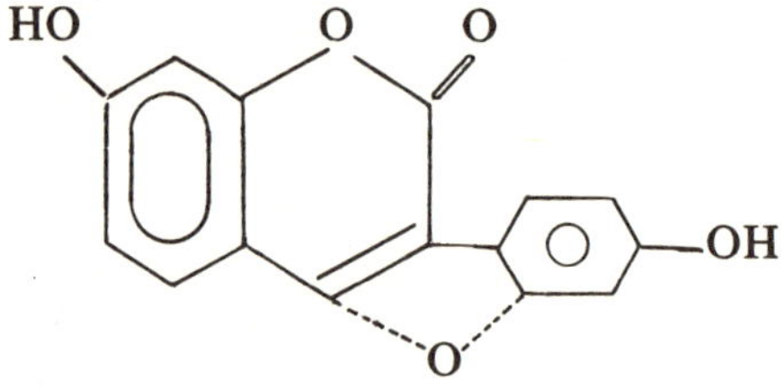

[Structure of Colimestrol]

Colimestrol

Another compound which has a basic isoflavone structure and is found in feces of chickens and other animals which metabolize isoflavones is eguliol.

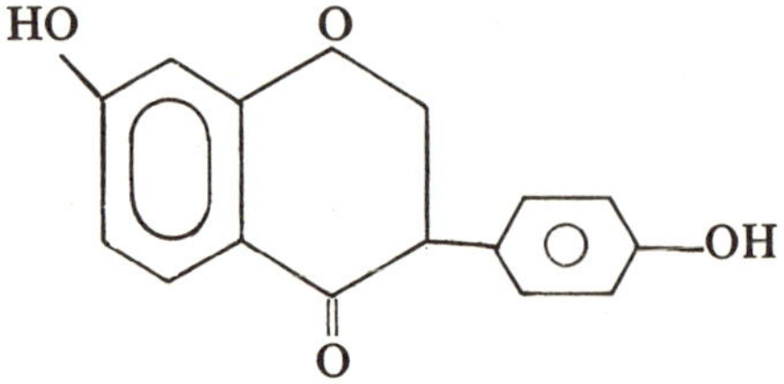

[Structure of eguliol]

The above compounds are effective only estrogenically, but they do not influence growth.

Miroestrol is a very orally effective plant estrogen obtained from the roots of *Puerarin anirifica*, Siamese Climbing Plant. Its structure is different from isoflavones, but the distance between its OH groups is similar to that of estradiol which gives it its estrogenic activity.

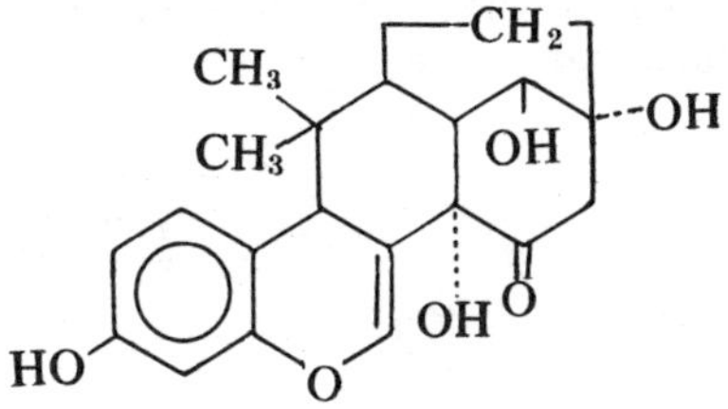

PROGESTERONE

Progesterone maintains the uterus. In 1903, Frankel removed the corpus leuteum from rabbit ovaries, which resulted in abortion.

Progesterone allows a tremendous polyferation in the endometrium.

There is estrogen surge first to condition the uterus; then there is a tremendous growth or polyferation caused by progesterone and resulting in the development of the placentomata.

Separation of Progesterone

1. By gas-liquid chromatography (GLC).
2. Protein-binding assay.

Synthesis of Progesterone

1. From the ovaries (granulosa cells of the ovaries).
2. From the adrenal cortex.
3. From the placenta.

Synthetic Pathway

Acetate ⟶ Cholesterol ⟶ Pregnenolone ⟶ Progesterone ⟶ to blood, to target organs (uterus, mammary glands, hypothalamus, pituitary).

After progesterone has its effect on the target organ, it is then metabolized through conjugation with sulfuric or glucuronic acids, or progesterone may undergo total reduction into pregnanediole, which is a main metabolite in human.

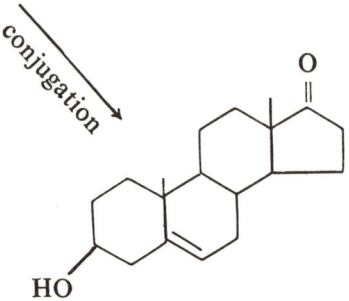

The half-life of progesterone in the blood of a nonpregnant cow is 34 minutes, and is 21 minutes in the pregnant cow.

Measuring the secretion rate through estimation of progesterone coming out of the ovaries, a nonpregnant cow's secretion rate is estimated as 46 mg/day per 500 kg cow, compared to a rate of 116 mg/day/1000 kg cow at six months of pregnancy.

In women, the secretion rate during the follicular phase is estimated as 5 mg/day, in luteophase as, 32 mg/day, and in the third trimester of pregnancy as 335 mg/day. Thus, although the human is only 1/8 of the weight of cow, it is producing three times as much progesterone. Thus, cows are low in steroid hormone secretion in relation to body weight, while humans are high. Cows, however, are not low in their thyroid hormone secretion compared to their body weight.

The human male secretes 5 mg/day progesterone and also secretes some estrogen.

Women, for their part, also secrete testosterone.

There are beneficial functions of testosterone in women as well as beneficial functions of estrogen and progesterone in men. Perhaps the beneficial function is mainly through action on the nervous system.

Action of Progesterone

Progesterone transforms an already estrogen-primed endometrium to a secretory tissue that will provide nutrients to the fertilized ovum, thus, by increasing the vascularity of the endometrium, its glycogen secretion and content along with other nutrients are increased.

Progesterone causes further development of the breasts by increasing the alveolar growth, and since progesterone is secreted immediately after ovulation, it will cause a body-temperature increase at this time.

CHORIONIC GONADOTROPIN (HCG)

A peculiar type of hormone produced by humans is human chorionic gonadotropin (HCG) which is produced by pregnant women in the cells of chorion of the fetal placenta.

HCG secretion starts at about forty days of pregnancy. It stays fairly high from 40-90 days, but its level peaks at 60-70 days of pregnancy.

HCG acts back on ovary and keeps the corpus leuteum functioning (producing progesterone only) and causes the uterus to maintain fetal growth.

HCG spills from the blood to the urine in pregnancy and is used as test for pregnancy.

HCG is a glyco-protein molecule of 30,000 molecular weight, and has a hexose moiety for its biological activity. If there is no pregnancy in humans, HCG becomes like LH in action.

In the early period of pregnancy, the ovary requires HCG to maintain corpus leuteum. After this period is over, HCG is not needed.

If a woman is not pregnant and she is given FSH, follicles will develop *in utery;* then if she is given HCG, ovulation will occur.

PREGNANT MARE SERUM GONADOTROPIN (PMSG)

Its origin is the endometrial cups of maternal placenta of the mare, and was discovered in the serum of pregnant mares. It is present in the blood, but not in urine. It is found at forty days of pregnancy, and peaks at 60-70 days. It disappears at 120-150 days of pregnancy.

PMSG stimulates the ovary to produce a new batch of follicles. Many of these regress, and some may form a new corpus leuteum which is a new source of progesterone for maintenance of pregnancy in mares.

At 150-180 days of pregnancy in the mare, its ovary becomes inactive. From here, the placenta takes over for production of estrogen and progesterone.

Neither HCG nor PMSG is a pituitary hormone. They both originate from pregnancy tissues, and are like the pituitary hormones in function, HCG being like LH in its action on intertissue stimulation.

Assay for PMSG is similar to that of FSH.

HORMONE OF THE TESTES

Androgen is the one major category, and is secreted by the Leydig cells dispersed in the interstitial space in and around the seminephrous tubules.

ACTION OF ANDROGEN

1. Anabolic
2. Androgenic

As anabolic, its effect is mainly on muscle, where it increases muscle mass and nitrogen retention. This anabolic influence is selective where stimulation is made only on certain muscles, such as the levator ani muscle.

Although we may succeed in increasing muscle mass through the use of androgen, if too much androgen is given, it will cause a feedback effect where damage to testes and accessory sex organs will result.

If androgen is given to an animal, it acts on the brain and causes aggressiveness. If the male is castrated, however, a more docile type of animal is the result.

Testosterone causes aggressiveness and brings about the onset of puberty in the male.

Androstenedione, which is a much weaker male hormone than testosterone, is present in animals in the prepuberty period at higher levels than testosterone. It is present much in higher levels in the fetus, 9 : 1 compared to testosterone.

At prepuberty, the ratio goes down to 1 : 1, and goes up again to 9 : 1 in the adult. Changes in the enzymatic system probably is the controlling factor in these shifts back and forth.

The reason for the high level in the fetus may be in order to lay down accessory sex organs.

The newborn human has large testes at first, and they get smaller later on until reaching adolescence. Before adolescence, the role of androstenedione is more metabolic than androgenic. At adolescence in the male, there is a large increase in androgen (as testosterone) which exerts effects on hair follicles, sweat glands, beard, hair pattern, and pubic hair.

Androgen influence in the female is metabolic.

FUNCTIONS OF TESTES

1. To produce gametes.
2. To provide substance that provides stimulus in getting sperm to female (process of fertilization).

Leydig cells produce the male hormone, and they are resistant to damage.

Testicles, in most species, produce estrogen besides testosterone.

Stallion, mare, dog, human, pig testicles all produce estrogen in the sertoli cells.

The bull testicles do not produce estrogen; however, the adrenal gland secretes some.

Temperature changes affect spermatogenesis rapidly. If, for example, the scrotum of a bull is warmed for a time, several weeks later we may observe abnormal spermatozoa. This lasts till the whole spermatogenesis cycle is over.

Although heat and x-rays will cause damage to spermatogenesis, Leydig cells are usually not affected.

Thus, the testes, besides causing spermatogenesis or gametogenesis at the seminephrous tubules, also cause steroidgenesis at the Leydig cells.

FSH is the gametogenic hormone. It stimulates the seminephrous tubules for spermatogenesis, while LH stimulates the Leydig cells for production of testosterone.

The male hormone is testosterone. It is the most active product of the testes. It acts on the organ which produced it, i.e., on the gonads, and it also acts on accessory sex organs.

The prostate gland is one organ that all species have in common, while the seminephrous vesicles are not common to all species. For example, dogs do not have seminephrous vesicles.

Rats and rodents have the most complicated accessory sex glands and have a complex prostate system of lateral and dorsal parts.

In humans, the big gland is the prostate vesicle, which is the vehicle of transport of the sperm into the female.

There is preejaculatory fluid in some species (before ovulation) which functions in clearing up the urethra by flushing it. The semivesicles, however, deliver the bulk of fluid and contain substances for the nourishment of the sperm.

Dogs and boars have large prostates but no vesicles. Dogs put out no fructose in their ejaculate for the nourishment of sperm.

The prostate gland puts out an alkaline fluid, the role of which is unknown in maintaining fertility in the male.

There are usually no sperms in the seminal vesicles. The seminal vesicles empty sperms into the urethra, the prostate, the vas deferens, and, at the time of ejaculation, the fluids mix just before ejaculation. The fluids may change the sperms' metabolism, but they have no effect on fertility.

ADRENAL OUTPUT OF SEX HORMONES

Although in smaller quantities, the adrenal cortex can produce all the gonadal hormones depending on variations in the enzyme concentrations which direct the synthesis in a certain direction.

FEEDBACK CONTROL OF GONADAL HORMONES (MALE)

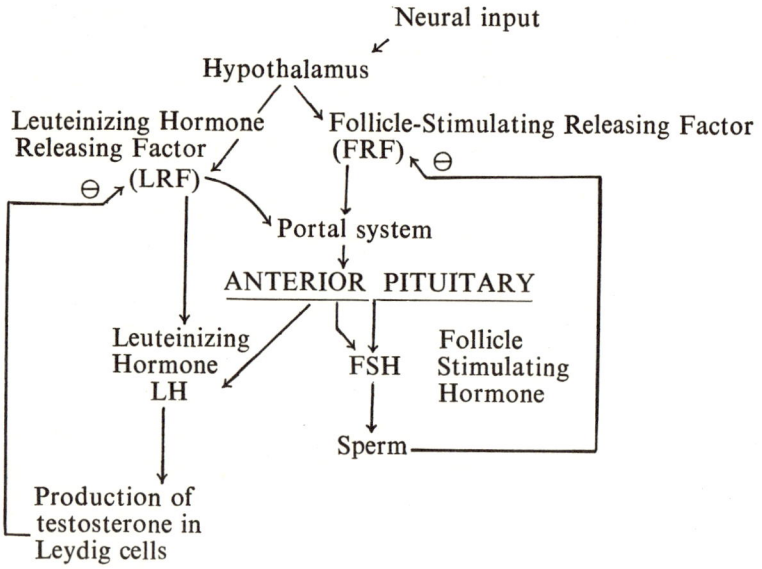

In the above scheme, the Leydig cells produce testosterone through the aid of follicle-stimulating hormone, which, in the case of male, is essential for normal sperm development. LH causes the production of testosterone in the Leydig cells. Excess testosterone may cause a feedback inhibition on the levels of FRF and LRF.

Estrogen, too, will cause a feedback on the level of FRF and LRF.

The male system is a stable, nonoscillating, self-controlling system.

FEEDBACK CONTROL OF GONADAL HORMONES (FEMALE)

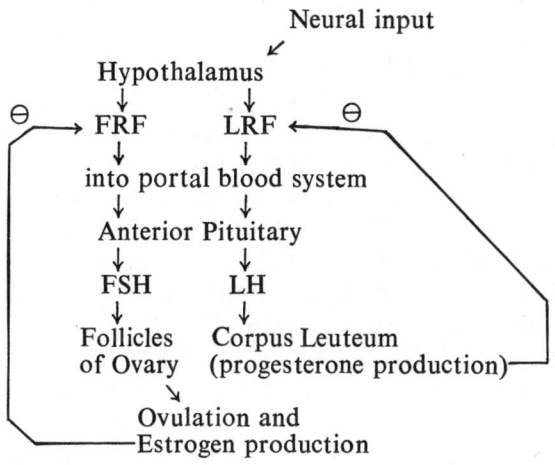

The above female system is similar to the male system except that it is oscillatory and, unlike the male's, is unstable. It, thus, seems that the activation of an already inhibited hypothalamus induces the onset of puberty in both male and female. At puberty, the hypothalamus is no longer inhibited and it starts to release FRF and LRF which stimulate the production of FSH and LH by the anterior pituitary. These, respectively, stimulate sperm production and gonadal growth in the male, and follicle production and gonadal growth in the female as well as the normal secondary sex characteristics.

There is no active gonadal tissue if puberty does not occur. Such an individual in both sexes is a "eunuch." A eunuch, as well, can be the result of a defect in any area of the system.

A eunuch may demonstrate a very tall stature with long extremeties due to the absence of testosterone or estrogen which promote growth but in the meantime limit bone growth through causing epipheseal plate closure. He or she is also very narrow in stature, and the female shows an absence of the familiar hip development.

CAUSES OF INFERTILITY

Perhaps as many as 15% of married couples have problems bearing children. This infertility may be contributed by either or both sexes. In the case of the male, the major problem is usually not caused by hormonal deficiency but rather by a duct obstruction. A subfertile male has a sperm count less than 20,000 per ml. (minimum count needed for fertility. The normal count considered as 500,000 per ml.). If properly trained, a subfertile male may be able to fertilize his wife. In case of the female, the infertility problem is usually associated with hormonal manifestations, such as a small or nonexistent surge of LH which may result in infertility. This may be corrected by administering FSH for ten days and then LH. The short lifetime of the corpus leuteum (which maintains pregnancy) is another cause of infertility in the female, where pregnancy gets terminated early.

THE PILL

1. Sequential: where the female takes estrogen for 14-16 days, and then progesterone for 5-8 days. In this case, the surge in LH is blocked and the normal pattern is disturbed.
2. Combined: where the female takes combined estrogen and progesterone for 20-21 days. Here, too, the LH surge is blocked, but is accompanied by an abnormal histology.
3. Endometrium: where, under control of progesterone, the fertilized ovum is not implanted.
4. Sperm mobility: sperm mobility may be decreased with progesterone, which in this case will cause a more viscous secretion by the uterus endometrial glands.

ACTIVITY OF STEROIDS

Progesterone is the most active steroid on the cellular level.

For estrogens, estradiol is the most active.

For androgens, it is thought testosterone is the most active form at the cellular level, but it is questionable whether it is testosterone itself or its conversion into the dihydrotestosterone form.

ASSAYS FOR SEX HORMONES

1. Assay based on urinary by-products, where testosterone concentration usually is shown by the 17-ketosteroid level in urine. The estriol level is an index of estrogen concentration, and the pregnanediol level in the urine is an indication of the progesterone metabolism. This is used as indication of pregnancy or ovulation.

2. Levels in plasma assay are becoming more popular through the radioimmuno and protein-binding methods.

REFERENCES

Averill, R. L. W. 1966. The hypothalamus and lactation. *Brit. Med. Bull.* 22:261.

Curet, L. B.; Mann, L.; Abrams, R.; Crenshaw, M. C.; and Barron, D. H. 1970. Effect of estrogen on arterial blood levels of alpha-amino nitrogen. *J. Appl. Physiol.* 28:1.

Drill, V. 1966. *Oral contraceptives.* New York: McGraw-Hill Book Company.

Halasz, B. 1968. Hypothalamic control of reproduction and lactation. *Proc. Int. Union of Physiol. Sciences* 6:204.

Hytten, F. E., and Leitch, I. 1964. *The physiology of human pregnancy.* Philadelphia: F. A. Davis Company.

Lawton, I. E., and Sawyer, C. H. 1970. Role of amygdala in regulating LH secretion in the adult female rat. *Amer. J. Physiol.* 218:622.

Lloyd, C. W., and Weisz, J. 1966. Some aspects of reproductive physiology. *Ann. Rev. Physiol.* 28:267.

Locke, M. 1968. *Control Mechanisms in developmental processes.* New York: Academic Press, Inc.

Masters, W. H., and Johnson, V. E. 1966. *Human sexual response.* Boston: Little, Brown and Company.

McCann, S. M. 1962. A hypothalamic luteinizing-hormone-releasing factor. *Amer. J. Physiol.* 202:601.

———. 1963. Recent studies on the regulation of hypophysial luteinizing hormone secretion. *Amer. J. Med.* 34:379.

McLaren, A. (ed.) 1968. *Advances in reproductive physiology.* New York: Academic Press, Inc., Vol. 3.

Nalbandov, A. V., and Cook, B. 1968. Reproduction. *Ann. Rev. Physiol.* 30:245.

Parkes, A. S. 1956–1966. *Marshall's physiology of reproduction.* 3rd ed., 3 volumes. Boston: Little, Brown and Company.

Pincus, G. 1965. *The control of fertility.* New York: Academic Press, Inc.

Rogers, J. 1963. *Endocrine and metabolic aspects of gynecology.* Philadelphia: W. B. Saunders Company.

Saunders, F. J. 1968. Effects of sex steroids and related compounds on pregnancy and on development of the young. *Physiol. Rev.* 48:601.

Savard, K. 1968. Steroid synthesis in the ovary. *Proc. Int. Union of Physiol. Sciences* 6:243.

Sharman, A. 1956. *Reproductive physiology in the post-partum Period.* Baltimore: The Williams & Wilkins Co.

Short, R. V. 1967. Reproduction. *Ann. Rev. Physiol.* 29:373.

Chapter 35

ADRENAL CORTEX HORMONES

INTRODUCTION

The adrenal cortex is essential for life, and, if removed without the exogeneous intake of its hormones, death occurs in about a week. Its necessity for life lies in its maintenance of mineral balance. The adrenal cortex hormones also regulate carbohydrate metabolism.

STRUCTURE OF THE ADRENAL CORTEX

The adrenal cortex is located on the superior pole of each kidney. The two of them weigh about eight grams in man. Ninety percent of the gland is the cortex and ten percent is the medulla. The cortex has its origin from the mesoderm, while the medulla comes from a neural origin.

ZONES AND THEIR SECRETION

1. Zona glomerulosa is the outer zone and secretes aldosterone—a mineralocorticoid. Zona glumerulosa is not under ACTH control.

2. Zona fasciculata is the middle zone and secretes the glucocorticoids and sex hormones.

3. Zona reticularis is the inner zone, and secretes glucocorticoids and sex hormones, or it may be a layer of cells differentiating into the other two outer layers.

REGENERATION AND HYPERTROPHY OF THE ADRENAL CORTEX

The adrenal cortex has a most remarkable characteristic of regeneration and hypertrophy. If it is destroyed but a small bud is left, this small bud will regenerate and hypertrophy to the original gland. Also, if one is removed, the other will hypertrophy and completely compensate for the one removed.

There are also the accessory adrenal tissues or the "adrenal cell rests" which do not function at normal circumstances but, if needed, may proliferate and become functional in compensating for the adrenal glands.

BLOOD SUPPLY

The adrenal cortex has a very rich blood supply which flows at a very rapid rate.

NERVE SUPPLY OF THE ADRENAL CORTEX

There is no direct innervation of the adrenal cortex. The control is only humoral.

BIOSYNTHESIS OF THE ADRENAL CORTEX HORMONES

Control of steroid biosynthesis in the adrenal cortex is through humoral agents only, where adrenocorticotrophic hormone (ACTH) plays a major role. Angiotensin II, high plasma sodium, and low potassium also promote steroidgenesis in the adrenal cortex.

The first step in synthesis of corticosteroids from cholesterol is by cleavage of the 6-carbon side chain of cholesterol by the enzyme demolase and oxidation at carbon-20, thus converting it to pregnenolone. Once cholesterol leaves the fat droplet and goes to the mitochondria for oxidation—once this point is passed—movement is very rapid into synthesis of the rest of corticosteroids. The rate-limiting step, thus, is the conversion of cholesterol to pregnenolone. Once pregnenolone is made, it moves back out to the smooth endoplasmic reticulum (SMR) where progesterone is synthesized. The steroid moves back into the mitochondria and has the 11-carbon position hydroxylated for the final synthesis of corticosterone.

The first and rate-limiting step will go under ACTH stimulation and, except for a short initial phase, it requires protein synthesis. See Figure 1.

If there is a defect in the 21-hydroxylase enzyme that converts 17α-hydroxyprogesterone to α-hydroxy-11 deoxycorticosterone, this will result in the build-up of 17α-OH progesterone, and the steroid synthesis in this case will take the route toward androstenedione and, consequently, the production of androgens, resulting in a virilization effect. In such a case, the route to synthesizing cortisol is blocked and cortisol is not produced. On the other hand, lack of plasma cortisol level will stimulate ACTH release from the anterior pituitary due to the absence of the negative feedback that would otherwise be exerted by cortisol on ACTH production.

This rise in ACTH will push further the synthetic pathway toward the androgenic hormone production.

A defect in the 11β-hydroxylase enzyme production will also result in blocking cortisol synthesis, the build-up of androgens, and an increase in masculinity.

The adrenal cortex secretes five male sex hormones of moderate strength, and also secretes estrone and estradiol as shown in Figure 1.

METABOLISM OF GLUCOCORTICOIDS

All the hormones of the adrenal cortex are passed from the circulation to the liver where the first reduction of C-4 occurs, followed by a second reduction of the ketone group at C-3 to a hydroxyl. A third reduction occurs at the keto group at C-20 to hydroxyl. The tetrahydro-form is then conjugated to glucuronic acid at the C-3 position which occurs in the liver. This conjugation with glucuronic acid renders the steroid water-soluble. The bulk of the adrenal steroids are excreted in the urine.

BINDING TO PROTEINS IN CIRCULATION

90-95% of the corticosteroids in plasma are bound to a specific protein. This specific protein is an a_1-globulin (corticosteroid-binding globulin, CBG). 5-10% of the corticosteroids are bound to albumin. The bound form is in equilibrium with the free form in blood plasma, and there is a shuttle between the two forms according to the rate of usage of the free form. The bound form of the hormone does not get conjugated or solubilized for excretion by the liver, thus, binding to protein is a protection to its presence in plasma. Moreover, the specificity with which the hormones bind with protein is very high, which has been a tool in the assay of these hormones through radioassay and radioimmunoassay.

Aldosterone, however, is mainly bound to albumin instead of globulin. Moreover, the extent of its binding with protein is not as high as the other corticosteroids, being only 70% bound, and the protein specificity is not as great as for the other corticoids.

HORMONES OF THE ADRENAL CORTEX

Aldosterone

Its secretion rate is 150–300 μg/day. Low Na^+ increases renin secretion causing increased aldosterone. Aldosterone has a very short half-life as it is completely removed from the hepatic circulation in a single passage. This also means that the hormone intake cannot be oral.

The hemiacetal on C 18 of aldosterone is important as it reacts with tissue receptors.

Aldosterone is the main mineralocorticoid. It is produced by the outer zone (the glomerulosa) and is not under the influence of ACTH. Aldosterone is the most potent mineralocorticoid. It maintains Na^+ through its reabsorption in the distal tubule of kidney and excretes K^+. Adrenalectomized subject would survive if provided with 1% NaCl in his diet daily. In order to produce a compound with much more activity than aldosterone, Ciba has put the synthetic compound "dexamethazone" in the market, which is the most potent glucocorticoid developed.

Dexamethazone

Control of Aldosterone Secretion

The juxtaglomerulus apparatus produces renin. Renin stimulates the conversion of angiotensinogen to angiotensin-I (a deca peptide). Another converting enzyme in the blood cleaves the last two amino acids of angiotensin-I, converting it to angiotensin II (an octa peptide). Angiotensin-II causes arterial constriction, and, thus, an increase in arterial blood-pressure and consequently the release of aldosterone. Aldosterone goes through the blood to the kidney and causes Na^+ retention. This is accompanied by water retention too, and there is an increase in the extracellular fluid. This increase in the extracellular fluid results in an increase in the arterial blood pressure in the renal arterioles which, in turn, causes reduction in renin secretion. A reduction in renin secretion means shutting off aldosterone production and a decrease in blood pressure. A decrease in blood pressure stimulates renin secretion and the cycle continues on.

Aldosterone activates the production of an enzyme which causes the movement of Na^+ across the cell membrane to other cells. A specific protein is synthesized through aldosterone's effect on DNA-m-RNA-template formation.

Actions of Aldosterone

1. On electrolyte and water metabolism; where aldosterone exerts its primary effect in promoting Na^+ reabsorption,

and secondarily, the passive reabsorption of water. The action of the hormone is exerted on the distal tubule, and at the tissue cell level, exchanging Na^+ and H^+ for K^+ removal. Aldosterone, thus, increases the water exchange.

2. Aldosterone exerts its action on sites other than the kidney, such as the salivary glands, skin, and intestines, where it causes a decrease in Na^+ excretion, and an increase of K^+ excretion in sweat, saliva, urine, and feces. Aldosterone may also act similarily on muscle cells, where it increases the intracellular Na^+ and decreases intracellular K^+. This may give rise to periodic paralysis.

3. Aldosterone causes an increase in renal filtration rate and blood flow.

4. As aldosterone increases the blood volume, it causes a blood pressure increase.

Cortisol

Its secretion rate is 15-20 mg/day in man. Three-fourths of this amount is secreted between midnight and the early morning. Practically all the glucocorticoids are characterized by having an OH group at the carbon-11 position. This OH on C-11 directs cortisol to the liver for metabolism. There is an equilibrium between cortisol and cortisone. This is of significance as cortisone, in this case, may be given to human patients. It then goes into equilibrium with cortisol, the effector agent, since cortisol is the dominant corticosteroid in humans.

Corticosterone

Its secretion rate is 2-5 mg/day in humans.
The steroids of the adrenal gland are built around corticosterone.

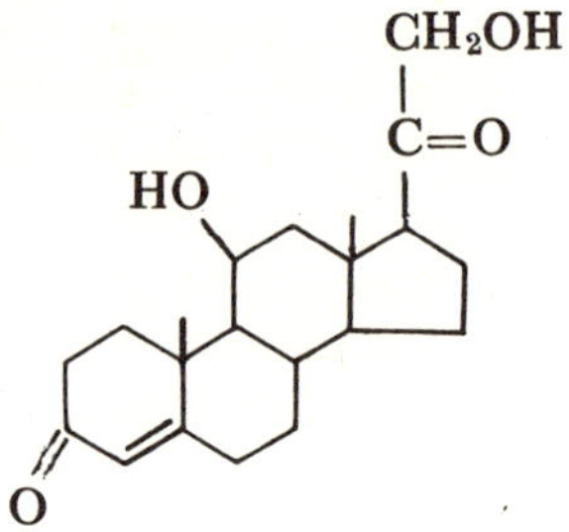

Corticosterone

Corticosterone is the glucocorticoid that lies in the middle of the synthetic pathway.

The OH group on C-11 distinguishes it from the 11-oxo-glucocorticoids.

The corticoid with the 17α-OH is cortisol, which has the highest glucocorticoid activity. If the H on C-11 of cortisol is removed, it gets converted to cortisone, which is much different in activity. If, instead, the oxygen is removed from C-11, 11-deoxy-corticosterone (DOC) is produced, which is a mineralocorticoid with a different target site from the C-11-oxygenated glucocorticoids, the maximum effect of which depends on the presence of the OH group at C-11. The OH group, especially in the β-position, gives the steroid much more potency than if it were a keto group at the same position.

Desoxycorticosterone (DOC)

Its secretion rate is 200–500 mg/day. Its effectiveness is thirty times less than that of aldosterone.

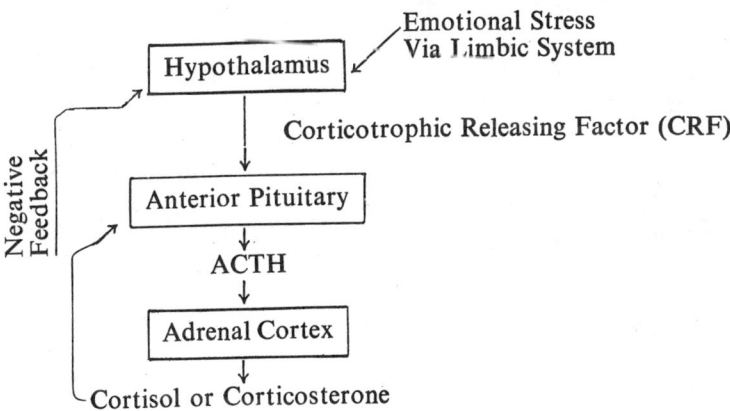

Desoxycorticosterone (DOC)

Control of Cortisol and Corticosterone Secretion

Cortisol and corticosterone cause a negative feedback on ACTH secretion on the level of the anterior pituitary and possibly on the hypothalamus level, as shown below:

```
                                    Emotional Stress
                                    Via Limbic System
       ┌──────────────┐           ↙
       │ Hypothalamus │ ←
       └──────────────┘
              │   Corticotrophic Releasing Factor (CRF)
              ↓
       ┌────────────────────┐
       │ Anterior Pituitary │
       └────────────────────┘
              ↓
             ACTH
              ↓
       ┌──────────────┐
       │ Adrenal Cortex │
       └──────────────┘
              ↓
       Cortisol or Corticosterone
```

(Negative Feedback)

At stress, ACTH secreted from special cells of the anterior pituitary travels through the afferent nerve fiber up the lateral spinothalamic tract to the spinal cord, to mid-brain, to the reticular formation, to the hypothalamus, where the release of corticotrophic-releasing factor (CRF) occurs from specific neurons in the hypothalamus.

The amygdala and hippocampus exert an influence on CRF release and on the hypothalamus in general. The amygdala has an excitatory influence on the hypothalamus, while the hippocampus is inhibitory of CRF release.

The amygdala and hippocampus exert their excitatory and inhibitory influence on the hypothalamus cells probably by lowering and raising threshold, respectively, of the neurons of the hypothalamus, thus changing the excitability of these neurons. The feedback probably is effective on the hypothalamus, limbic system, and the mid-brain levels. The mid-brain probably exerts an inhibitory effect on the secretion of the corticoids. Since the adrenal secretion is higher during sleep, where the activity of the reticular formation is operating at a reduced level, and since the reticular formation is in the mid-brain, consequently, the inhibitory influence on ACTH becomes depressed during sleep, and thus an increase in the corticoid's secretion results. In sleep, therefore, inhibition coming from the reticular formation system is depressed.

There is, on the other hand, a direct negative feedback of the corticosteroids on the anterior pituitary which reduces ACTH secretion. Secretion of ACTH may also be regulated through a direct action of ACTH itself on the brain through a negative feedback. Here, the rate of firing of the brain cells is changed.

The rate of ACTH secretion is determined by the magnitude of CRF secretion and by the magnitude of feedback from cortisol.

EFFECTS OF ADRENALECTOMY

Lack of the adrenal cortex hormones results in loss of appetite; delayed absorption in the intestines; nausia; stomach ulceration; decrease in blood volume, flow, and pressure; a decrease in Na^+, Cl^-, CO_3^-, and glucose, and an increase in K^+ and nonprotein nitrogen; and loss of resistance to stress or toxin.

MISCELLANEOUS FUNCTIONS OF THE GLUCOCORTICOIDS

1. Catabolism of protein to amino acids, to keto acids, to carbohydrate. They activate the transaminase enzymes.

2. Antiflammatory: cortisone-like steroids exert a corrective effect on a ruptured lysosomal membrane, thus stopping leakage of hydrolytic lysosomal enzymes, like the cathepsins, and the digestion of the tissue's protein which caused the inflammation.

SCHEME OF THE BIOSYNTHESIS OF THE ADRENAL CORTEX HORMONES (Figure 1).

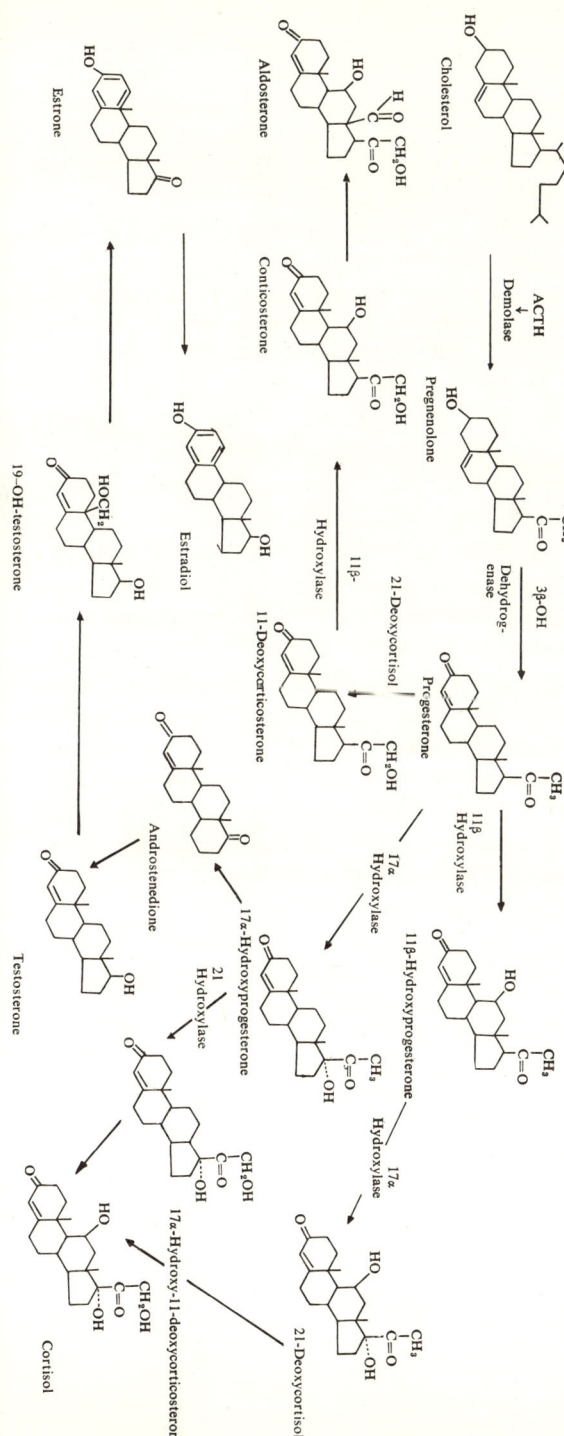

REFERENCES

Adler, S. 1970. An extrarenal action of aldosterone on mammalian skeletal muscle. *Amer. J. Physiol.* 218:616.

Applegren, L. 1967. Sites of steroid hormone formation. *Acta Physiol. Scand.* 71, Suppl. 301.

Blair-West, J.; Cain, M.; Catt, K.; Coghlan, J.; Denton, D.; Funder, J.; Nelson, J.; Scoggins, B.; Wintour, M.; and Wright, R. 1968. The control of aldosterone secretion. *Proc. Int. Union of Physiol. Sciences* 6:249.

Bransome, E. D., Jr. 1968. Adrenal cortex. *Ann. Rev. Physiol.* 30:171.

Ciba Foundation Study Group No. 27. 1967. *The human adrenal cortex.* Boston: Little, Brown and Company.

Cope, C. L. 1965. *Adrenal steroids and disease.* Philadelphia: J. B. Lippincott Co.

Denton, D. A. 1965. Evolutionary aspects of the emergence of aldosterone secretion and salt appetite. *Physiol. Rev.* 45:245.

Eisenstein, A. B. 1967. *The adrenal cortex.* Boston: Little, Brown and Company.

Fimognari, G. M.; Fanestil, D. D.; and Edelman, I. S. 1967. Induction of NRA and protein synthesis in the action of aldosterone in the rat. *Amer. J. Physiol.* 213:954.

Goodwin, F. J.; Knowlton, A. I.; and Laragh, J. H. 1969. Absence of renin suppression by deoxycorticosterone acetate in rats. *Amer. J. Physiol.* 216:1476.

Chapter 36

GASTROINTESTINAL HORMONES

INTRODUCTION

Unlike the salivary glands, other glands in the gastrointestinal tract respond to neural stimuli or to blood-borne hormones. These hormones may originate in the digestive tract or come from an endocrine system.

GASTRIN

It is produced in the duodenal mucosa and it stimulates gastric secretion. Gastric juice is composed of hydrochloric acid (HCl) and pepsin. HCl is secreted by the parietal cells of the stomach. Secretion of the gastric glands is caused by the presence of food in stomach, or stimulation of the vagus nerve itself which causes gastric secretion. If the vagus nerve is cut, we still get secretion in response to the presence of food in stomach.

Presence of partially digested food in the stomach acts on the antrium cells to produce a hormone which is carried in the blood to the parietal cells and causes the secretion of HCl and pepsin.

Gastric juice is a nonacid secretion from the antrium. If all vagus nerves are cut, this would not affect its secretion from the antrium to the duodenum. Also, an increase or decrease in blood flow has no effect. Histamine will cause an increase in gastric juice secretion.

Presence of food in the mouth, or thinking of it, will cause gastric secretion as well.

Vagus stimulation causes parietal cells to secrete HCl.

There is also a local nervous reflex for secretion of gastric juice which can be blocked by atropine or cocaine. Besides the antrium, gastrin is also produced from the delta cells of the islets of the pancreas.

Presence of fat in the small intestines inhibits gastrin secretion.

SECRETIN

Its source is the duodenal mucosa. It is secreted due to HCl presence in the stomach. It increases the secretion of the exocrine portion of pancreas and the production of enzyme-rich pancreatic juices.

Secretin resembles glucagon in structure and is made up of 27 amino acid residues. Secretin also stimulates the hepatic cells of the liver for secretion of bile juice.

ENTEROGASTRONE

It is produced from the duodenal part of the small intestines. The presence of fat is the stimulus for its secretion. The presence of sugar or acid in the small intestines may also stimulate its secretion. The presence of enterogastrone may inhibit the secretion of gastrin.

CHOLECYSTOKININ (PANCREOZYMIN)

Secreted due to fat presence in the duodenum, it is composed of 33 straight-chain amino acid residues, the last five of which are similar to the last five amino acids of gastrin. It also has a sulfated ester group which causes similar activity to that of gastrin.

This hormone causes the contraction of the gall bladder and the release of bile pancreatic juice which is rich in enzymes (notably the lipases, which aid in metabolism of fat), and bicarbonates.

While secretin inhibits gastrin's action, cholecystokinin has an additive effect on gastrin. Thus, gastrin and cholecystokinin may act on the same type of receptor cells.

VILLIKININ

Secreted from the duodenal mucosal cells due to the presence of chyme in the small intestines, it stimulates movement of intestinal material.

ENTEROCRININ

Secreted from cells of the duodenum due to presence of chyme, it is carried in the blood and brings about intestinal secretion.

REFERENCES

Ciba Foundation Symposium. 1962. *The exocrine pancreas: normal and abnormal function*. Boston: Little, Brown and Company.

Code, C. F. 1968. Hormonal inhibition of gastric secretion. *Proc. Int. Union of Physiol. Sciences* 6:194.

Cooke, A. R. 1967. The glands of Brunner. In *Handbook of physiology*. Baltimore: The Williams & Wilkins Co., Sec. VI, Vol. II, p. 1087.

———. 1969. Potentiation of acid secretion in dogs. *Amer. J. Physiol.* 216:968.

———, and Grossman, M. I. 1968. Comparison of stimulants of antral release of gastrin. *Amer. J. Physiol.* 215:314.

Davenport, H. W. 1967. Physiological structure of the gastric mucosa. In *Handbook of physiology*. Baltimore: The Williams & Wilkins Co., Sec. VI, Vol. II. p. 759.

Farrar, G. E., Jr., and Bower, R. J. 1967. Gastric juice and secretion: physiology and variations in disease. *Ann. Rev. Physiol.* 29:141.

Gregory, R. A. 1965. Secretory mechanisms of the digestive tract. *Ann. Rev. Physiol.* 27:395.

———. 1968. Chemistry of gastrin. *Proc. Int. Union of Physiol. Sciences* 6:188.

Grossman, M. I. 1961. Gastrin: new life for an old hormone. *Gastroenterology* 40:149.

———. 1967. Neuronal and hormonal stimulation of gastric secretion of acid. In *Handbook of physiology*. Baltimore: The Williams & Wilkins Co., Sec. VI, Vol. II, p. 835.

———. 1968. Hormonal regulation of gastrointestinal function. *Proc. Int Union of Physiol. Sciences* 6:187.

Jacobson, E. D.; Swan, K. G.; and Grossman, M. I. 1967. Blood flow and secretion in the stomach. *Gastroenterology* 52:414.

Johnson, L. R., and Grossman, M. I. 1968. Secretin: the enterogastrone released by acid in the duodenum. *Amer. J. Physiol.* 215:885.

———. 1969. Effects of fat, secretin and cholecystokinin on histamine-stimulated gastric secretion. *Amer. J. Physiol.* 216:1176.

———. 1970. Analysis of inhibition of acid secretion by cholecystokinin in dogs. *Amer. J. Physiol.* 218:550.

Nakajima, S., and Magee, D. F. 1970. Influences of duodenal acidification on acid and pepsin secretion of the stomach in dogs. *Amer. J. Physiol.* 218:545.

Nakajima, S.; Nakamura, M.; and Magee, D. F. 1969. Effect of secretin on gastric acid and pepsin secretion in response to various stimuli. *Amer. J. Physiol.* 216:87.

Uvnas, B. 1968. Gastrin and wagus. *Proc. Int. Union of Physiol. Sciences* 6:189.

INDEX

Carbohydrates, 44
  aldonic acid, 55
  analytical procedures for detection of, 68
  degree of sweetness of, 62
  enzymatic action on starch, 61
    enzymatic browning of, 71
    obtimum temperature range for phenolases, 74
    possible mechanism of enzyme inhibition, 74
    possible mechanism of melanoidin formation, 75
    properties of the phenolase enzymes, 74
  gums and mucilages, 65
  metabolism of, 85
    anabolism, 85
    catabolism, 85
    control mechanisms, 90
    glycolytic system, 85
  monosaccharides, 44
  mucopolysaccharides, 57
    chitin, 58
    chondriotin sulfate, 58
    hemicellulose, 59
    hyaluronic acid, 57
  non-enzymatic browning, 76
    aldol condensation, 82
    carmilization, 83
  oligosaccharides, 51
    cellibiose, 51
    lactose, 52
    maltose, 54
    sucrose, 53
    trehalose, 53
  pectic substances, 62
    pectic acid, 63
    pectin, 63
  pentose shunt, 94
    anaplerotic reactions, 98
    metabolism of galactose, 100
  polysaccharides, 59
  prevention of Mailard browning, 84
  saccharic acid, 56
  tricarboxylic acid cycle, 91
    energy production, 94
  trisaccharides, 54
  types and sources in the American diet, 66
  uronic acid, 56
  various forms of glucose in solution, 50

Cereal chemistry, 205
  gas production in dough making, 206
  gas retention in dough, 207
  introduction to, 205
  vitamins and minerals content, 206

Flavonoids, 171
  biological activity of, 178
  bitterness of, 178

changes that occur during processing, 180
classification of, 171
flavanone, 174
  anthocyanins, 174
  anthoxanthins, 174
  catechols and tannins, 174
general properties of, 177
occurrence of, 175
tannins, 178

Flavors in foods, 182
  introduction to, 182
  in various foods, 186
  off-flavor (development), 187
  off-flavor (milk), 188
  off-flavor (vegetable), 189
  sensory evaluation of, 185

Fruits, 192
  chemical composition of, 192
  classification of, 193
  general changes during cooking of, 195
  introduction to, 192
  structure of 193
  turgor of, 195
  volatile sulfur compounds of, 197

Hormones, 343
  adrenal cortex hormones, 443
    binding to proteins in circulation, 446
    biosynthesis of the adrenal cortex hormones, 444
    blood supply of, 444
    effects of adrenalectomy, 452
    introduction to, 443
    hormones of the adrenal cortex, 447
      action of aldosterone, 448
      aldosterone, 447
      control of aldosterone secretion, 448
      control of cortisol and corticosterone secretion, 451
      corticosterone, 449
      cortisol, 449
      deoxycorticosterone, 450
      metabolism of glucocorticoids, 446
      miscellaneous functions of the glucocorticoids, 453
    nerve supply of the adrenal cortex, 444
    regeneration and hypertrophy of the adrenal cortex, 444
    scheme of biosynthesis of the adrenal cortex hormones, 453
    structure of the adrenal cortex, 443
    zones and their secretion, 443
  adrenocorticotropic hormone (ACTH), 365
    assay of, 369
    biological activity of, 366
    chemical structure of, 365
    mechanism of action of ACTH on cell membrane, 366
    secretion and feedback of, 369
  anterior pituitary hormones, 360
    histology of pituitary gland, 360
    introduction to, 360
    major groups of anterior pituitary hormones, 361
  antidiuretic hormone (ADH), 357
    assay of, 358
    control of secretion of, 358
    functions of 358
      antidiuresis, 358
      effect on ACTH, 358
      pharmacological effect, 358
      water retention, 358
    synthesis and storage of, 357
  catecholamines, 343
    action of catecholamines on adipose tissue, 354
    action of catecholamines on various organs and tissues, 348
    biochemical action of, 349

biochemical mechanism of action of epinephrine, 350
hyperglycemia, 350
lipolysis, 351
biosynthesis of catecholamines, 346
control of secretion of, 351
distribution of urinary metabolites of epinephrine, 353
estimation of, 353
hormones of the adrenal medulla, 343
internal negative feedback, 352
metabolism of catecholamines, 352
origin, storage and release of, 345
physiological effects of adrenal medullary hormones, 347
receptors, 348
regulation of catecholamines secretion, 344
secretion and release of catecholamines, 345
gastrointestinal hormones, 455
cholecystokinin, 457
enterocrinin, 457
enterogastrone, 456
gastrin, 455
introduction to, 455
secretion of, 456
villikinin, 457
glucagon, 416
action vs. insulin, 416
function of, 416
glucagon and lipolysis, 418
introduction to, 416
gonadotropic and sex hormones, 419
action of androgens (anabolic, androgenic), 435
activity of steroids, 441
adrenal output of sex hormones, 437
assay of, 441
protein-binding assay, 441
radioimmunoassay, 441
urinary by-products, 441

biosynthesis and metabolism of sex hormones, 423
causes of infertility, 440
chorionic gonadotropin, 434
feedback control of gonadal hormones (female), 439
feedback control of gonadal hormones (male), 438
follicle stimulating hormone, 423
functions of testes, 436
fertilization, 436
production of gametes, 436
hormone of the testes, 435
hormones of the ovary, 424
action of estrogen, 425
bioassay of, 427
biological effects of, 427
estrogen, 424
plant estrogens, 429
solubility of, 427
source of the naturally occurring in the body, 426
synthetic estrogens, 428
types of, 428
naming of, 421
pregnant mare serum gonadotropin, 434
progesterone, 431
action of, 433
separation of, 431
synthesis of, 431
steroid biosynthesis, 422
steroids, 419
structures of, 421
the pill, 440
growth hormone, (GH), 371
assay of, 374
bioassay, 374
infusion of insulin, 374
plasma amino acids level, 374
radioimmunoassay, 374
tibia test, 374
biological effects of growth hormone, 371
effect on bone growth, 372
effect on fat metabolism, 372

effect on glucose phosphorylation, 371
  effect on potassium retention, 372
  growth hormone and lipolysis, 372
  increase of amino acid uptake, 371
 chemical structure, 371
 secretion and feedback control of growth hormone, 373
  amino acid level and influence on secretion, 373
  arginine level, 373
  catecholamines level, 373
  hypoglycemia, 373
  psychic factors, 373
  testosterone level, 374
  thyroxine level, 374
insulin, 404
 action on cell membrane, 406
 bioassay of, 412
 biological action of, 406
  effect on intermediate metabolism, 406, 407
   inhibition of enzymes synthesis, 406
   permeability of glucose, 406
   stimulation of enzymes synthesis, 406
 factors that influence insulin's release, 406
 influence of insulin on skeletal muscle cells, 408
 insulin and carbohydrate metabolism, 407
 insulin and fat metabolism, 409
 insulin and protein metabolism, 410
 introduction to, 404
 miscellaneous actions of insulin, 410
  on chondriotin sulfate, 410, 411
  on membrane potential, 411
  on sulfhydryl groups of muscle, 410

 orally effective agents for control of diabetes, 412
 organs not affected by insulin, 411
 symptoms and causes of diabetes melletis, 411
 synthesis of, 405
 types of, 413
oxytocin, 356
 actions of, 356
  milk ejection, 356
  naturesis, 357
  sperm transport and ejaculation, 357
 assay of, 357
  biological, 357
  milk ejection, 357
 release of, 356
 synthesis and storage of, 356
parathyroid hormone, 398
 control of calcium metabolism, 399
 function of the parathyroid gland, 399
 interaction of parathyroid -Ca- Vit. D, 400
  phosphate excretion, 401
  plasma membrane permeability, 400
  removal of Ca from bone to blood, 400, 401
  synthesis of carrier protein, 400
 introduction to, 398
prolactin, 376
 assay methods, 377
 biological effects of, 377
 control of secretion of, 378
 introduction to, 376
 metabolism of, 377
 structure of, 376
thyroid hormones, 380
 characteristics of thyrotoxicosis, 391
 control of release of thyroid hormones, 386
 cretinism, 394

examples of goitrogenic compounds, 391
formation of the hormone, 383
function of the hormone, 387
goitrogenic compounds, 391
graves' disease, 393
Hashimoto thyroiditis, 393, 394
introduction to, 380
myexedema, 393
LATS, 394
plumner's disease, 393
relation of thyroxine to other hormones, 389
  effect on gonads, 390
  relation to parathyroid gland, 390
  relationship with adrenal cortex hormones, 390
  specific action of T_3 and T_4, 389
  structure of the thyroid hormones, 382
  theories of the thyroid hormones action, 388
  action of deficiency of in young, 388
  T_3, 388, 389
  T_4, 388, 381, 389

Lipids, 5
  autoxidation of unsaturated fatty acids of, 27
    carbonyls, 30
    peroxide test, 29
    TBA test, 29
  classification and functions of, 5
    complex forms, 11
    fatty acids, 7
    non-saponifiable, 6
    saponifiable, 6
    steroids, 7
  definition of, 5
  digestion and absorption of, 17
  factors that cause inhibition of oxidation of, 30
  factors that cause oxidation of, 30
  fat-soluble vitamins, 12
    vitamin A, 12, 13, 14
    vitamin D, 14, 15
    vitamin E, 15, 16
    vitamin K, 16
  fatty acids of, 7
  metabolism of, 34
    biosynthesis of cholesterol, 42
    β-oxidation of fatty acids, 34-38
    fatty acids synthesis, 38-40
    role of carnitine, 36-38
    synthesis of triglycerides, 40-42
  physical properties of, 19
    crystalization, 19, 20
    double bond rearrangements 20, 21
    hydrogenation, 21, 22
    hydrolysis of and relation to foods, 23
    pyrrolysis of, 21
    terpenoids, 22, 23
  possible ways of reducing oxidation of, 31
    antioxidants, 31, 32
    mechanism of action of antioxidants, 31, 32
    synergists, 32, 33
  processing of, 24
  refining of, 24
  sources of oxidation of, 33

Milk, 200
  enzymes of, 203
  introduction to, 200
  non-combustible ash content of, 203
  other vitamins content of, 203
  proteins content of, 114-116
  vitamin A, D, E and K of, 201

Minerals, 315
  introduction to, 315
  the essential minerals, 319
    border-line elements, 319
    micro-elements, 319

trace minerals and the biological system, 315-318
calcium, 319
   deficiency symptoms of, 320
   factors affecting its requirement of, 321
   human daily requirement, 320
   requirement of, 320
   role of, 319
   source of, 319
   toxicity symptoms of, 320
chromium, 336
   functions of, 336-337
cobalt, 334
   biochemical functions of, 334
   deficiency symptoms of, 334, 335
   requirement of, 334
   source of, 334
   toxic level of, 334
   toxicity symptoms of, 335
copper, 327, 316
   activation of enzymes, 327
   deficiency symptoms of, 328
   functions of, 327, 328
   requirements of, 328
   sources of, 328
   toxicity symptoms of, 329
fluorine, 336
   functions of, 336
   requirement of, 336
   source of, 336
   toxic level of, 336
iron, 326, 315-319
   absorption of, 326, 327
   biochemical function of, 326, 327
   functions of, 326
   iron-copper relationship, 327
   requirements of, 326
   sources of, 326
magnesium, 323, 315-319
   Ca-P-Mg relationship, 324, 325
   deficiency symptoms of, 324
   functions of, 324
   role in enzyme activation, 324, 325
   role in hypernation, 325

manganese, 331, 315-319
   biological role of, 331
   enzymes activation of, 331
   functions of, 331, 332
   Mn-biotin relationship in metallo-enzymes, 332
molybdenum, 332, 315-319, 325, 326
   enzyme activation of, 333
   functions of, 332, 333
   requirement of, 333
   sources of, 333
   toxic level of, 333
   toxicity symptoms of, 333
phosphorus, 322, 315-319, 321
   biochemical functions of, 322
   sources of, 322
   toxicity symptoms of, 322
potassium, 323
   function and requirement of, 323
   toxicity symptoms of, 323
selenium, 335
   biochemical function of, 335, 336
   deficiency symptoms of, 335, 336
   functions of, 335, 336
   requirement of, 336
   selenium-Vit. E relationship, 335, 336
   source of, 335
   toxic level of, 335
   toxicity symptoms of, 335, 336
sodium, 322, 315-319
   biochemical functions of, 322
   deficiency symptoms of, 323
   toxicity symptoms of, 323
zinc, 329, 315-319, 325, 326
   biochemical functions of, 329, 330
   deficiency symptoms of, 330, 331
   enzyme activation, 329-330
   requirement of, 329
   sources of, 329
   toxic level of, 331
   toxicity symptoms of, 331

Proteins, 104
  amino acids in foods, 104
  amino and keto acids interconversions, 155, 156
  acid base characteristics of amino acids, 109
  cereal proteins, 121
    albumines of, 121
    amino acids of, 121
    globulins of, 121
    glutelins of, 121
    prolamines of, 121
    proteoses of, 121
  classification of proteins, 111
    albuminoids of, 112
    albumins of, 111
    globulins of, 111
    glutelins of, 111
    histones of, 112
    prolamines of, 112
    protamines of, 112
    simple proteins, 111
  Conjugated proteins, 112
    chromoproteins of, 113
      derived proteins of, 113
    glycoproteins of, 113
    lipoproteins of, 113
    nucleoproteins of, 112
    phosphoproteins of, 113
  deamination reactions of amino acids, 152, 153
  decarboxylation of some amino acids, 157, 158
  dehydrogenation reactions, 154
  egg composition, 116
    shell of, 116
    shell membrane (inner) of, 117
    shell membrane (outer) of, 116
    vitelline of, 117
    yolk of, 117
  egg proteins, 116
  egg white proteins, 117
    avidin of, 120
    conalbumin of, 118
    flavoprotein of, 120
    lysozymes of, 118, 119
    minor proteins of, 120
    ovalbumin of, 117
    ovoinhibitor of, 120
    ovomucin of, 119
    ovomucoid of, 119
  enzymes in meat, 141
    enzymes of bacteria and fungi, 141, 142
  formation of creatine, 168, 169
  formation of peptide linkages between amino acids, 109
  glycine, 162, 163
  important reactions of, 158-165
  interconversion of sulfur amino acids, 165, 166
  introduction, 104
  Lysine, 163
  meat curing, 139, 140
    function of smoking of meats, 139-141
    tenderness of, 140, 141
  meat emulsions, 142-146
    definition of, 143
    percent of soluble proteins of, 145, 146
  meats, 124
    amino acid content of, 131
    composition of cooked meat, 130
    composition of skeletal muscle, 126
    connective tissue of, 126, 127
    connective tissue membranes of skeletal muscle of, 127
    connective tissue proteins of, 126
    lipids component of muscle of, 129
    meat color of, 133-138
      green pigments of, 136, 137
      structure of myoglobin and hemoglobin of, 133, 134
    meat salts, 139
    muscle contraction, 128, 129
    protein composition of skeletal muscle, 126
    rigor mortis, 132
    specialized connective tissue, 130

types of muscles, 125
types of proteins of skeletal muscle, 126, 127
water extractives, 129
metabolism of amino acids, 146
  amino acids requirement, 150, 151
  digestion and absorption of, 151, 152
  essential amino acids, 150
  general reactions of, 152-161
  making use of nitrogen, 146, 147
  nitrogen fixation, 147, 148
  processes of NH_3 fixation, 148-150
  transamination reactions of, 152, 153
milk proteins, 114
  casein fractionation, 115
  fractionation of milk proteins, 114
  role of casein in cheese-making, 115
proteins in the egg yolk, 120
relationship between urea and TCA cycles, 168, 169
shape and structure of proteins, 113
  fibrous, 113
  globular, 113
tryptophan, 106, 107
urea cycle, 166-168
valine, 104, 105
vegetable proteins, 123
wheat proteins, 122
  composition of hard wheat, 122
  morphological composition of, 122
  glutin composition, 122-123
  glutin modification, 122-123
  glutin protein, 122, 123
  glutin subdivision, 123
  soluble proteins, 123
  albumins, 121
  globulins, 121

Vegetables, 192
  chemical composition of, 192
  classification of, 193
  general changes during cooking of, 195
  introduction to, 192
  structure of, 193
  turgor of, 195
  volatile sulfur compounds of, 197

Vitamins, 213
  vitamin A, 213
    absorption and uptake of, 216
    biological function of, 217
    factors in conversion of carotenes to, 215
    introduction to, 213
    involvement in vision, 219
    role in bone, 220
    role in glycogen synthesis, 221
    summary of deficiency of, 223
    summary of excessive symptoms of, 224
    symptoms of deficiency of, 222
    tests for vitamin A, 216
    vs. thyroxine, 221
  vitamin C, 251
    ascorbic acid and adrenal functions, 256
    biosynthesis of, 259
    effect of ascorbic acid on metal ions, 255
    factors affecting ascorbic acid requirement, 258
    introduction to, 251
    other biochemical functions of, 254
    oxidation-reduction system of 253
    relationship between ascorbic acid and cholesterol, 257
    symptoms of scurvy, 258
    various functions of, 257
  vitamin D, 227
    absorption of, 228, 231
    deficiency symptoms of, 233
    functions of, 230

introduction to, 227
mechanism of Ca-transport of, 231
relationship with other chemicals, 233
relationship with other hormones, 233
structure of various forms of, 229
sources of, 232
summary of deficiency symptoms of, 233
toxicity symptoms of, 234
vitamin E, 236
absorption and utilization of, 237
factors determining its intake, 241
introduction to, 236
sources of, 237
structure of various forms of, 237
summary of metabolic functions of, 240
symptoms of deficiency of, 241
various forms of, 236
vitamin K, 244
antagonists of, 248
deficiency symptoms of, 248
factors affecting poor clotting time of vit. K deficiency, 248
intrinsic system of vit. K action, 247
introduction, 244
various forms of, 246
biotin, 283
biochemical functions of, 285
in carbohydrate metabolism, 285, 286
in lipid metabolism, 286
in propionate metabolism, 285, 286
in purine synthesis, 287
in urea synthesis, 286
deficiency symptoms of, 288
enzymes activated with, 287, 288
introduction to, 283

possible mechanism of binding with Co_2, 284
role in protein synthesis, 288
sources of, 285
structure of, 284
summary of roles in biochemical systems, 287
amino acid and nucleic acid metabolism, 287
carbohydrate metabolism, 287
lipid metabolism, 287
B_{12}, 296
absorption, 298
co-enzyme functions of, 298
biosynthesis of methyl groups, 298, 299
nucleic acid metabolism, 300
propionate metabolism, 299
deficiency symptoms of, 301
factors affecting B_{12} requirement, 300
general functions of, 298
introduction to, 296
production of methionine, 300
relationship between B_{12} and folic acid, 301
structure of, 296
summary of B_{12} deficiency symptoms, 301
in chicks, 302
in man, 301, 302
in ruminants, 302
in swine, 302
synergistic action with THF, 301
folic acid, 303
co-enzyme forms, 304
examples of reactions catalyzed with, 307
functions of, 306
catabolism of histidine, 306
epileptic patients and folic acid, 306
folic acid and B_{12}, 306
folic acid and the sulfanamide drugs, 306
glycine, serine and methionine, 306

one carbon metabolism, 306
purine and thymine synthesis, 306
pyrimidines and glycolytic enzymes' synthesis, 306
relationship with biotin, 306
introduction to, 303
structure and forms, 304
symptoms of deficiency of, 309
niacin, 275
deficiency symptoms of, 278
function of the co-enzymes, 275
introduction to, 275
source of, 276
structure and forms of, 276
pantothenic acid, 279
deficiency symptoms of, 282
introduction to, 279
structure of, 279
summary of functions of, 281
types of reactions catalyzed with, 280
head to tail condensation, 281
nucleophillic attack, 280, 281
pyridoxine (B_6), 289
deficiency symptoms of, 295
introduction to, 289
metabolism of, 294
possible mechanisms of reactions of, 291
role of B_6 in, 291
decarboxylation of amino acids, 291, 292
racemization of amino acids, 291
transamination of amino acids, 291
tryptophan metabolism, 294
various forms of, 290
riboflavin (B_2), 271
deficiency symptoms of, 274
enzymes activated by FAD, 274
enzymes activated by FMN, 274
functions of, 273
introduction to, 271
metallo-flavoproteins, 274
role in electron transport, 273
structure and forms of, 271
thiamin (B_1), 263
deficiency symptoms of, 269
functions of, 264
introduction to, 263
possible mechanism of decarboxylation reactions, 268
possible mechanism of non oxidative decerboxylation, 267
specific reactions catalyzed by TPP, 265
carbon transfer reactions, 265
non-oxidative decarboxylation, 265
oxidative decarboxylation, 265
transketolase reactions, 266
structure of, 264

Water in foods, 1
adsorbed, 1
as temperature stabilizer, 2
bound form, 1
characteristics as life sustaining system, 1
effects on foods, 3, 4
free form, 1
heat conductance of, 2
high surface tension of, 2
hydrate forms of, 1
hydrogen ion activity of, 2, 3
imbibed form of, 1
latent heat of fusion of, 1, 2
latent heat of vaporization of, 1, 2
solvent property of, 2
some unique properties of, 2
specific gravity of, 2
specific heat of, 1, 2
structure of, 2, 3
transparency to light of, 2